INTRODUCTION TO THE MODERN THEORY OF METALS

Alan Cottrell

THE INSTITUTE OF METALS
LONDON 1988

Book 403
Published by

The Institute of Metals
1 Carlton House Terrace
London SW1Y 5DB

and

The Institute of Metals North American Publications Center
Old Post Road, Brookfield, VT 05036, USA

First Published 1988
Reprinted 1991

British Library Cataloguing in Publication Data

Cottrell, Alan
Introduction to the modern theory of metals.
1. Metals
I. Title
669
ISBN 0–904357–97–X

Library of Congress Cataloging in Publication Data

applied for

Made and printed in Great Britain

Contents

Preface

The electron theory of metals has taken great strides forward since the 1940s when the classic introductions for metallurgists were written by Hume-Rothery and Raynor*. When they were writing, in the immediate postwar years, there was a wide gap between the picture of a metal provided by the theoretical physicists and the general view held by metallurgists. Hume-Rothery was led to write his book, in his words 'in the hope of providing a bridge by means of which the student of metallurgy might be led to an understanding of the general ideas on which the electron theories of metals and alloys are based.' Recent decades have seen many advances in the theory, some of which have completely overturned the ideas of the 1940s, and the gap now seems to have become wider than ever. It is thus for this reason that I have written the present book with exactly that same purpose, here in the 1980s, that motivated Hume-Rothery and Raynor in the 1940s.

This book deals only with metals (and alloys); nothing is said about semiconductors, insulators, ionic crystals and so on, beyond what is necessary to place metals in context and to mark out the boundaries of the field. However, the class of metals is now seen to be very much larger than was envisaged in the 1940s. It goes well beyond the traditional metals on the left-hand side of the Periodic Table; and the indications are that almost anything becomes a metal when its atoms are brought close enough together. Indeed, the best conductor of electricity at the temperature of liquid nitrogen is now no longer a traditional metal, but an oxide. It is possible that the same may one day be true at room temperature.

I have tried to get as quickly as possible to the things that have happened since the days of Hume-Rothery and Raynor's texts, and so I have merely summarized the earlier parts of the theory - somewhat sketchily - in the introductory Appendices, hoping always that the reader will have a lively recall of the foundations of the subject as admirably presented by Hume-Rothery and Raynor. Because of their interest to metallurgists and materials scientists, I have given particular emphasis in the main body of the book to the cohesive properties of metals, including various aspects of heat of formation, crystal structure, elastic properties, alloying behaviour, surfaces and defects, even

*Hume-Rothery, W., *Atomic Theory for Students of Metallurgy*, The Institute of Metals, London (1946); updated by W. Hume-Rothery and B. R. Coles, 1969.
Raynor, G.V., *An Introduction to the Electron Theory of Metals*, The Institute of Metals, London (1947).

though these are in general not the properties that can be dealt with most readily by the theory.

The modern theory is inevitably intensely mathematical, and it is impossible to avoid going into some of this if one wants to understand why metals are as they are. Nevertheless, I have tried to keep the mathematical treatment as simple as possible (indeed, oversimplified in all too many places), and relegated more detailed workings to the Appendices. Another problem is that the number of entities that require to be represented by mathematical symbols is rather large, and some symbols have to stand for more than one (e.g. T for both temperature and kinetic energy). I hope that the context is always sufficiently clear for there be no possibility of confusion, but I have also provided a list of the main symbols and what they represent.

Viewing the wood rather than the trees, it may be useful here to give a broad, qualitative picture of the metallic state as suggested by the modern theory. One of the biggest changes concerns that most elementary question: what is a metal? The distinction between metals and insulators based on Brillouin zones has steadily lost ground, since Mott's 1949 paper, to the more 'chemical' view, long championed by Pauling but recently seen to have a much older, half-forgotten pre-quantum basis. A material becomes a metal when its atoms are brought sufficiently close together for the energy cost of transferring a valency electron from one atom to another (as a result of the electrostatic repulsion from the valency electrons already in this other atom) to be more than offset by the energy that would be gained by 'liberating' the electron from its atomically localized state into a 'free-electron' state in which it can wander through the entire volume of the material.

The old free-electron idea of a metal thus survives, although with a different foundation. Moreover, what was an assumption in the old theory – that a free electron in a metal behaves as if the strong electrostatic fields from both the atomic cores and the other free electrons are almost non-existent – is now properly understood. The atomic cores have little effect (apart from providing a nearly uniform field of 'smeared out' positive charge) because, by the Pauli exclusion principle, the electrons in the atomic cores repel the free electrons with an intensity that in ideal metals almost perfectly balances the electrostatic attraction of the positively charged cores. As a result, the electrostatic fields of the atomic cores behave as if they were very weak 'pseudopotentials', and the simple wave-mechanical picture of free electrons, which goes back to the earliest days of the quantum theory of metals, is then quite applicable.

The apparent freedom of the free electrons from one another's influences is a more subtle and difficult problem, but the qualitative explanation is simple. Each electron moves about, not as a 'bare' particle but as a 'quasiparticle', a combination of the electron itself with an 'exclusion zone' or 'correlation hole', a small region around it which is usually empty of other electrons. Each such 'clothed' electron is in effect electrically neutral since it consists of a negative electron and a positive hole; these neutral quasiparticles move around as almost independent entities.

The cohesion of 'simple' non-transition metals such as sodium, magnesium and aluminium results from the electrostatic attraction between the negative electrons and the positive atomic cores, although the kinetic energy of the electrons, due to quantization and the Pauli principle, greatly complicates the behaviour. The energy of this electrostatic attraction strongly favours the close-packed structures: f.c.c., c.p.h. and b.c.c. The resistance to elastic shear stems mainly from the change in this electrostatic energy when a close-packed crystal structure is distorted out of its symmetrical form.

The Jones theory of alloys has long familiarized metallurgists with the idea that the energy of the free electrons can be decreased by the perturbing effect of Brillouin zone boundaries. Although it is small compared with the major volume-dependent energy term, this band-structure energy plays a significant part in, for example, producing the less simple crystal structures of gallium, indium and mercury, the large axial ratios of zinc and cadmium, and the small elastic shear resistance (and hence, probably, low melting point) of aluminium. When the large, volume-dependent part of the total energy is separated out, the remainder can be interpreted in terms of 'bonds' acting between the atoms along the lines between their centres. While this is useful for dealing with some of the more complicated structural problems such as lattice defects, it does not justify the naive method of estimating structural energies by counting the numbers of 'bonds'.

Because of the dominating influence of their partly filled d shells, transition metals are best viewed as gigantic, unsaturated, covalently bonded molecules. The atomic d orbitals of neighbouring atoms overlap slightly, with either the same signs (bonding overlaps) or opposite signs (antibonding overlaps), and each quantum state of the material consists of one complete pattern of such overlaps throughout the whole structure. As a result, the d electrons exist in an energy band of such states, determined by the numbers of bonding and antibonding overlaps. The cohesion comes from the band being only partly filled by electrons, with more in the low-energy, mainly bonding states than in the high-energy, antibonding ones. This cohesion is greatest when the band is half-filled, although in the first long period there are complications caused by ferromagnetism and other effects. A significant feature is that, for the b.c.c. lattice, the d band is almost split into a low-energy half and a high-energy half, with a bridging region in between where the density of states is low. The low-energy half can hold all the electrons in the earlier members of the transition series and so many such elements show a strong preference for the b.c.c. structure.

It is ironic that copper, silver and gold, Hume-Rothery and Raynor's 'ideal' metals for working out the rules of alloying, are very far from simple. They are, as it were, 'transitional' between the simple metals and the transition metals and so cannot be satisfactorily treated by either the free-electron or the covalent-bond approximation. Although they have filled d shells, the energies of their d electrons lie close to those of the valency electrons and as a result there is in the solid strong 'hybridization' between the d states and the free-electron states which contributes appreciably to the moderately good cohesion of these metals. The old picture, in which the filled d shells make the atoms of these metals resemble hard billiard balls held together by the free electrons, has not survived however. By hybridizing, the d electrons help to pull the atoms together, and it is the kinetic energy of the free electrons that mainly provides the 'billiard ball' hardness.

The dominating factor in the formation of alloys between true metals is the Hume-Rothery size factor. The electron theory has justified the simple theory that deals with the size factor in terms of the elastic energy needed to expand or shrink the atomic volumes to fit the different atoms together. Many complexities of alloy structure reflect the tendency to provide sites of different atomic volumes in order to accommodate small and large atoms with the least strain. The electrochemical factor is also important, but the electron theory shows that (at least for the simpler alloys) this factor expresses itself mainly through the band-structure energy rather than the electrostatic energy of differently charged ions, which is a small effect.

The simple form of the Jones theory of alloys suffered a great shock in 1957 when Pippard showed, by experiment, that the Fermi surface already touches the zone boundary in pure copper. The theory showed signs of recovery with Heine's subsequent demonstration that what matters is not the actual shape of the Fermi surface, but the spherical approximation to it as assumed in the simpler theories. Various other complexities have appeared though, and this, perhaps the earliest, application of electron theory to alloys remains in an unsettled state. More generally, the complexities of alloy theory, particularly for transition metal alloys, have led to various semi-empirical theories and classification schemes, the latest of which, developed by Pettifor, has proved most successful in bringing greater rationality to this complex subject.

The energy barrier at the surface of a metal, which imprisons the free electrons and which has many important practical aspects (e.g. in thermionics, contact potentials, adsorption and catalysis), is fairly well understood. This understanding is based partly on the idea of an electrical double layer formed as a result of the tendency of the free electrons to spread a little way beyond the atomic surface of the metal, and partly on the concept of the image force, which results from the way in which the free electrons just inside the surface correlate their positions when another electron is brought towards this surface from outside. The recent forms of the electron theory have enabled good estimates of surface energy to be made, but many challenges still remain, for example to explain the reconstruction of the outermost layer of atoms in a metal surface into a two-dimensional crystal pattern different from that inside.

The electron–electron interactions responsible for, among other things, correlation holes and image forces achieve a most spectacular effect in the superconductivity that occurs in many metals at low temperatures. The basic process is that, by means of such an interaction, the electrons loosely associate in pairs (known as 'Cooper pairs') in which form they behave in a fundamentally distinct way, quantum mechanically; more like the atoms in superfluid helium. In metals and alloys this association takes place through the interaction of electrons with the lattice vibrations, or phonons. One electron attracts the nearby lattice atoms together, electrostatically; these close atoms then in turn attract another electron, and the two electrons behave as if they were attracted to each other and thereby form a Cooper pair. The BCS (Bardeen, Cooper, Schrieffer) theory of superconductivity shows that the phonon interaction is always likely to be weak, and this is confirmed by the fact that even in the most favourable alloys the thermal energy, even at only 25 K, is sufficient to break up the pairs and destroy the superconductivity. As the reader will know, a whole new aspect of this subject has recently and unexpectedly opened up with the discovery of oxides which remain superconducting to very much higher temperatures. Although the theory of these high-temperature superconductors is, as yet, far from established, it is already clear that the mechanism is unlikely to be a BCS-type theory, which is limited to low temperatures. There is thus the possibility that materials which remain superconducting at room temperature will be discovered.

Alan Cottrell
Cambridge, July 1987

Acknowledgements

It is a pleasure to thank various friends and colleagues for their help during the preparation of this book. In particular, Dr R. W. Cahn read the manuscript and offered many helpful suggestions. Dr J. A. Charles made the initial contacts with the Institute of Metals, as a result of which I was encouraged to begin the book, and I have enjoyed several stimulating discussions with Dr P. Edwards about the metallic state of matter. Professor D. Hull kindly made available to me the facilities of the Department of Materials Science and Metallurgy, Cambridge University.

List of symbols

Å	Ångström
a	lattice spacing; atomic spacing; eigenvalue
a_H	orbital radius
a_0	Bohr radius
$\mathbf{a}$	unit lattice vector
$\mathbf{a}^*$	reciprocal lattice vector
A	general operator
A	constant; parameter; weighting factor; coefficient in Thomas–Fermi equation; affinity energy
b	atomic spacing; eigenvalue
$\mathbf{b}$	unit lattice vector
$\mathbf{b}^*$	reciprocal lattice vector
B	general operator
B	constant; parameter
c	constant; speed of sound; atomic concentration
c_{11}, c_{12}, c_{44}	elastic constants
c/a	axial ratio of hexagonal crystal
$\mathbf{c}$	unit lattice vector
$\mathbf{c}^*$	reciprocal lattice vector
$\mathrm{c}(l \times m)$	centred surface reconstruction net
C	constant
d	spacing of crystal planes; interatomic distance
d	atomic d state symbol
$\mathrm{dd}\sigma$, $\mathrm{dd}\pi$, $\mathrm{dd}\delta$	overlap energies of d orbitals
d_ε, d_γ	symmetry classes of d orbitals
e	electrical charge of electron
E	energy (particular forms of which are identified by various subscripts and superscripts)
E^*	chemical potential
E_F	Fermi energy
$\mathscr{E}$	electric field
f	atomic f state symbol
$f(q)$	atomic form factor; Fourier transform of $f(x)$
$f(x)$	general function of x
F	electric field strength

$\boldsymbol{g}$	reciprocal lattice vector
G	Green's function
h	Planck's constant ($\hbar = h/2\pi$); direction cosine; Miller index
H	magnetic field strength
H	Hamiltonian operator
i	general variable and subscript
i	$=\sqrt{-1}$
I	ionization energy; electron–electron interaction energy; intensity of scattered beam
j	electrical current; general variable and subscript
k	wavenumber; direction cosine; Miller index
k_B	Boltzmann's constant
k_F	wavenumber at Fermi surface
$\boldsymbol{k}$	wave-vector ($k = \lvert\boldsymbol{k}\rvert$)
K	constant; normalization factor; bulk modulus of elasticity
l	position coordinate; angular momentum quantum number; direction cosine; Miller index; mean free path
$l \times m$	surface reconstruction net
$\boldsymbol{l}$	general lattice vector
L	length of metal; angular momentum
L_m	latent heat of melting
m	mass of electron (or other particle)
m^*	effective mass of electron
m_l	atomic quantum number
m_s	spin quantum number
M	mass of atom or ion
M_m	mth moment of density of states distribution
n	any integer; principal atomic quantum number; number of particles per unit volume; electron density
$n(E)$	density of states per atom
N	number of particles (electrons per atoms), total or per unit volume (particular examples of which are identified by subscripts)
$\mathrm{N_A}$	Avogadro's number
N_k	number of states with wave-vector $\boldsymbol{k}$
N_P	number of atoms per unit area of a plane
$N(E)$	density of states (i.e. number of quantum states per unit volume in a given unit energy range)
$N(0)$	density of one-electron states at Fermi surface
p	momentum; probability of occupation of a state; probability of a transition
p_l	number of nodes in wavefunction
p	atomic p state symbol
$\mathrm{p}(l \times m)$	primitive surface reconstruction net
$\boldsymbol{P}$	spin polarization vector
q	wavenumber ($= \lvert\boldsymbol{q}\rvert$)

$\boldsymbol{q}$ difference between wave-vectors of two states
$\boldsymbol{Q}$ wave-vector
r radius (particular examples of which are identified by subscripts); distance of electron ($= |\boldsymbol{r}|$)
$\boldsymbol{r}$ position vector
R internuclear spacing; radial component of wavefunction; molar refractivity
s spin
s atomic s state symbol
S entropy; overlap integral; structure factor
t time
t_{2g} symmetry class of d orbitals
T temperature; kinetic energy
T_m melting point
$u(x), u(\boldsymbol{r})$ lattice site factor in Bloch wavefunction
U electrical field of atom; electrostatic repulsive energy between electrons in same atomic orbitals
v volume; pseudopotential of atom
v_s velocity of sound
v^0 unscreened pseudopotential of atom
$\boldsymbol{v}$ velocity vector
V potential energy (particular forms of which are identified by subscripts); perturbing potential; matrix element; molar volume; electron–phonon interaction energy
V_{ps} pseudopotential
w number of arrangements of electrons in quantum states; width of d band
$w(q)$ pseudopotential form factor; Fourier coefficient of $v(r)$
W depth of potential energy well; number of arrangements of electrons in quantum states; amplitude of total scattered beam; energy of superconducting state
$W(q)$ Fourier coefficient of $V(r)$
x general variable; position
y general variable; position
z general variable; position; number of nearest neighbours
z_d number of electrons per atom in d band
Z nuclear charge number; charge number of ionic core; number of metallic electrons per atom (metallic valence)

α polarizability; ratio of wavenumber to Fermi wavenumber; structure constant equivalent to Madelung constant; matrix coefficient in tight-binding theory
β Pauli repulsion, in pseudopotential; matrix coefficient in tight-binding theory
γ screening factor; interfacial energy; stacking fault energy; angle of shear

Γ centre of Brillouin zone
δ general variation symbol; Kronecker (Dirac) delta; molecular orbital symmetry
δ_l effective angle of phase shift
Δ spread of a variable
2Δ superconducting energy gap
ε energy measured relative to Fermi level; elastic strain
ε_g symmetry class of d orbitals
ε_{xc} exchange and correlation energy of a uniform electron gas
ξ superconducting coherence length
η angle of phase shift; close-packing factor
θ angle; Bragg angle; phase of wavefunction; angular measure of wavenumber
Θ angular component of wavefunction
Θ_D Debye temperature
κ dielectric constant
$\kappa(q)$ dielectric function
λ undetermined multiplier; wavelength; screening radius; superconducting penetration depth
λ_{TF} Thomas–Fermi screening radius
μ undetermined multiplier; elastic shear modulus
ν frequency of vibration
π molecular orbital symmetry
ρ probability density; number of particles per unit volume; probability of occupation of a quantum state; charge density
σ molecular orbital symmetry; cross-sectional area; surface energy
ϕ wavefunction; electrostatic field potential
ϕ^* electronegativity parameter
Φ work function
$\Phi(r)$ effective ion–ion interaction
$\Phi(q)$ Fourier transform of $\Phi(r)$
χ perturbation characteristic; chemical scale factor
$\psi, \psi(x), \psi(\mathbf{r})$ general wavefunction
ψ^* complex conjugate of ψ
Ψ time-dependent wavefunction
Ω volume
Ω_a volume per atom; volume of Wigner–Seitz cell
ω angular frequency ($= 2\pi\nu$); volume of crystal cell
ω_p frequency of plasma oscillation
! factorial symbol
∂ partial differential symbol
∇^2 $\frac{\partial^2}{\partial x^2} + \frac{\partial^2}{\partial y^2} + \frac{\partial^2}{\partial z^2}$
$\langle x \rangle$ average value of x
$\langle \psi |$ bra form of wavefunction, i.e. ψ^*

$\vert\psi\rangle$	ket form of wavefunction, i.e. ψ
$\langle\phi\vert V\vert\psi\rangle$	matrix element of perturbation V between states ψ and ϕ, in Dirac notation

PHYSICAL CONSTANTS AND UNITS

1 Å (angstrom) = 10^{-10} m
1 mol (mole) = $6{\cdot}03 \times 10^{23}$ particles
1 ev (electron-volt) per particle = 96 500 J mol^{-1}
1 T (tesla) = 10^4 G (gauss)
e = charge of electron = $1{\cdot}6 \times 10^{-19}$ C (coulomb) = $4{\cdot}8 \times 10^{-10}$ electrostatic units
h = Planck's constant = $6{\cdot}626 \times 10^{-34}$ J s
m = mass of electron = $9{\cdot}1 \times 10^{-31}$ kg
k_B = Boltzmann's constant = $1{\cdot}38 \times 10^{-23}$ J K^{-1}
1 eV = $k_B T$ at T = 11 605 K
a_0 = Bohr radius = $0{\cdot}529 \times 10^{-10}$ m
1 Ry (Rydberg) = 13·6 eV

1

What is a metal?

1.1 METALS, INSULATORS AND SEMICONDUCTORS

The first question for a student of metallurgy – 'What is a metal?' – has a long and surprising history of answers. The 'modern' answers begin with the classical simplicity of Drude and Lorentz, in the period 1900–1916, followed in the 1930s by the quantum-mechanical subtleties of the Sommerfeld, Bloch, Wilson and Brillouin theories, and then in 1949 by Mott's drastic reappraisal which in recent years has led back to classical simplicity, although with some extremely sophisticated developments.

The most direct answer of course is simply to identify metals by their properties. Metals differ enormously from non-metals in their electrical conductivities, those of copper, silver and aluminium, for example, being some 10^{20} times larger than those of insulators such as diamond, mica and aluminium oxide. High electronic conductivity is not a sufficiently precise criterion, however, because many semiconductors are fairly good electronic conductors at or above room temperature. Nevertheless, unlike metals, they owe their conductivity to thermal activation and so become insulators when cooled to absolute zero. In a true metal the electrical resistivity remains finite (or, in a superconductor, falls to zero) as the temperature approaches absolute zero. Closely related to this is the fact that in semiconductors the resistivity falls when the temperature is increased, whereas in metals it rises.

We thus have a working definition of a metal. But this does not tell us what a metal is. Before going on to the theory of metals, we first see how large the class of metals is. Most of the elements in the Periodic Table (*see* Fig. 1.1) are metals, the non-metallic group being essentially confined to the upper right of a diagonal line running from boron to astatine. The dual status of tin, which can exist as either a metal (white tin) or a non-metal (grey tin) is familiar. Under high pressures, as in the interior of the planet Jupiter, it is likely that all elements are metals. In the laboratory, hydrogen (the major constituent of Jupiter) has now been made metallic by compressing it at low temperature to the density of water. Iodine is a metal at pressures greater than 160 kbar. Conversely, experiments on normally metallic elements such as Hg, Cs, Rb and K heated to the liquid–vapour transition at high temperatures ($> 2\,000\,°C$) have shown that, at sufficiently low densities, a transition to a non-metallic state occurs. At normal densities some elements which are marginally non-metallic when crystalline, such as Si, Ge, Se and Te, become metallic when melted.

IA	IIA	IIIA	IVA	VA	VIA	VIIA	VIIIA			IB	IIB	IIIB	IVB	VB	VIB	VIIB	
H																	He
Li	Be											B	C	N	O	F	Ne
Na	Mg											Al	Si	P	S	Cl	Ar
K	Ca	Sc	Ti	V	Cr	Mn	Fe	Co	Ni	Cu	Zn	Ga	Ge	As	Se	Br	Kr
Rb	Sr	Y	Zr	Nb	Mo	Tc	Ru	Rh	Pd	Ag	Cd	In	Sn	Sb	Te	I	Xe
Cs	Ba	La	Hf	Ta	W	Re	Os	Ir	Pt	Au	Hg	Tl	Pb	Bi	Po	At	Rn
Fr	Ra	Ac															

Ce	Pr	Nd	Pm	Sm	Eu	Gd	Tb	Dy	Ho	Er	Tm	Yb	Lu
Th	Pa	U	Np	Pu	Am	Cm	Bk	Cf	Es	Fm	Md	No	Lw

Fig. 1.1 The Periodic Table, showing the dividing line between metals and non-metals

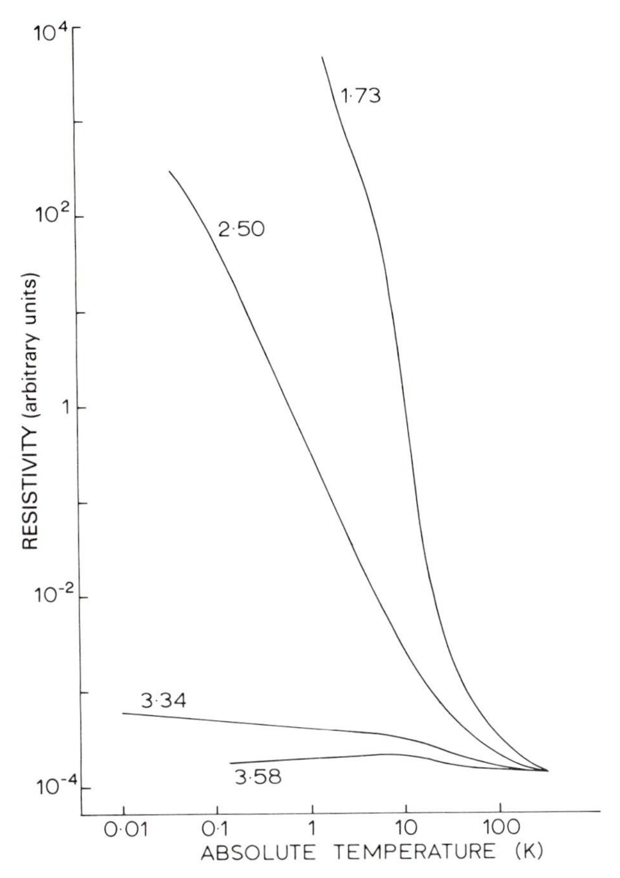

Fig. 1.2 The electrical resistivity of phosphorus-doped silicon as a function of temperature; the content of electron donor atoms, in units of 10^{24} m^{-3}, is indicated on each curve (Sasaki 1980, Edwards and Sienko 1983)

Some non-metals become metallic when impure. Figure 1.2 shows the typical transition to the metallic state that occurs when a semiconductor is sufficiently heavily doped with an impurity of different valency. In this particular example the addition of phosphorus to silicon produces the transition at a concentration of additional valency electrons of about 3×10^{24} m^{-3}. Below this concentration the resistivity climbs to a high value when the temperature falls towards absolute zero (0 K), as is usual for a semiconductor; above it, the resistivity approaches a finite value at 0 K, not much different from that at room temperature.

Liquid ammonia becomes a metal when small amounts of some metals, particularly alkali metals, are dissolved in it, a behaviour which resembles that of heavily doped semiconductors. Very dilute ammonia solutions have a bright blue, non-metallic appearance, but above a certain range of concentration they look like molten bronze and conduct electricity well. Some chemical compounds are metallic in the solid state; examples are Fe_3O_4 and the 'vanadium bronzes' (e.g. $Na_xV_2O_5$, where $0{\cdot}2 < x < 0{\cdot}6$) and complexes such as $K_2Pt(CN)_4Br_{0{\cdot}3}$. Several metallic organic compounds are now known, for example iodine-doped polyacetylene $(CHI_{0{\cdot}25})_n$ and TTF–TCNQ (tetra-thiofulvalenium-7,7,8,8-tetracyano-*p*-quinodimethanide). These are both chain structures and act as one-dimensional metals, conducting well along the chain axis.

1.2 FREE ELECTRONS

The theory of metals began with Drude (1902) and Lorentz (1916), who suggested that metals contain *free electrons* – electrons which have become separated from their parent atoms and are free to wander through the entire body of the material as a kind of gas or plasma. The theory was extremely successful. It immediately explained the high electrical conductivity, as the mobile electrons could easily drift towards the positively charged end of the metal in an applied electrical field. Their similar movements to and fro, in rapidly oscillating applied electromagnetic fields, explained the typical optical properties of metals, i.e. opacity and lustre. The good thermal conductivity was also explained in terms of the redistribution of the free electrons in a temperature gradient.

The free-electron theory has since been seen to provide a general understanding of the metallurgical properties of metals. Bereft of their free electrons, the atoms in a metallic crystal act as positively charged spheres, and the cohesion of the material comes from the electrical attraction of these positive charges to the 'sea' of negative electrons flowing among them. It follows that, as long as they keep close to the free electrons, these atoms are not very sensitive to their positions relative to one another. Hence metals tend to have small elastic shear moduli (as compared with their moduli of compression); also they often exist in more than one crystal structure and tend to have somewhat low melting points. Since there are no directional chemical bonds, only an isotropic attraction of positively charged spheres to a negatively charged fluid, metals generally crystallize in simple close-packed structures. They also cohere together readily in alloys, and in welded and soldered joints, over wide ranges of composition. Their close-packed crystal structures provide crystallographic planes and directions suitable for slip; this geometric property, together with the low shear modulus, makes metals such as copper and aluminium ductile, soft and very resistant to cleavage fracture.

Early direct evidence for the existence of free electrons in metals was provided by Tolman and Stewart (1917). They showed that accelerating metals sharply created brief pulses of current of just the kind that would be expected if free electrons were thrown towards one end by such movements.

Despite its many successes, the classical free-electron theory faced some severe difficulties, particularly the problem of specific heat. When its temperature was increased the free-electron gas was expected to absorb heat, much like any gas heated at constant volume, but in practice there was no sign of any such effect. The problem could not be solved by supposing that only a few of the valency electrons in a good metal are free, because the measured electrical conductivity showed that in a metal such as silver there is about one free electron per atom, making the free-electron gas extremely dense.

The problem was solved by Sommerfeld (1928), who applied quantum mechanics to the theory and showed that the high electron density leads to a low specific heat. (A summary of elementary quantum mechanics is given in Appendices 1 and 2.) To be able to describe a gas by classical kinetic theory it must be possible, in principle, to observe the individual particles of the gas. Thus if N is the number of free electrons per unit volume, we have to be able to locate the position of any electron to within an error $(\Delta x)^3$ that is smaller than N^{-1}. But, by Heisenberg's uncertainty principle (*see* Appendix 1), a particle cannot be located with an exactitude Δx unless its momentum exceeds a minimum value Δp, given by

$$\Delta p \, \Delta x \approx h \tag{1.1}$$

where h is Planck's constant ($6{\cdot}626 \times 10^{-34}$ J s). Since the kinetic energy E of the particle is given by $E = p^2/2m$, where p is the momentum and m the mass, the electron gas cannot behave according to classical kinetic theory unless it is heated to such a temperature that its thermal energy per electron exceeds the value

$$E_{\mathrm{F}} \approx \tfrac{1}{2}(\Delta p)^2/m \approx \tfrac{1}{2}h^2 N^{2/3}/m \tag{1.2}$$

The exact result (*see* Appendix 3) is

$$E_{\mathrm{F}} = \frac{h^2}{8m}\left(\frac{3N}{\pi}\right)^{2/3} \tag{1.3}$$

which for silver corresponds to a temperature of 60 000 K. At ordinary temperatures the motion of the free electrons is completely dominated by quantum effects. Even at 0 K these electrons are in motion inside the metal at speeds of about 10^6 m s^{-1}, and they occupy a band of quantized kinetic energy levels from a *ground energy level* of almost zero up to a maximum at E_{F}. This state of affairs remains much the same at temperatures up to the boiling point of the metal, so that very little heat energy is absorbed by the free electrons.

Despite its successes and Sommerfeld's improvement, the electron theory of metals remained very incomplete in 1928. Why were the electrons free? And why were they not free in non-metals? (A significant step towards answering these questions had already been taken by Goldhammer (1913) and Herzfeld (1927), but this made use of some very different ideas and was largely overlooked until the 1970s; we consider it later.) After Sommerfeld's work, and with the general development of quantum mechanics, the main objective in 1928 was to give the theory of metals a thoroughly quantum-mechanical treatment.

1.3 ELECTRON WAVES

In quantum mechanics (or wave mechanics) an electron is of course represented by a wave. This wave undulates continuously throughout both space and time, although in the important class of equilibrium states in which the electron has a *stationary state* of motion the time coordinate can usually be omitted. The symbol for the electron wave is then $\psi(x)$ for one space direction x, or $\psi(\boldsymbol{r})$ in three dimensions with $\boldsymbol{r}$ as the vector coordinate. The amplitude of the wave at any point x is then given by $\psi(x)$, or by $\psi^*(x)$, which is the complex conjugate of ψ, for the *wavefunction* is a mathematically complex quantity (i.e. $\psi = a + ib$ and $\psi^* = a - ib$, where a and b are real numbers and $\mathrm{i} = \sqrt{-1}$). The wavefunction is the fount of all information about the dynamical state of the electron. For example, the probability $\rho(x)\,\mathrm{d}x$ of finding the electron in the region between x and $x + \mathrm{d}x$, along the x axis is given by

$$\rho(x) \equiv |\psi(x)|^2 \equiv \psi^*(x)\,\psi(x) \equiv a^2 + b^2 \tag{1.4}$$

Other observables are deduced by *operating* on $\psi(x)$. For example, the *momentum* $p(x)$ is given by

$$\frac{1}{\psi}\frac{\hbar}{\mathrm{i}}\frac{\partial\psi}{\partial x} \tag{1.5}$$

where $\hbar = h/2\pi$, and the energy E is given by

$$\frac{1}{\psi}\mathrm{H}\psi \tag{1.6}$$

where H is the *Hamiltonian* operator, given in one dimension by

$$\mathrm{H} = \frac{1}{2m}\left(\frac{\hbar}{\mathrm{i}}\frac{\partial}{\partial x}\right)^2 + V(x) = -\frac{\hbar^2}{2m}\left(\frac{\partial^2}{\partial x^2}\right) + V(x) \tag{1.7}$$

$V(x)$ being the electrostatic potential energy of the electron at the point x. On rearrangement, this expression for the energy operation becomes the familiar Schrödinger equation in its time-independent one-dimensional form:

$$\frac{\mathrm{d}^2\psi}{\mathrm{d}x^2} + \frac{2m}{\hbar^2}(E - V)\psi = 0 \tag{1.8}$$

The condition for a free electron is V = constant, and we can take this constant as zero, in which case equation (1.8) is a simple equation for sinusoidal waves. The wavefunction for a free electron is then conveniently written as

$$\psi(x) = C\exp(\mathrm{i}kx) \tag{1.9}$$

where C is a constant and k is the *wavenumber* (in three dimensions it is $\mathbf{k}$, the *wave-vector*), which is related to the wavelength λ of the wave by

$$k = 2\pi/\lambda \tag{1.10}$$

Since $\exp(i\theta) = \cos\theta + i\sin\theta$, this ψ gives a constant probability along the x axis:

$$\begin{aligned} \rho(x) &= C^2(\cos kx - i\sin kx)(\cos kx + i\sin kx) \\ &= C^2 = \text{constant} \end{aligned} \tag{1.11}$$

as is to be expected since there is no non-uniformity of V to attract a free electron to any particular part of the line. The constant C^2 is determined by the condition that the electron must be *somewhere* on the whole line of x, i.e. the integral of $\rho(x)$ over the whole of the x line must equal unity. We assume that the units of ψ and ρ are chosen so as to satisfy this requirement; this is known as *normalization*. The momentum and energy of this electron are, from the above operations,

$$p(x) = \hbar k \tag{1.12}$$

and

$$E(k) = \hbar^2k^2/2m \tag{1.13}$$

The energy thus increases as the square of the wavenumber, for a free electron, as is indicated in the 'E, k parabola' of free-electron theory.

In a finite piece of metal not all values of k are possible, because the wavefunction has to become zero beyond the boundaries of the piece, or – in the case of a closed loop of wire – because ψ has to join up smoothly all round the loop. Each allowed value of k serves as a discrete *quantum number* and it defines one *quantum state* which, by the *Pauli exclusion principle*, can hold not more than two electrons, which must have opposite electron spins. The characteristic *band* of occupied quantum states results from this. If there are N free electrons in the metal, then the $\frac{1}{2}N$ states of lowest energy are occupied, with two electrons in each, up to the highest state, at the *Fermi level*, the energy of which is given by equation (1.3).

1.4 EFFECT OF A PERIODIC FIELD

The concepts of wave mechanics enabled Bloch in 1928 to attack the central puzzle, the existence of free electrons. To represent in the simplest possible way the real state of affairs inside a crystal, he replaced the V = constant of the free-electron theory by a periodically varying V, in accordance with the electrostatic attraction of the electrons to the positive ions, assumed to be arranged in a perfectly periodic lattice. Bloch's solution in one dimension (*see* Appendix 4) is

$$\psi(x) = u(x)\exp(ikx) \tag{1.14}$$

which of course represents the free electrons of the V = constant theory in the limit $u(x) \rightarrow C$. In the general case, in which $V(x)$ is repeated exactly at every

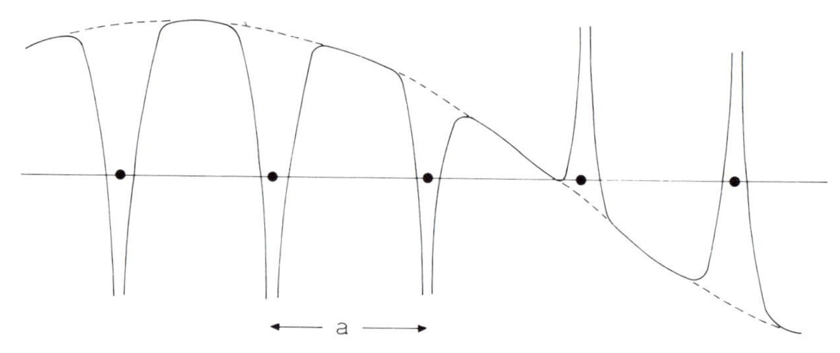

Fig. 1.3 A partial representation of a Bloch function along a line of atomic centres

lattice site, so also is $u(x)$ a repetitive lattice function. That is, if a is the lattice spacing along the x-axis, then

$$V(x + a) = V(x) \tag{1.15}$$

$$u(x + a) = u(x) \tag{1.16}$$

Within each lattice site, $u(x)$ owes its form to the variation of $V(x)$ across that site, and where $V(x)$ locally is like the field inside a single atom then $u(x)$ resembles an atomic wavefunction. Thus the Bloch function simply combines the two features we would expect it to: an atomic-like function, the same at every site, and an overall sinusoidal 'free-electron' function extending throughout the crystal.

Figure 1.3 shows what one part, e.g. $u(x) \cos kx$, of a Bloch function might look like. Here the 'carrier' wave, $\cos kx$, is the same as a free-electron wave, but it is modulated at each lattice site by the function which multiplies it, this $u(x)$ being exactly the same at each site. *The periodic lattice field, equation (1.15), thus neither scatters nor destroys the free-electron wave, but merely modulates it.* This was perhaps the most important single result of the quantum theory of metals. The Bloch wave retains the key property of the free-electron wave, of extending continuously and repetitively throughout the whole of the space concerned, i.e. throughout the whole of the crystal. The freedom of the electron to move through the entire solid is thus unaffected by the periodic lattice field; the electron is still in this sense free, even though $V(x)$ is of the form given by equation (1.15).

1.5 THE ZONE THEORY

In the words of Wilson (1980), recalling his work in the 1930s, 'Bloch's theory had proved too much. Before Bloch it was difficult to explain the existence of metals. Afterwards it was the existence of insulators that required explanation.' In 1931 Wilson extended the theory by showing that the bands of filled quantum states in crystals could be either of two general types: partially filled or completely filled (or completely empty). Only the first type allows the free electrons to act as *conduction electrons* and so make the solid a metal.

The basis of the effect (*see* Appendix 4) is that the periodic lattice field, given by equation (1.15), creates *energy gaps*, or *band gaps*, in the band of allowed quantum states – certain ranges of energy in which there are no Bloch states. These gaps occur at critical values of k, of which the primary one is

$$k = \pm \pi/a, \quad \text{i.e.} \quad \lambda = 2a \tag{1.17}$$

where a is the lattice spacing. Although the $V(x)$ field of the perfect lattice does not scatter a Bloch electron it can reflect it, and equation (1.17) expresses the condition for this, in the 'one-dimensional' example of an electron with wavenumber k pointing along an x axis perpendicular to the reflecting planes of spacing a. In fact we recognize equation (1.17) as a particular form of *Bragg's law of reflection* (of electrons, X-rays, and so on). Consider an electron moving in the positive x direction with the Bloch function given by equation (1.14). Here the positive value of k indicates this positive direction of motion. However, if equation (1.17) is satisfied, the electron will sooner or later be reflected into the reverse motion, i.e. towards $-x$ and with $+k$ replaced by $-k$ in equation (1.14). In this reverse motion it is again eventually reflected, and acquires $+k$ again; and so on, to and fro. The complete motion of the electron is thus represented not by the single Bloch function of equation (1.14), but by a function which gives equal place to $+k$ and $-k$. With suitable normalization, there are two functions of this type:

$$\psi_1(x) = \frac{1}{2}u(x)[\exp(ikx) + \exp(-ikx)] = u(x)\cos kx \tag{1.18}$$

and

$$\psi_2(x) = \frac{1}{2i}u(x)[\exp(ikx) - \exp(-ikx)] = u(x)\sin kx \tag{1.19}$$

These are *standing waves* and represent the fact that, when k satisfies the Bragg condition, the electron on average moves neither forward or backward through the reflecting planes, since it is as likely to have $-k$ as $+k$.

Figure 1.4 shows this behaviour, along a line of lattice sites with the periodic potential $V(x)$. The cosine function is a maximum at the centres of these sites, i.e. where $V(x)$ is lowest. Thus the standing wave function of equation (1.18) gives a maximum electron density (cf. equation (1.4)) in the regions of lowest electrostatic potential energy, and so the electron in this state has lower energy than the value, given by equation (1.13), of a truly free electron with this same k value. Similarly, the sine wave of equation (1.19) has a higher energy than a free electron of the same k. Thus, if the k value of a Bloch electron is gradually increased through the critical value given by equation (1.17), the energy of the electron jumps discontinuously from a lower (cosine) value to a higher (sine)

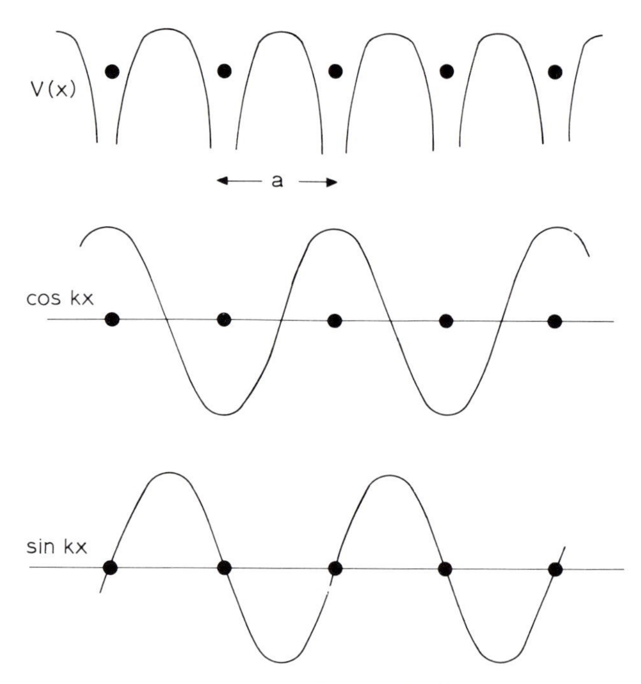

Fig. 1.4 Standing waves through a line of lattice sites with periodic potential $V(x)$

value, and the E, k parabola is thereby split into segments or *allowed bands*, as shown in Fig. 1.5, separated by energy gaps such as AB.

As shown in Appendix 4, the number of quantum states in each allowed energy band is equal to the number of atomic sites. Hence, in a monovalent element such as sodium the band containing the valency electrons is half-filled, with two electrons of opposite spins in each of the states in the lower half of the energy range of the band, up to the highest occupied states at the *Fermi level* or *Fermi surface*. In a macroscopic sample (e.g. 10^{23} states) the *density of states* in the band is extremely high, and so there are empty quantum states of imperceptibly greater energies above the Fermi level. Being so near in energy, they can easily be taken up by electrons in the Fermi surface when these electrons are accelerated by a small applied voltage. It is thus easy, with such a voltage, to *unbalance* the electron distribution by transferring some electrons, in the region of the Fermi level, from $-k$ states into $+k$ states. There are then more electrons moving in the $+k$ direction than against it, so the net effect of this imbalance in the positive and negative flows is that an electronic current is flowing: the material is a metal.

How, then, could a divalent element such as calcium be a metal? Its two valency electrons per atom ought to fill the band precisely, in which case the electrons would face the large energy gap, AB in Fig. 1.5, and so be prevented from unbalancing their Fermi distribution under small applied voltages. The

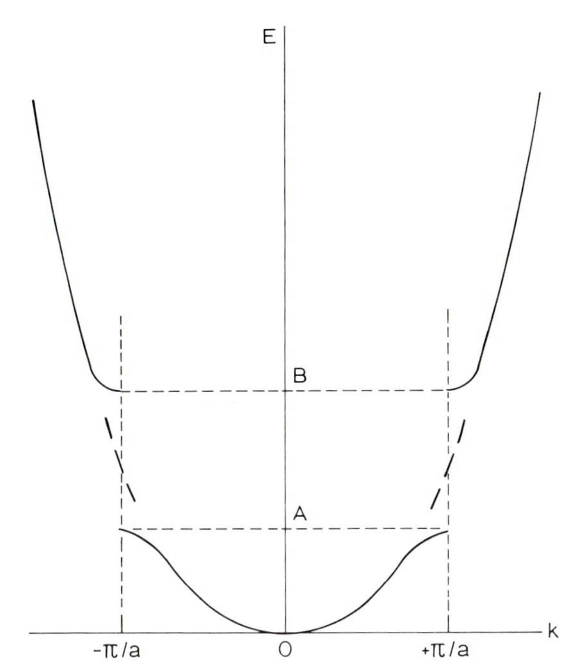

Fig. 1.5 Discontinuities produced in the *E*, *k* parabola by Bragg reflection, at $k = \pm\pi/a$, giving an energy gap AB

resolution of this problem lies in the three-dimensionality of the solid. We recall that k is really a vector, $\mathbf{k}$, having direction as well as magnitude. An allowed energy band contains $\mathbf{k}$-states for every direction of motion through the crystal, and the lattice structure fixes the critical k values – and the associated energy gaps – *differently* in the different crystallographic directions of this motion. The map of this variation in a 'k-space' diagram is the *Brillouin zone* of the structure. The variation of k with angle is familar from Bragg's law,

$$\lambda = 2a \sin\theta \tag{1.20}$$

(for a first-order reflection), where θ is the angle of incidence to the reflecting planes.

If the energy gap (AB in Fig. 1.5) is small, it is possible for those states of highest energy in one allowed band (i.e. states in the far corners of the Brillouin zone) to have energies *above* those at the bottom of the allowed band above the energy gap. The two bands, or zones, then 'overlap' in energy and must give partial filling since, by the principle of filling states of lowest energy first, it is impossible to fill the lower-energy band without beginning to fill the next one; the material is thus a metal. By contrast, if the energy gap is sufficiently large this crystallographic effect may be too small to bridge the gap and in this case, when

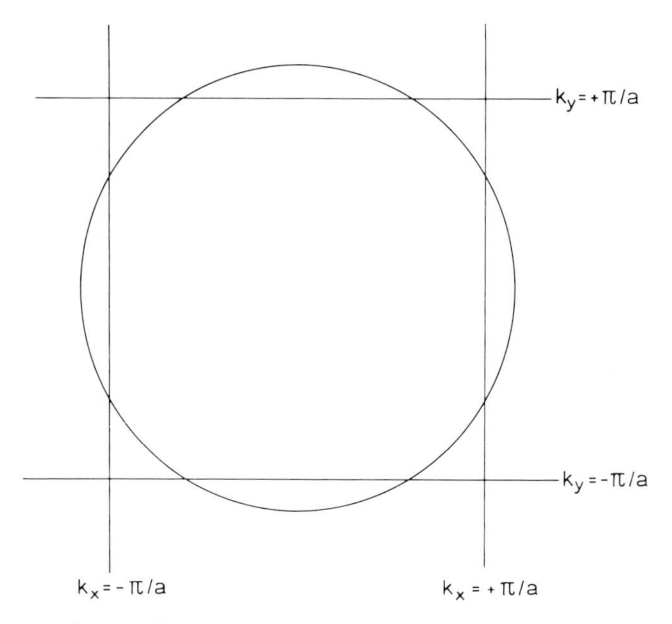

Fig. 1.6 The first Brillouin zone for a two-dimensional square lattice and an inscribed Fermi surface of free electrons in a divalent material

the number of electrons is sufficient to fill a band exactly, the material is a non-metal.

Figure 1.6 shows the first Brillouin zone, i.e. the lines corresponding to the Bragg reflections at $k_x = \pm \pi/a$ and $k_y = \pm \pi/a$, for a two-dimensional square lattice of spacing a. If the band gap is really small, the electrons are *nearly free* and the Fermi surface can be approximated by the *Fermi sphere* (or circle, in two dimensions) of free-electron theory in which, as in equation (1.13), the energy depends only on the magnitude of $\boldsymbol{k}$, not on its direction. Inside the circular Fermi surface of Fig. 1.6 the quantum states are filled by electrons; outside it they are empty. We see that the condition for a metal, i.e. partial filling of a band or zone, is necessarily satisfied in this *nearly-free electron* (NFE) limit.

The concept of Brillouin zones thus appeared in the 1930s to have brought the wave-mechanical theory to final success in its attempt to explain the existence of metals and insulators. In fact for many years this was the standard theory, presented as such in all the textbooks. However, there remained some grounds for doubt. On the conceptual side, the theory led to an absurd conclusion in a hypothetical limit. Starting with an ordinary sodium crystal, imagine that we are able somehow to expand it enormously, while preserving the crystal structure intact, until the atoms are far apart, say a = 1 km. According to the above argument the crystal should still be a metal! In fact, of course, once the lattice spacing exceeds a few angstroms the free electrons can save energy by condensing onto the atoms, one electron per atom, so forming a crystal of neutral atoms with no free electrons. On the practical side, examples became known of crystalline substances (e.g. NiO) which, according to the theory, ought to be metallic, but in fact are insulators.

The defect of the theory turned out to be that, whereas it deals fully with the distribution of electrons in $\boldsymbol{k}$-space, it virtually ignores their distribution in *real* space. Also, related to this, it is a *one-electron theory* in the sense that the particular form of the Schrödinger equation, on which it is based, is for a single electron only. The influence on any one of all the other electrons is either ignored or contained implicitly in the *static* potential, $V(x)$, of this particular Schrödinger equation. Such a simplification was hardly surprising because the rigorous Schrödinger equation for more than one electron is virtually insoluble. The practical effect of making the one-electron approximation was that *correlation effects* in the mutual positions of the electrons were ignored in the first instance. The problem of correlation has since proved to be a difficult aspect of the more advanced quantum theories, even though some important solid-state phenomena such as superconductivity owe their very existence to it.

The theories to which we shall now turn, although less fundamental than the above, give much greater prominence to correlation effects. They may be regarded as *chemical* theories of the solid state because their starting assumption is not the free electron but the *chemical bond*, localized and directed in space. This approach owes much to the pioneering work of Pauling (1938).

1.6 BONDS AND BANDS

We start with a chemist's picture of atoms bonded together into a crystalline solid. Specifically, consider a crystal of pure silicon. This, like the diamond form of carbon, germanium or grey tin, is made from quadrivalent atoms, each of which forms four *electron-pair, covalent* bonds with four neighbours. We can represent such a bond structure two-dimensionally as in Fig. 1.7(a) in which each line represents a two-electron bond between the joined atoms. The

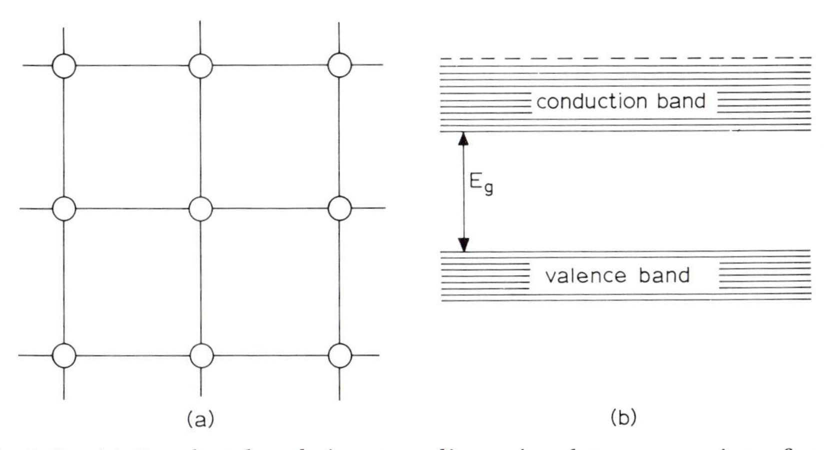

Fig. 1.7 (a) Covalent bonds in a two-dimensional representation of a silicon crystal; (b) the corresponding energy band picture

energies of these bound electrons in their molecular orbital quantum states can be indicated as in Fig. 1.7(b) by a completely filled *valence band*, of constant energy throughout the solid. These quantum states exist in a (narrow) band of energy levels because the atomic wavefunctions, from which the bond orbitals derive, partially overlap and perturb one another in the solid state. Alternatively, in a Bloch picture, we can think of the valence band as a narrow, filled, allowed energy band above which there is an energy gap E_g.

In order to understand E_g in the bond picture, we can imagine trying to dislodge one of the electrons from its bond. The energy E_g required to pull it away is generally large because of the stability of the covalent bond and because the dislodged electron cannot form a covalent bond anywhere else in the (otherwise perfect) crystal, since the bonding states are all saturated (i.e. the valence band is full). The electron, unbonded, thus wanders through the material in a band of levels which starts at a level E_g above the valence band. The electron is in fact free, and the band in which it resides is the *conduction band*. In escaping from its covalent bond it leaves behind an unfilled bond, an *electron hole*, and this hole can contribute to electrical conductivity by hopping from bond to bond. The valence band has, to a very slight degree, become only partially filled by the removal of the electron and, as a partially filled band, is also able to provide a slight electrical conductivity.

Suppose now that one of the silicon atoms is replaced by a pentavalent phosphorus atom. Four of the valency electrons of this atom enter into the covalent bond structure of the crystal, as shown in Fig. 1.8, but the fifth is unbonded and remains electrostatically attracted to the single net positive charge of this ion. The situation is quite similar to that of a hydrogen atom, in space, where again a single electron is electrostatically attracted to the single positive charge of the proton nucleus, with the potential at a distance r given by

$$V(r) = -e^2/r \tag{1.21}$$

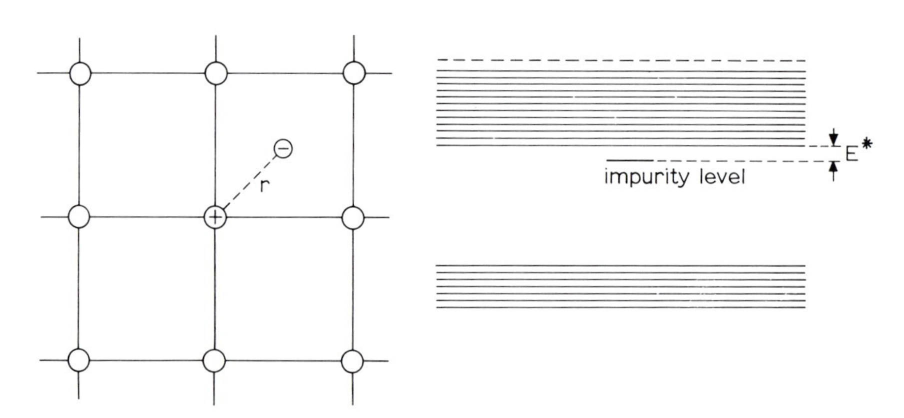

Fig. 1.8 **In a silicon crystal, the fifth electron of a phosphorus impurity atom is loosely bound, and lies at an energy just below the conduction band**

The solution of Schrödinger's equation with this potential shows that, in the lowest energy (i.e. *ground*) state, the electron is *bound* (i.e. confined) to the nucleus in a wavefunction or *orbital*

$$\psi = C \exp(-r/a_0) \tag{1.22}$$

where C is a normalization constant and a_0 is the *Bohr radius*:

$$a_0 = \hbar^2/me^2 = 0{\cdot}529 \times 10^{-10}\,\mathrm{m} \simeq 0{\cdot}53\,\text{Å} \tag{1.23}$$

The (negative) energy level of this state, relative to the (zero) energy of an electron in space, is the *Rydberg*:

$$1\,\mathrm{Ry} = me^4/2\hbar^2 = 13{\cdot}6\,\mathrm{eV} \tag{1.24}$$

where eV is the *electron-volt*. The Bohr radius and the Rydberg are often used as *atomic units* of length and energy, to simplify the equations of quantum mechanics. Thus, in atomic units the one-dimensional Schrödinger equation (equation (1.8)) becomes

$$\frac{\mathrm{d}^2\psi}{\mathrm{d}x^2} + (E - V)\psi = 0 \tag{1.25}$$

The fifth electron of the phosphorus atom does not, of course, move in an empty space about the point positive charge, but in a space already filled with the electrons of the covalent bonds and the *core* electrons of the deeper energy levels inside the atoms. The problem of electron *correlations* thus challenges the theory again. However, it is found that a good approximation can be obtained by simply regarding all the other electrons as merely providing another kind of 'space', which differs from free space by two parameters, a *dielectric constant* κ and an *effective mass* m^* of the electron. The first of these comes from the *polarizability* of the medium, a result mainly of the flexibility of the bond orbitals. The electrons in these orbitals are pulled slightly towards the positive charge of the phosphorus ion and so partially *screen* or *shield* this charge from the fifth electron. As a result the electrostatic interaction given by equation (1.21) is weakened to

$$V(r) = -e^2/\kappa r \tag{1.26}$$

The replacement of the real mass of the electron, m, by m^* in the Schrödinger equation takes quite good account of the periodic lattice field of the medium, particularly for a large orbital. This approximation is related to the Bloch theory, for in this theory the effect of the periodic field, near the bottom of the band, is simply to give the E, k parabola (*see* Fig. 1.5) there a different curvature, which is the same as if the electron were free but had a different mass. This simplification, known as *renormalization*, enables the periodic component of the total field V to be dropped from the Schrödinger equation, leaving only the simple *Coulomb* term as given by equation (1.26). We thus obtain a hydrogen-like ground state, but with the parameters

$$a^* = (m/m^*)\kappa a_0 \qquad (1.27)$$

$$E^* = (m^*/m\kappa^2)\ \mathrm{Ry} \qquad (1.28)$$

In silicon $m/m^* \simeq 4$ and $\kappa \simeq 12$, which gives $a^* \simeq 25$ Å and $E^* \simeq -0{\cdot}025$ eV. (The observed value of E^* is $-0{\cdot}045$ eV.) We note the very large orbital and weak binding; the fifth electron is very nearly a free electron. In fact at room temperature there is sufficient heat energy ($k_BT \simeq 0{\cdot}025$ eV) to excite the electron into the conduction band, so enabling ionization to occur and the electron to escape, as a free electron, into distant parts of the medium. Phosphorus doping thus converts silicon into an *n-type impurity semiconductor.*

The opposite process, brought about by the addition of trivalent atoms, e.g. Al, Ga or In, produces localized impurity levels just above the valence band. Thermal excitation of valence electrons into these states leaves behind electron holes in the valence band, the movement of which through the medium gives *p-type* impurity semiconduction.

1.7 THE MOTT TRANSITION

Suppose that the number, N per unit volume, of phosphorus atoms in silicon is gradually increased (*see* Fig. 1.2). At first, while N is very small, these atoms are so far apart as to behave as if isolated. At 0 K the fifth electron of each is then bound to its parent atom, in a localized impurity state, and the substance is an insulator (i.e. $N < 3 \times 10^{24}$ in Fig. 1.2). We arrive at Mott's conclusion that a *small* number of dissociated electrons and electron holes cannot exist at absolute zero (Mott 1949). As N grows larger, however, and the outermost regions of the localized atomic wavefunctions (equation (1.22) with a_0 replaced by a^*) begin to overlap, a major new effect appears which, as first recognized by Mott, can cause a transition to the metallic state. In addition to the screening provided by the valence bond electrons (equation (1.26)), the fifth electrons, by correlating their mutual positions, also become able to *screen* one another from the positive charges of the phosphorus ions. Moreover, their large weak binding orbitals are highly deformable and so can polarize extensively, thus giving a strong screening effect.

It then becomes possible for electrons to transfer from one such orbital to the next, as was recognized by Pauling. The electrons are beginning to become free and to provide electrical conductivity. The effect is self-reinforcing since, as electrons become free and start to move about, the material becomes too good a conductor to permit long-range electrical fields to exist inside it. In other words, the screening becomes much more effective, which enables yet more electrons to become free. The process of freeing electrons thus develops sharply and 'catastrophically', at a critical value of N, and turns the material into a metal. This is the *Mott transition* from an insulator to a metal, at 0 K.

Mott originally gave a semiclassical calculation of the transition, based on a *screened-charge* 'Thomas–Fermi' interaction potential

$$V(r) = -\frac{e^2}{\kappa r}\exp(-\gamma r) \tag{1.29}$$

where

$$\gamma^2 = \frac{4m^* e^2(3N/\pi)^{1/3}}{\kappa\hbar^2} \tag{1.30}$$

The radius of the screening sphere round a donor atom is thus about γ^{-1}. The condition for metallic conductivity is then that γ^{-1} should be smaller than the orbital radius a_H ($= a^*$ in the present case). This leads to Mott's condition

$$N_c^{1/3} a_H \simeq 0{\cdot}25 \tag{1.31}$$

for the critical value $N = N_c$ at the metal-non-metal transition. Although this treatment is now regarded as an oversimplification, and equation (1.29) has been replaced by other expressions, equation (1.31) has nevertheless proved extremely successful in locating the boundary between metals and non-metals. It has been confirmed experimentally for values of N_c ranging over ten orders of magnitude (Edwards and Sienko 1978, 1983).

1.8 THE HUBBARD TRANSITION

Mott's theory opened the way to many later treatments. An important one is due to Hubbard (1963, 1964a, b). Imagine that we have a 'crystal' of, for example, atomic hydrogen or sodium, with an enormous lattice parameter, so that all the valency electrons are bound to their parent (neutral) atoms. Remove a valency electron from one atom and attach it to another. An *ionization energy* I has to be spent to do this, but a (smaller) *affinity energy* A is regained when the electron joins the other atom. We suppose that the electron, with opposite spin, joins the valency electron already present in the outermost orbital of this other atom. There is a Coulomb repulsion e^2/r between these two electrons, at an average separation r in this orbital, i.e. $I - A = e^2/r$. It is expected that $r \simeq a_H$. The calculation in fact gives

$$I - A = \frac{5}{8}\frac{e^2}{a_H} \tag{1.32}$$

When the atoms are very far apart this is the only contribution to the energy change brought about by the transfer. It is positive, usually a few electron-volts, and creates the *Hubbard energy gap* which preserves the atomically neutral, non-conducting state of the material at large lattice spacings.

We have to remember the Bloch principles, however. Even though the Bloch energy bands are narrowed down to single atomic energy levels at these large lattice spacings, the transferred electron can equally well go to *any* of the other atoms of the crystal, and so is in a Bloch state. It can thus go into the state at the bottom of its band. Similarly, the hole it leaves behind can transfer to any atom, and so it too is in a Bloch state. If the lattice parameter is now reduced, so that these bands begin to broaden, the electron and hole can both go into those states in their bands which give the crystal lowest energy. The gain in energy resulting from this effect offsets the increase indicated by equation (1.32), and the Hubbard theory shows that, at a certain lattice spacing, it offsets it completely. Ionization can then take place freely, at no energy cost, and the material turns into a metal. The detailed theory of this *Hubbard transition* leads to the criterion

$$N_c^{1/3} a_H \simeq 0{\cdot}2 \tag{1.33}$$

which is remarkably close to the value given by equation (1.31), considering that the physical argument is quite different.

1.9 THE TRANSITION AS A POLARIZATION CATASTROPHE

The development, since Mott's 1949 paper, of theories of the metal–non-metal transition in terms of electron–electron correlations has led to the realization (Berggren 1974, Edwards and Sienko 1978, 1983) that a much older and half-forgotten pre-quantum theory (Goldhammer 1911, Herzfeld 1927) was closely related to them. The basic idea of the Herzfeld–Goldhammer theory was that the transition from insulator to metal is brought about by a *polarization* or *dielectric catastrophe*. We return to the hypothetical crystal of neutral one-electron atoms with an atomic spacing so large that each valency electron remains bound to its atom, so that the material is an insulator. However, we now think – quite classically – of this binding as an elastic 'spring' between the valency electron and its atom. The strength of the binding is indicated by the stiffness of the spring and hence, in classical theory, by the high frequency ν of oscillations of the electron on its spring. The insulating state is represented by stiff springs which allow the electrons to move only slightly away from their equilibrium positions in the neutral atoms; that is, their *polarizability* is small. A small electric charge introduced into the medium will not produce much displacement of these electrons, and so its field will be only weakly screened. The dielectric constant κ is small, not much larger than unity.

We now envisage a progressive weakening of the spring constant. The electrons can make larger displacements, the natural frequency of electronic oscillations falls, the material becomes more polarizable and its dielectric constant increases. Eventually, when the spring constant falls to zero there is no restoring force pulling an electron back to its parent atom. The vibrational frequency ν has dropped to zero: the electron no longer oscillates but moves always forward, leaving its parent atom behind and passing onward through the material. In the classical theory of dielectrics this is the *polarization catastrophe*; in our language it is the transition to the metallic state.

The classical relation for the dielectric behaviour of an insulator of the type we are considering (i.e. non-polar) is the *Clausius–Mossotti relation*

$$\frac{\kappa - 1}{\kappa + 2} = \frac{4\pi}{3} N\alpha_0 \tag{1.34}$$

which relates the dielectric constant κ to the *static polarizability* α_0 of an isolated atom, where N is the number of atoms per unit volume in the medium. This polarizability, which is a measure of the elastic 'softness' of the spring, is the induced *dipole moment* (i.e. the charge times the distance by which it is displaced) of the polarized atom, divided by the applied electric field that induces it. In the crystal (or liquid or dense gas) the polarizing field is not only an externally applied one but also the combined effect of the electric fields due to the induced dipoles of all the other atoms. This brings in a *cooperative polarization* which leads to the catastrophe when the atoms are closely crowded together (i.e. when N becomes large in equation (1.34)). We see that when N reaches a critical density

$$N_c = \frac{3}{4\pi\alpha_0} \tag{1.35}$$

the catastrophe is brought about by the requirement that $\kappa - 1 = \kappa + 2$ in equation (1.34), i.e. $\kappa \rightarrow \infty$.

Following custom we write $N = N_A/V$ in equation (1.34), where N_A is Avogadro's number and V is the molar volume. The relation then becomes

$$\frac{\kappa - 1}{\kappa + 2} = \frac{R}{V} \tag{1.36}$$

where

$$R = \frac{4\pi}{3} N_A \alpha_0 \tag{1.37}$$

is the experimentally measured *molar refractivity* (at low density).

As Edwards and Sienko (1978, 1983) have emphasized, the criterion for the metal–non-metal transition can then be expressed remarkably simply in terms of two measurable properties of free atoms, their *molar refractivity* and *molar volume*. For non-metals $R < V$, and for metals $R > V$. A remarkably large number of elements, compounds and conditions can be successfully classified on this basis (Edwards and Sienko, 1978, 1983), including the compressions required to bring hydrogen, solid rare-gas elements and alkali halides into the metallic state; the expansions necessary to make the alkali metals non-metallic; and the critical concentrations for the polarization catastrophe and onset of the metallic state in heavily doped semiconductors.

It is known in physical chemistry that the molar refractivity R of a pure species in the form of a pure, dilute gas depends upon the actual volume of its atoms or molecules. The factor R/V thus measures the ratio of the volume of the atom in a dilute gas and in the *condensed* state (i.e. crystal, liquid or highly compressed gas). The *size* of the atom, when isolated or surrounded by close neighbours, is thus the key factor which determines whether the state is metallic, on this interpretation of the Herzfeld–Goldhammer theory.

Edwards and Sienko have also pointed out that the simple criterion $R = V$ can be closely linked to the Mott criterion. The polarizability of hydrogen-like states is known to be of the form

$$\alpha_0 = \frac{9}{2} a_H^3 \tag{1.38}$$

and from this the condition for the catastrophe (equation (1.34)) then becomes

$$N_c^{1/3} a_H = 0{\cdot}376 \tag{1.39}$$

which is not greatly different from the value given by equation (1.31).

1.10 DISORDERED MATERIALS

The assumption of a crystalline structure in the above discussion was made merely for convenience. Once we no longer depend on Bragg reflections and Brillouin zone boundaries for differentiating between metals and non-metals, much of the distinction between crystalline and non-crystalline arrangements of the atoms becomes irrelevant to the theory of the transition. Indeed, the example of this transition represented in Fig. 1.2 is for a system in which the donor atoms (phosphorus) are arranged randomly throughout the host lattice (silicon).

There is one major effect, however, which occurs only in disordered structures – *Anderson localization* (Anderson 1958). Consider a model system in which the disorder is not in atomic positions, but in the *spread W* of the (random) depths of the potential wells at the sites of a lattice (*see* Fig. 1.9). This could represent, for example, a multi-component, random, substitutional solid solution. Suppose that the lattice spacing is large enough for the atomic valency wavefunctions to overlap only slightly, so that the lattice wavefunctions can be regarded approximately as atomic wavefunctions added together. This is the approximation of the *linear combination of atomic orbitals* (LCAO) or the *tight-binding method* (*see* Appendices 2 and 4). If the localized atomic wavefunctions are to combine well, so as to form Bloch states extending throughout the structure, thus giving the possibility of free electrons and metallic properties, the functions must not only overlap significantly. They must also be compatible, in two ways. First, they must have compatible symmetries (*see* Appendix 2); we shall suppose that this is so. Second, for the reason given in Appendix 5, their energy levels

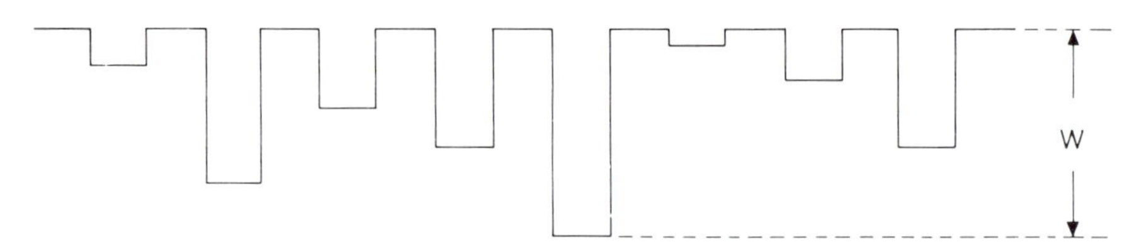

Fig. 1.9 A lattice of potential energy wells of random depths, with spread *W*

must be fairly similar. In a perfect crystal, where $W = 0$, this third condition is automatically satisfied.

Suppose, however, that W is large. The localized atomic wavefunction at one of the deepest wells cannot then combine satisfactorily with the other states, because in general the nearby wells are much shallower, whereas states in wells of similar depths are too far away to overlap. Localized, non-conducting states can thus exist when W is large.

Anderson's theory proves this conclusion. It is a kind of generalization of the theory of bound impurity states in a semiconductor. Thus, as long as the average spacing of the phosphorus atoms in silicon is sufficiently large, all the fifth electrons are in localized states, bound to their present phosphorus atoms, and there is no conductivity. In the present case a sufficiently large spread W ensures that there are some widely spaced, very deep levels which can bind electrons. In fact, these *Anderson states* exist, like the impurity states in semiconductors (Fig. 1.8), at the fringes or *tails* of the conduction and valence bands. Even though these two bands may, in a disordered structure, 'overlap' in energy in the sense that there is a set of Anderson states in the gap between them, there is nevertheless no conductivity, at 0 K, when the energy level of the Fermi surface falls in this gap, because all these Anderson states are localized.

Anderson's conclusion is based on one-electron theory and so does not depend on electron correlation effects, unlike the above theories of the metal–non-metal transition. It is closely related to the theory of the *electrical resistance* of metals, resistance which is brought about by the scattering of conduction electrons at irregularities in a crystal lattice. In the standard theory of electrical resistivity the scattering is expressed in terms of a *mean free path* for a conduction electron, between successive collisions with irregularities. It is known, from both observation and theory, that in normal metals the mean free path is long, of the order of 100 lattice spacings even at room temperature. Under such circumstances a conduction electron maintains its state of uniform motion, undisturbed, over sufficiently long sections of path to justify this motion being defined in terms of the wave-vector $\mathbf{k}$, which is in this sense a 'good' quantum number. But there are materials with much shorter mean free paths (e.g. less than 2 lattice spacings, in liquid barium), and for these such a definition of motion is inappropriate. In highly disordered materials, where the potential varies strongly from one site to the next and the mean free path is of the order of the atomic spacing, the traditional free-electron type of theory is inapplicable and an analysis of the Anderson type is necessary. The general condition for Anderson localization to occur is that

$$W/B > C \tag{1.40}$$

where B is the width of the allowed energy band when $W = 0$, and $C \simeq 2{\cdot}2$. In a normal liquid metal the Fermi level does not fall in the tail of a band, so the states at the Fermi surface are conducting states, similar to those of free electrons. But in an Anderson insulator the Fermi level falls in the band gap region, where the states are localized. The only conduction possible in these circumstances is by a process of 'hopping' in which an electron occasionally jumps, with the aid of thermal fluctuations, from one localized site to another. A full review of such processes is given by Mott and Davis (1979).

REFERENCES

Anderson, P. W., *Phys. Rev.*, **109,** 1492 (1958).
Berggren, K.-F., *J. Chem. Phys.*, **60,** 3399 (1974).
Edwards, P. P., and Sienko, M. J., *Phys. Rev. B,* **17,** 2575 (1978); *Int. Rev. Phys. Chem.*, **3,** 83 (1983).
Goldhammer, D. A., *Dispersion und Absorption des Lichtes*, Teubner, Leipzig (1911).
Herzfeld, K. F., *Phys. Rev.*, **29,** 701 (1927).
Hubbard, J., *Proc. R. Soc. A,* **276,** 238 (1963); *Ibid.*, **277,** 237 (1964a); *Ibid.*, **281,** 401 (1964b).
Mott, N. F., *Proc. Phys. Soc. A,* **62,** 416 (1949).
Mott, N. F., and Davis, E. A., *Electronic Processes in Non-Crystalline Materials*, 2 edn, Clarendon Press, Oxford (1979).
Pauling, L., *Phys. Rev.*, **54,** 899 (1938).
Sasaki, W., *Phil. Mag. B,* **42,** 725 (1980).

2

Why are metallic electrons so free?

2.1 THE PROBLEM

As it moves through a metal a conduction electron is continuously exposed to intense and rapidly varying electrostatic forces. Its electrostatic potential plunges to $-\infty$ at each atomic nucleus, and climbs up again, in a complicated way, in the field of the core electrons which surround each nucleus. Finally, there are the fluctuating forces resulting from its interactions with all the other conduction electrons moving through the metal. The simple theory of metals has swept all this complication aside by its two main assumptions: that of *freedom*, in which the intense field of the lattice ions is either assumed to be completely smoothed away, as in the free-electron theory, or represented by the gently undulating periodic field of the nearly-free electron (NFE) theory; and that of *independence*, in which each conduction electron is assumed to be governed by a one-electron Schrödinger equation, independent of its interactions with the other conduction electrons.

Despite these drastically simplifying assumptions, the theory of metals has been extremely successful. Moreover, the experimental methods developed in the 1960s by Pippard and others for measuring the Fermi surfaces in metals have shown that in some metals, for example aluminium and lead, the actual surfaces are remarkably similar to the simple Fermi spheres of the free-electron theory. The assumptions evidently have a much better scientific basis than mere simplifying approximations - they seem to represent some real physical characteristics of the electronic structure of metals. The challenge to the theory has thus been to explain these characteristics and so to put the theory of metals on a more sound - even if essentially unchanged - basis. This challenge has been successfully met, as we shall show in this chapter and the next.

2.2 THE WIGNER–SEITZ METHOD

The first step towards understanding the behaviour of the conduction electrons in the field of the ionic cores was taken by Wigner and Seitz (1933, 1934) when they introduced the *cellular method* for calculating exact wavefunctions in metals. The basic idea is the Wigner–Seitz cell (*see* Appendix 4). For this, the metal crystal is imagined to be divided up into equal polyhedra, constructed by

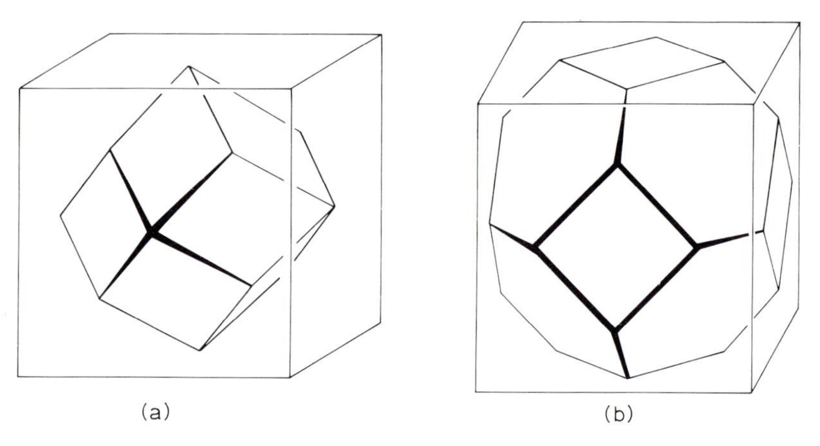

Fig. 2.1 The Wigner–Seitz cells in (a) f.c.c. and (b) b.c.c. crystals

drawing lines between the centres of neighbouring atoms (and next-nearest neighbours in crystals such as BCC) and then bisecting each such line with an orthogonal plane. Figure 2.1 shows the Wigner–Seitz (WS) cells of FCC and BCC structures.

The intention was to solve Schrödinger's equation for a single atom, but with the boundary conditions of the free atom replaced by those on the surface of the WS cell, surrounding this atom, which are appropriate to the structure of the crystal. The sharp edges of the WS cell are mathematically awkward, however, and so Wigner and Seitz, recognizing that in close-packed structures such as FCC and BCC the cells are almost spherical, made the approximation of replacing the polyhedron by an equivalent Wigner–Seitz sphere of the same volume $\boldsymbol{\Omega}_a$, with a radius r_s:

$$\frac{4\pi}{3} r_s^3 = \boldsymbol{\Omega}_a \tag{2.1}$$

The boundary condition on the *ground state* Bloch function ψ_0 (i.e. for which $k = 0$ and the energy is lowest) is then easily expressed. This ground state function is exactly repeated, without change of phase, from cell to cell, i.e. $\psi_0(\mathbf{r} - \boldsymbol{l}) = \psi_0(\mathbf{r})$, and so must be both continuous and of zero slope

$$\partial\psi_0/\partial r = 0 \tag{2.2}$$

at the boundary of the sphere, $r = r_s$. The whole problem then has spherical symmetry and so requires only the radial part of the Schrödinger equation

$$\frac{1}{r^2}\frac{\mathrm{d}}{\mathrm{d}r}\left(r^2\frac{\mathrm{d}\psi}{\mathrm{d}r}\right) + \frac{2m}{\hbar^2}(E - V)\psi = 0 \tag{2.3}$$

for an s-like state. The potential $V(r)$ of the atomic core is chosen from a Hartree–

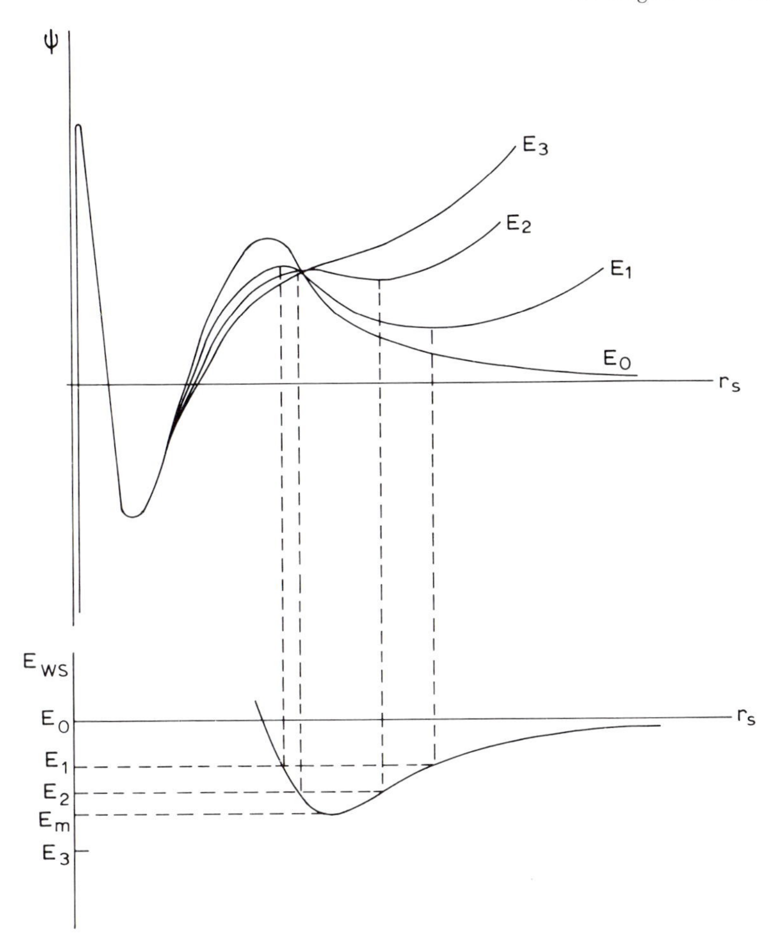

Fig. 2.2 Wavefunctions at various energy levels in the Wigner–Seitz method

Fock calculation (*see* Appendix 2), or from a later development known as the *quantum defect* method in which $V(r)$ is constructed semi-empirically by fitting atomic core functions to the observed energy levels of the free atom. Given $V(r)$, the radial equation with the WS boundary condition can then be solved numerically.

Figure 2.2 shows schematically some solutions for various values of the electron energy $E = E_{WS}$. The part of ψ deep inside the atom is hardly affected by the relatively small changes in E_{WS}. The free atom has a wave with energy E_0, which satisfies equation (2.2) at only one point outside the core. At a slightly lower energy, E_1 or E_2, there are two such points where the wave is horizontal, and from the set of all such points the energy curve is constructed. As the energy is reduced, from E_0 to E_1 to E_2 and so on, the two points converge and eventually

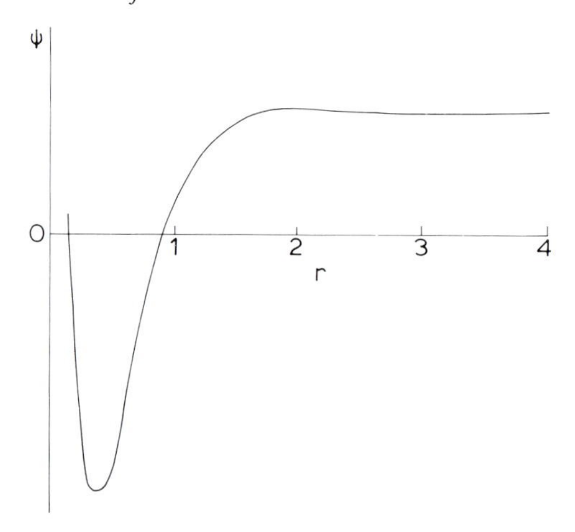

Fig. 2.3 The Wigner–Seitz ground state wave function for sodium (r in atomic units, i.e. 0·53 Å)

unite at the minimum E_m of the energy curve. Wavefunctions of lower energy, e.g. E_3, do not satisfy equation (2.2) at any point outside the core.

This E_{WS} curve gives the energy of the ground Bloch state as a function of the WS atomic radius $r = r_s$. Moving inwards from large r_s, the curve falls at first as the contraction of the WS sphere brings the electron more into the low-V inner regions of the atom. Because this is an electrostatic attraction, the (negative) energy changes roughly as r^{-1} in this region. At closer spacings the energy rises, fairly steeply, since the kinetic energy increases as the electron is squeezed into a WS sphere of ever smaller volume, as follows directly from the Heisenberg uncertainty principle (*see* Appendix 1). When r_s is extremely small, equation (2.2) can be satisfied only by bending the *inner*, *atomic*, part of the wavefunction in order to make the slope horizontal there; the resistance to this bending gives the steep increase in E_{WS} at these small r_s values.

Figure 2.3 shows the $k = 0$ wavefunction of sodium obtained by the Wigner–Seitz method. Its flatness at the larger values of r is striking. Because $\Omega_a \propto r^3$, ψ is almost constant throughout 80% of the total volume of the metal. In this constancy it is very like the ground state wavefunction ($k = 0$) of free-electron theory. The Wigner–Seitz method thus provided the first, albeit limited, theoretical justification of this theory.

2.3 MORE ADVANCED THEORIES

There have since been many attempts to take account of the atomic potentials in the wavefunctions of metallic electrons, leading up to the major development known as the *pseudopotential theory*. The Wigner–Seitz solution (Fig. 2.3) opened the way to some of these, based on the *muffin-tin potential*, (Fig. 2.4) in which the periodic field of the lattice ions is pictured as a mainly flat, i.e. $V = constant = 0$, 'free-electron' field into which is inset a lattice of small spheres, inside each of

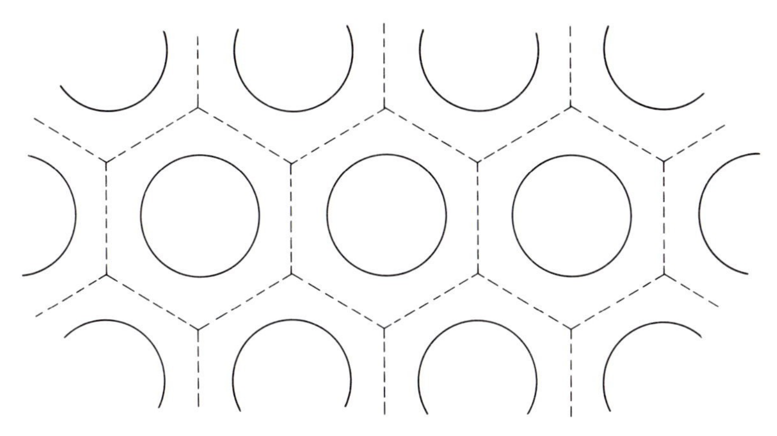

Fig. 2.4 The muffin-tin potential, in which V is atomic-like inside the spheres and constant outside them

which V is an atomic potential such as the Hartree–Fock potential.

Since, from Fig. 2.3, the wavefunction in the flat parts between the 'muffins' can be well represented by a free-electron plane wave, $\exp(i\boldsymbol{k} \cdot \boldsymbol{r})$, i.e. the three-dimensional generalization of equation (1.9), it should be possible to construct a complete solution, also valid *inside* the muffins, by making a sum of several such waves, using various k values. This has proved impracticable, however, because the rapid variation of V in the central regions of the atoms requires ψ to bend sharply to and fro there, which can be reproduced only by using a large number of short-wave components in the sum.

In his *augmented plane wave* (APW) method Slater circumvented this difficulty by making a sum of waves which are plane between the muffins and atomic-like inside them. This required finding a solution of Schrödinger's equation, inside a muffin, which joined smoothly onto the plane wave outside. Several such APWs were then summed to make a Bloch wave with arbitrary coefficients, the values of which were then found from the requirement that the energy be a minimum, using the method of variations (*see* Appendix 5). Although requiring heavy computing, the method has been used to calculate the band structures of several metals.

Another way of managing the muffin-tin potential is the *KKR method*, proposed by Korringa, Kohn and Rostoker, which uses *Green's functions* in place of the sum. Green's functions stem from Huyghen's idea, in the earliest days of classical wave theory, that a complete wave may be regarded as the sum of innumerable wavelets which spread out from every point of the medium irradiated by the wave. The KKR method is developed *self-consistently*: the complete wave $\psi(0)$ at the point $\boldsymbol{r} = 0$ is regarded as the sum of all wavelets arriving at this point from every other point $\boldsymbol{r}$ of the medium; and the amplitude of each wavelet is made proportional to that of the complete wave $\psi(\boldsymbol{r})$ at the point $\boldsymbol{r}$ and to the strength of the potential $V(\boldsymbol{r})$ which scatters the wave at $\boldsymbol{r}$ into its wavelets.

The quantum-mechanical version of this is expressed as

$$\psi(0) = \int G(\mathbf{r}) V(\mathbf{r}) \psi(\mathbf{r}) \, \mathrm{d}\mathbf{r} \tag{2.4}$$

where the Green's function is

$$G(\mathbf{r}) = -\exp(\mathrm{i}\kappa r)/4\pi r \tag{2.5}$$

with $r = |\mathbf{r}|$ and

$$\kappa^2 = \begin{cases} 2mE/\hbar^2 & (E > 0) \\ -2mE/\hbar^2 & (E < 0) \cdot \end{cases} \tag{2.6}$$

where E is the energy. The Green's function thus measures the diminution in amplitude of the scattered wavelets with increasing distance r from a source of unit strength. In the KKR method the emission of a spherical scattered wave from a single atom of the crystal is first considered. The complete wave impinging upon this atom, from which that scattered wave is created, is then constructed self-consistently from the set of all such spherical scattered waves arriving from the other atoms of the crystal. A form of equation (2.4) is thus constructed in which the Green's function is summed over all atoms. From it, trial wavefunctions are constructed and the method of variations is used to optimize the final form of the wavefunction. Although it requires heavy computation, the KKR method has proved to be slightly less laborious than the APW method, and it gives similar results.

2.4 ORTHOGONALIZED PLANE WAVES

An important method which does not require the muffin-tin approximation was introduced by Herring (1940), and led eventually to the pseudopotential theory. As we have seen from the Wigner–Seitz solution, the actual wavefunction of a metallic electron is rather like that of a free electron in the regions between the lattice ions, but oscillates in an atomic-like manner inside each ion. These oscillations represent to a large extent its response to the presence of the inner electrons in the core orbitals of the atom. Herring's idea was to simplify the problem of constructing such a wavefunction by using the wavefunctions of these core electrons – already known from atomic theory – in a most direct way to introduce atomic-like oscillations into the otherwise plane-wave free-electron solution. In the limit of an infinitely expanded lattice in which each atom becomes isolated, these oscillations have to reduce to the appropriate atomic orbital. In sodium metal, for example, which has one metallic electron per atom outside the core of 1s, 2s and 2p electrons, the Bloch state of the metallic electrons reduces in the expanded limit to the 3s orbital of the isolated sodium atoms.

Herring's key point was that these atomic limit states of the metallic Bloch function are the least energetic atomic states which are *orthogonal* to the quantum states of the core electrons. We recall (*see* Appendix 2) that the structure of a free atom is built up from atomic orbital states, each of which is characterized by its pattern of *nodal* surfaces across which ψ changes sign; and that each nodal pattern, designated 1s, 2s, 2p, 3s, 3p, 3d and so on, can hold a limited number of electrons as determined by the Pauli principle. Mathematically, the atomic wavefunctions belonging to different patterns are *orthogonal* to one another. Two wavefunctions ψ_A and ψ_B are said to be orthogonal when

$$\int \psi_A^* \psi_B \, dv = 0 \tag{2.7}$$

where the integral is taken over the whole volume (*see* Appendix 1). For example, the (spherical) 2s state is orthogonal to the (spherical) 1s state because it has one spherical node which divides it into two parts – an outer hollow sphere and an inner sphere – which have opposite signs, whereas 1s has no node and so has the same sign everywhere. Each elementary contribution to $\psi_{2s}^* \psi_{1s}$ from inside the node is matched by one of opposite sign outside, so that the whole integral is zero. As a second example, $2p_x$ is orthogonal to 1s because it has a nodal plane, through the centre of the atom perpendicular to the x axis, which divides it into two equal lobes – one positive, the other negative. Again, each positive $\psi_{2p_x}^* \psi_{1s}$ on one side of this nodal plane is matched by a corresponding negative one on the other side.

The first step in Herring's *orthogonalized plane wave* (OPW) method is to link together the quantum states of the core electrons belonging to the various atoms into the form of a Bloch wave, with a common k value, running through the whole crystal (equation (1.14)). Consider for simplicity a one-dimensional chain of lithium atoms, and let $\phi(x)$ be the wavefunction of the 1s core state of an atom at the origin, $x = 0$. Then the core state of the atom at l is $\phi(x - l)$. The Bloch function ϕ_c is a sum of all these atomic states centred at the various atomic sites, l, with the k wave factor attached:

$$\phi_c(x) = \sum_l \exp(ikl)\phi(x - l) \tag{2.8}$$

This is an extreme example of a *tight-binding Bloch function* (*see* Section 1.10 and Appendix 4). It simply represents a line of separate $\phi(x)$ atomic orbitals, with each of which is associated the appropriate phase orientation according to the value of its $\exp(ikl)$ factor, as shown in Fig. 2.5.

Next, the free-electron plane-wave function is introduced, with the *same* k as the core function wave:

$$\phi_0 = \exp(ikx) \tag{2.9}$$

and finally the OPW ψ_k for this k is constructed by subtracting the core wave from the free-electron wave, as shown in Fig. 2.5, and given by

$$\psi_k = \phi_0 - a_c \phi_c \tag{2.10}$$

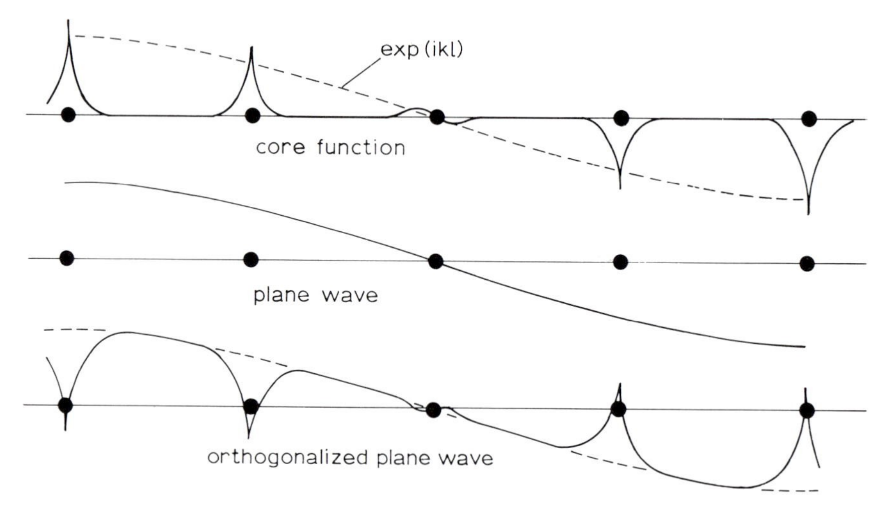

Fig. 2.5 Construction of an orthogonalized plane wave by subtracting a core function wave from a plane wave

In this the coefficient

$$a_c = \int \psi_c^* \phi_0 \, dv \tag{2.11}$$

gives the orthogonalization, as can be seen by multiplying equation (2.10) through by ϕ_c^* and integrating to obtain

$$\int \phi_c^* \psi_k \, dv = \int \phi_c^* \phi_0 \, dv - \left(\int \phi_c^* \phi_0 \, dv\right)\left(\int \phi_c^* \phi_c \, dv\right)$$
$$= 0 \tag{2.12}$$

taking the core functions as normalized so that $\int \phi_c^* \phi_c \, dv$ is unity. Thus the OPW ψ_k is orthogonal to the core functions. This orthogonality can be seen in Fig. 2.5 in the pairs of points, either side of each atomic site, where the OPW function crosses the origin. These represent a one-dimensional section of the spherical node of the OPW round each atomic site. In the regions between the cores, where $\phi_c \rightarrow 0$, we have $\psi_k \simeq \phi_0$, approximating as required to a free-electron wavefunction.

This ψ_k is for a *single k* and a *single* core state ϕ_c. In general, lattice ions have several core states. For example, the potassium core is of the form $(1s)^2$ $(2s)^2$ $(2p)^6$, i.e. there are two electrons in 1s, two in 2s and six in 2p. In such cases the OPW is constructed to be orthogonal to *all* the core states. This generalization of equation (2.10) is simply

$$\psi_k = \phi_0 - \sum_c a_c \phi_c \tag{2.13}$$

where the sum is over all the various core states of the atom.

For many values of k, a single OPW is usually sufficient to represent the exact wavefunction of the metallic electron fairly accurately. However, in sensitive regions of k-space, particularly near Brillouin zone boundaries, a more accurate expression is generally needed. This is obtained by forming a sum of OPWs, by adding to k in each term a $g = \pm 2\pi n/a$ (i.e. a *reciprocal lattice vector* **g** in three dimensions, *see* Appendix 4) where n is an integer, which simply transfers k to the equivalent point on the opposite side of the Brillouin zone. The total wavefunction is then taken to be

$$\sum_g b_{k-g} \psi_{k-g} \tag{2.14}$$

and the coefficients b_{k-g} are determined by the method of variations to give this total function the least possible energy for the k in question (*see* Appendix 5).

2.5 THE PSEUDOPOTENTIAL

Let V and ψ in equation (1.8) be the *exact* potential and wavefunction of a metallic electron. Both then have difficult features which make them very different from their free-electron counterparts. The potential V varies in a complicated way in every ionic core and plunges to $-\infty$ at its centre; and the wavefunction, as well as showing the exp (ikx) form of the free-electron theory, oscillates strongly in every core. Now, replace this exact ψ in the Schrödinger equation by an OPW ψ_k, which is a good approximation to it. Then, from equation (2.13) and using the Hamiltonian H from equation (1.7), we obtain

$$\mathrm{H}\phi_0 - \sum_c a_c \mathrm{H}\phi_c = E\phi_0 - E\sum_c a_c\phi_c \tag{2.15}$$

Since the Schrödinger equation for the core electrons is of the form $\mathrm{H}\phi_c = E_c\phi_c$, where E_c is the energy of an electron in core state ϕ_c, then equation (2.15) can be rewritten as

$$(\mathrm{H} + V_\mathrm{R})\phi_0 = E\phi_0 \tag{2.16}$$

where

$$V_\mathrm{R}\phi_0 = \sum_c (E - E_c) a_c \phi_c \tag{2.17}$$

Thus, in place of the original potential V in equation (1.8), we now have an *effective potential*

$$V_{ps} = V + V_\mathrm{R} \tag{2.18}$$

This is the *pseudopotential*. The strongly-varying and negative V is largely cancelled by the positive V_R. When the core functions are well chosen, this cancellation is almost exact. Then V_{ps} is weak, and gives an equation very similar to the Schrödinger equation of NFE theory. The transformation from equations (1.8) and (2.15) to (2.16), which was introduced by Phillips and Kleinman (1959), is a most useful device because it gets rid of the strongly varying part of the potential by the very same process of getting rid of the rapidly varying (atomic core) part of the metallic electron wavefunction. Therefore the state of these electrons can justifiably be represented by an NFE type of theory which uses the pseudopotential V_{ps} and the corresponding *pseudo-wavefunction*, very similar to ϕ_0, in place of V and ψ.

This answers the question asked at the beginning of this chapter. In physical terms, the reduction of the strongly varying lattice field to the gently undulating periodic field of the NFE model is justified because there are *two* interactions between a metallic electron and the ionic cores: the strong electrostatic interaction, represented by V; and the effect of the Pauli principle, which forbids the metallic electrons from occupying the already full core states and so forces them into the equivalent of higher atomic states. In effect, the core electrons repel the metallic electrons through the Pauli effect and this repulsion is recognized by the V_R term which opposes and largely cancels the V attraction.

True to its name, the pseudopotential is not a proper potential. The role of a true potential in the Schrödinger equation is a strictly *local* one, i.e. its effect, as an operator acting at any point on a wave function, is simply to multiply this function by the local value of the potential at the point in question. In contrast, V_{ps} is *non-local*. The pseudopotential also contains the energy (equation (2.17)), so V_R in equation (2.16) depends on the solution E of this equation and thus has to be found by a self-consistent procedure. It is also not uniquely defined since other core wavefunctions, with the same k, can be added without changing the NFE solution. Nevertheless, its introduction has been one of the greatest developments in the electron theory of metals, mainly because it has validated the NFE theory and thus given fresh impetus to the further development of this theory. Such developments often make use of approximations based directly upon the concept of the pseudopotential.

2.6 SCATTERING THEORY

Another approach to the whole problem is to examine the ability of the crystal ions to *scatter* a metallic electron passing through them. Because of the steep fall of V at the centre of each ion, a strong scattering might be expected, but the electrical conductivities of good metals show that such electrons enjoy long *mean free paths* between collisions with lattice ions, about 100 atomic spacings at room temperature and much longer at low temperatures. The freedom of such electrons is thus also a comparative freedom from scattering, which again suggests that the effective V of the lattice field is much weaker than the actual V. The analysis of metallic electrons in terms of their scattering behaviour is closely related to the KKR and APW theories.

Why is a lattice ion so poor at scattering a Bloch wave in a metal? As a preliminary, we can note two contributory factors. First, because a metal is a good conductor it does not allow long-range electrostatic fields to exist inside it; its mobile electronic charges adjust their mutual positions so as to screen or shield out any large electropositive centres. This electron correlation effect, which we shall discuss further in Chapter 3, is an additional reason why NFE behaviour is to be expected and goes beyond the arguments of the simple pseudopotential theory.

The second factor is the *Ramsauer effect* of atomic physics: electrons of certain energies pass through inert gases such as neon and argon with remarkably little scattering. As we shall see, the reason is that at one of these critical energies the distortion of the wavefunction of such an electron caused by the scattering centre, is just sufficient to 'pull' another node into the scattering centre; the wavefunction *outside* this node then appears as an undistorted function, almost as if there were no scattering centre.

The scattering of metallic electrons by the (spherically symmetric) $V(r)$ of a lattice ion is expressed in terms of the *phase shifts* of *partial waves* (for the case where $V(r)$ drops off faster than r^{-1} as the distance r from the scattering centre increases). The theory is developed as an *asymptotic behaviour* at large r, for which the wavefunction of the electron simplifies to the form

$$\psi \approx \exp(\mathrm{i}kx) + \frac{f(\theta)}{r}\exp(\mathrm{i}kr) \qquad (2.19)$$

The first term represents the incoming electron as a plane wave, with wavenumber k, which is approaching the scattering centre at the origin along the x axis. The second term is the outgoing wave of the electron after being *elastically* scattered (i.e. retaining the same k value) by $V(r)$ and it represents the electron leaving the ion along some radial direction r at an angle θ to the x axis. This term, being proportional to r^{-1}, conserves the number of such electrons since, when squared, it gives an outgoing electron density which satisfies the inverse square law. When the total wavefunction is squared, i.e. $\psi^*\psi$ is formed, there is a *cross-product* of the incoming and outgoing waves. This implies an *interference* between them, which is an important feature of the analysis.

The amplitude of the outgoing wave depends on the angle of scattering θ, through $f(\theta)$, which has to be found by the method of partial waves. For this, we rewrite ψ as a sum of partial waves:

$$\psi(r,\theta) = \sum_{l=0}^{\infty} R_l(r)\,\Theta_l(\theta) \qquad (2.20)$$

which are products of purely radial functions $R(r)$, and purely angular functions $\Theta(\theta)$. The latter, which we do not need to consider further, are a form of standard harmonic functions (Legendre polynomials) appropriate when there is circular symmetry around the x axis. This representation of ψ is very similar to that used in the theory of the hydrogen atom (*see* Appendix 1), and l denotes the symmetry of the partial wave just as the angular momentum quantum number l in the hydrogen theory indicates the symmetry of the atomic orbitals. The total scattering is thus built up from s-wave scattering ($l = 0$), p-wave scattering ($l = 1$) and so on. We now put equation (2.20) into Schrödinger's equation and extract the radial term,

$$\frac{\mathrm{d}^2(rR_l)}{\mathrm{d}r^2} + \left[\frac{2mE}{\hbar^2} - \frac{2mV(r)}{\hbar^2} - \frac{l(l+1)}{r^2}\right] rR_l = 0 \qquad (2.21)$$

which at large distances, $r \rightarrow \infty$, simplifies to

$$\frac{\mathrm{d}^2(rR_l)}{\mathrm{d}r^2} + \frac{2mE}{\hbar^2} rR_l = 0 \qquad (2.22)$$

i.e. a simple harmonic equation giving a sine-wave dependence of rR_l on r.

Finally, the two forms of ψ, equations (2.19) and (2.20), have to be coalesced. When $V = 0$ there is no scattered wave and the procedure in this case is to express the incoming plane wave, $\exp(\mathrm{i}kx)$, as a sum of spherical waves, centred on $r = 0$, with a sine-wave dependence on r. This is a standard problem in the theory of Bessel functions and it leads in the asymptotic limit ($r \rightarrow \infty$) to

$$R_l \simeq \frac{1}{r}\sin\left(kr - \tfrac{1}{2}l\pi\right) \qquad (2.23)$$

which is also a solution of equation (2.22), as required. Since equation (2.22) still applies when $V \neq 0$ the only way that $V(r)$ can change equation (2.23) is to introduce a *phase angle* η_l into the argument of the sine function. The general solution for R_l at large r is thus

$$R_l \simeq \frac{1}{r}\sin\left(kr - \tfrac{1}{2}l\pi + \eta_l\right) \qquad (2.24)$$

The *phase shift* η_l is zero when $V = 0$ and increases as V deepens.

Figure 2.6 shows schematically the effect of an attractive ($V < 0$) potential on the radial component R_l of the partial wave with k and l. The wave is 'pulled in' towards $r = 0$ by an amount η_l/k, which increases with increasing V, but when η_l reaches the value π a complete half-cycle of the wave is pulled in, with the effect that outside the node, at r_c, the wave is identical to the unscattered one ($V = 0$), apart from the reversal of its sign. Hence the scattering is zero for this particular V and η_l, and weak for nearby values. This is the basis of the Ramsauer effect and also of the explanation of the weakness of the $V(r)$ scattering in simple metals. The scattering does not continue to increase as V deepens, but alternately increases and decreases as more nodes of R_l are pulled into the ion and form more localized core states. The rapid oscillations of the wavefunction in the core do not add to the effective value of the η_l outside the core because this effective value is only what remains after the pulled-in nodes have been subtracted. To recognize this effect, we write η_l in the form

$$\eta_l = p_l\pi + \delta_l \qquad (2.25)$$

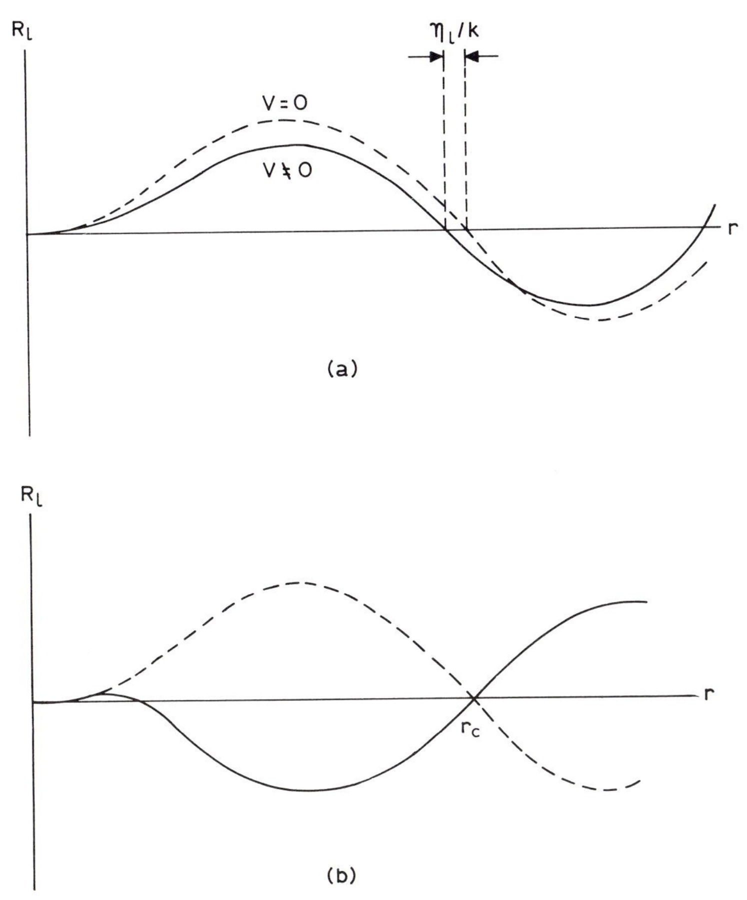

Fig. 2.6 (a) The radial component R_l of a partial wave is 'pulled in' towards the origin by an amount η_l/k by an attractive potential. (b) It is pulled in by an amount π, as in the Ramsauer effect

where p_l is the integer which counts the number of pulled-in nodes and is chosen so that δ_l, the *effective phase shift*, lies in the range $+\frac{1}{2}\pi > \delta_l > -\frac{1}{2}\pi$. For simple metals such as Na, Mg or Al, this effective phase shift is very small. Thus the scattering is small, the pseudopotential is weak and the general NFE approach is justified.

2.7 SCREENING

We have seen that the effect of the core potential V on a metallic electron, in a metal such as Na, Mg or Al, is small for two distinct reasons. First is the effect of the Pauli principle and the orthogonality of states in the cores, which lead to the weakness of the pseudopotential. Second is the screening effect of the other conduction electrons which build up a screening charge around each core, made up of small contributions from many such electrons as they each respond, slightly and transitorily, to the pseudopotential field, in their motions through the metal.

We discussed screening in an elementary way in Chapter 1. The theory of partial waves now allows us to discuss it more accurately, and also to consider a foreign atom which has an excess core charge of $+Ze$. Equation (2.24) shows that the screening of this charge by the redistribution of the conduction electrons is not highly concentrated in the atom itself, but is spread out into the neighbourhood in the form of concentric spherical *haloes*, according to the positions of the nodes in the R_l function. This was deduced for one particular direction of approach, along the x axis, of the incident electron, but in the metal electrons approach the scattering centre equally from all directions. This has the effect of averaging out the θ screening distribution evenly over all directions out of the atom, so that the θ dependence disappears; but the r dependence remains and, from equation (2.24), gives the concentric spherical haloes which are known in this context as *Friedel oscillations* (Friedel 1958). From equations (2.20), (2.23) and (2.24), the radial dependence of the density of the partial wave k, l is proportional to

$$\frac{1}{r^2}\left[\sin^2(kr - \tfrac{1}{2}l\pi + \eta_l) - \sin^2(kr - \tfrac{1}{2}l\pi)\right] = \frac{1}{r^2}\sin(2kr - l\pi + \eta_l)\sin\eta_l \quad (2.26)$$

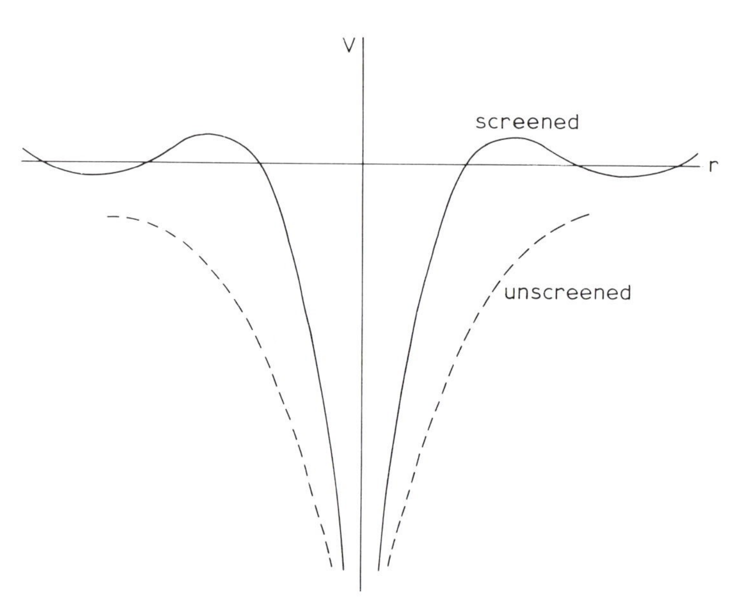

Fig. 2.7 The potential of a charged ion, unscreened and as screened by surrounding haloes in the density distribution of the conduction electrons

This is for one particular k. The total electron density in the haloes for the l-scattered wave is the sum of this over all k values up to k_F at the Fermi level. Hence, integrating the expression with respect to k, we obtain

$$\delta\rho_l \propto \frac{1}{r^3} \cos(2k_F r + \eta_l) \tag{2.27}$$

for this density. The cosine term gives the Friedel oscillations in the electron density around the scattering centre, and we note that this density decreases only as r^{-3} (unlike the exponential decrease of the Thomas–Fermi approximation). Figure 2.7 shows schematically the kind of screened potential produced by the charged ion and its accompanying haloes of screening conduction electrons.

The conductivity of a metal requires that, at sufficiently large distances, the excess charge Ze of an ion must be completely neutralized by electron screening. Any inadequacy in the screening associated with one value of l must then be made up by the screening associated with the other values. This requirement leads to the *Friedel sum rule*: that the total electronic charge pulled towards the ion by the motion of the conduction electrons passing through the neighbourhood achieves long-range neutrality when

$$Z = \frac{2}{\pi} \sum_l (2l + 1)\eta_l \tag{2.28}$$

where η_l is evaluated at the Fermi wavenumber k_F.

These Friedel oscillations play a significant part in the interactions between foreign atoms in metals and also in determining some of the physical properties of metals (*see* Chapter 5).

REFERENCES

Friedel, J., *Suppl. to Nuovo Cim.*, **7**, 287 (1958).
Herring, C., *Phys. Rev.*, **57**, 1169 (1940).
Phillips, J. C., and Kleinman, L., *Phys. Rev.*, **116**, 287, 880 (1959).
Wigner, E., and Seitz, F., *Phys. Rev.*, **43**, 804 (1933); *Ibid.*, **46**, 509 (1934).

3

Why are metallic electrons so independent?

3.1 THE PROBLEM

Despite its practical successes and despite the improvements outlined in Chapter 2, the electron theory of metals so far described has a profoundly unsatisfactory basis. It is a *one-electron* theory, i.e. the Schrödinger equation of the theory deals with a single electron moving in a fixed field of potential. There are no terms in it representing the energy of electrostatic interaction, e^2/r_{12}, where r_{12} is the distance between two (moving) electrons 1 and 2. If such terms were introduced, the wavefunction would have to take on a *many-electron* form $\psi(1, 2, \ldots, N)$, i.e. a function of the coordinates of *all* N electrons, which is very difficult to handle mathematically.

The one-electron theory has produced most of the characteristic features of the theory of metals: the plane-wave functions, energy levels and Fermi sphere of the free-electron theory; the Bloch functions, energy bands, Brillouin zones and Fermi surface of the NFE and tight-binding theories; and the concept of the pseudopotential. In all these theories the first step has been to find the wavefunctions and E, k relations from a one-electron equation. Only then have the N electrons been introduced into these already calculated quantum states, which have been filled up to the Fermi level in accordance with the Pauli principle. There has been no correlation in the mutual positions of the electrons in these theories, so that, quite apart from other consequences, the total energy of the system has been overestimated as a result of adding the electron–electron repulsion, since no allowance was made for the tendency of interacting electrons to avoid one another.

There are many important effects of electron–electron interaction. We have seen (Chapter 1) that in 'chemical' theories of the metallic state correlation is responsible for the most basic effect of all, the transition from metals to insulators; and that in the scattering theory (Chapter 2) the weakness of the effective lattice field in a metal is partly due to the conduction electrons mutually shielding one another from the core potentials. There are also major practical demonstrations of the effects of electron–electron interactions in metals: for example *ferromagnetism, plasma oscillations* and *superconductivity*.

The importance of the electron–electron problem extends far beyond the theory of metals. It is the key problem in the quantum theory of many-electron atoms and molecules, as discussed in Appendix 2. One of the starting points for the improvement of the theory of metals has in fact been to apply ideas already developed in atomic physics, in particular the self-consistent field methods of Hartree, Fock and Slater.

3.2 HARTREE–FOCK FREE-ELECTRON THEORY

In the electron–electron problem we obviously want to keep the ionic core effects as simple as possible. This leads to the *jellium model*, in which the positive charge of these cores is imagined to be smeared out uniformly throughout the metal, where its role is simply to neutralize the average negative charge of the metallic electrons and thus to hold everything together. The one-electron theory of this model is of course the free-electron theory in which a metallic electron has a plane-wave quantum state, $\psi = C \exp(\mathrm{i}\boldsymbol{k} \cdot \boldsymbol{r})$, and a *uniform* probability density, $\psi^*\psi$ = constant, thoughout the system.

The Hartree theory of jellium follows the general Hartree self-consistent method outlined in Appendix 2, in which one seeks to establish the behaviour of each electron in the *average* field of all the others. The interaction of any one electron (e.g. electron 1) with any other (e.g. electron 2) is thus simplified to the interaction of 1 with a *fixed* charge distribution, $e\psi_2^*\psi_2$, which represents electron 2; the interaction energy is

$$I = \int \frac{e^2}{r_{12}} \psi_2^*\psi_2 \, \mathrm{d}v \tag{3.1}$$

where r_{12} is the distance from electron 1 of the volume element $\mathrm{d}v$, which contains the element $e\,\psi_2^*\psi_2\,\mathrm{d}v$ of the charge cloud representing electron 2. Such integrals are summed over all the $N - 1$ electrons with which electron 1 interacts, using *assumed* wavefunctions, and the sum is then added to the V term in the one-electron Schrödinger equation for the wavefunction ψ_1 of electron 1. This equation is solved to find an improved version of ψ_1, which is then fed back, in turn, into the corresponding one-electron equations for $\psi_2, \ldots, \psi_N$, and the whole process is repeated until a self-consistent set of wavefunctions, $\psi_1, \ldots, \psi_N$, is obtained.

Usually this is a laborious method, even in single atoms where N is small, but in the jellium free-electron theory it becomes extremely simple because all the distributions such as $\psi_2^*\psi_2$ are *uniform* and so can be taken outside the integral over $\mathrm{d}v$. In fact, of course, since they precisely neutralize the equally uniform jellium charge everywhere, the result is $V = 0$ and the Hartree treatment of electron–electron interaction thus leaves the free-electron theory completely unchanged!

This is unsatisfactory, however. It turns out that there is *no* cohesion in the Hartree free-electron theory. This is because the uniform charge distributions 'allow' the electrons to approach one another closely, so that the (repulsive) electrostatic energy of the electrons is then far too large. The theory has in fact neglected correlation in the actual, as distinct from the average, positions of the electrons. It has neglected both the *charge* or *Coulomb correlation* which exists between all electrons because of their like charges, and also the *exchange* or *spin correlation* which exists between electrons of parallel spins as a result of the Pauli exclusion principle (Appendix 2).

Fock and Slater modified the Hartree method to take account of the exchange interaction. This exists because electrons are physically indistinguishable and so must exist in wavefunctions which remain unchanged (apart from a reversal of sign) if they exchange places. For two electrons A and B, with parallel spins, the combination of one-electron functions which represents both of them has the form

$$\psi(1, 2) = \psi_A(x_1)\psi_B(x_2) - \psi_B(x_1)\psi_A(x_2) \tag{3.2}$$

where x_1 and x_2 are their positions. The Hartree–Fock theory proceeds as before, but with charge distributions of the type $e\psi^*(1, 2)\psi(1, 2)$ in place of the simpler ones of the Hartree theory. This is an immense complication, not least because it makes the electron–electron term in the potential *non-local*; it cannot be expressed as a function $V(x)$ of x only, but is a function $V(x_1, x_2)$ of *two* position variables. Nevertheless, in the jellium model the states remain plane waves, enabling a solution to be obtained (see e.g. Ziman 1964). This can be shown to give the energy of an electron in the plane-wave state with wave-vector $\boldsymbol{k}$ as

$$E = \frac{\hbar^2 k^2}{2m} - \frac{e^2 k_F}{2\pi}\left[2 + \frac{(1-\alpha^2)}{\alpha}\ln\left|\frac{1+\alpha}{1-\alpha}\right|\right] \tag{3.3}$$

where $\alpha = k/k_F$ and k_F is the wavenumber of the Fermi surface. The energy is reduced by the second term, which is the *exchange energy* E_x, because the Pauli principle causes each electron to carry around with it a 'hole' in the free-electron fluid, i.e. a small volume which tends to be avoided by electrons of the same spin. This is the *exchange* or *Fermi hole*. The reduction of the energy below the Hartree free-electron value ($\hbar^2 k^2/2m$) saves the cohesion and in fact gives a fairly good value for the binding energy of an electron in a metal.

To obtain the total energy of all N electrons we would have to sum equation (3.3) over all states k up to the Fermi limit k_F, and allow for the two electrons, of opposite spins, in each state. The sum of the first term simply gives the total Fermi energy, $\frac{3}{5}NE_F$, of a free-electron gas. In summing the second term, i.e. the electron–electron interactions, there are two compensating factors to be noted. First, the sum has to be multiplied by two to allow for the two electrons of opposite spins in each state, as above. Second, the sum has to be divided by two because it is a sum of interactions between *pairs* of electrons and, in evaluating the interactions of every electron with all the others (as we did when summing the I of equation (3.1), each pair is included in the sum twice.

The theory is still not satisfactory, however. If we differentiate E in equation (3.3) with respect to k, the logarithmic factor appears in one of the resulting terms and becomes infinite when $k = k_F$. Since those electrons mostly responsible for metallic properties are located in the region of the Fermi surface, $k \approx k_F$, this infinity in the E, k slope is an unacceptable feature. It leads

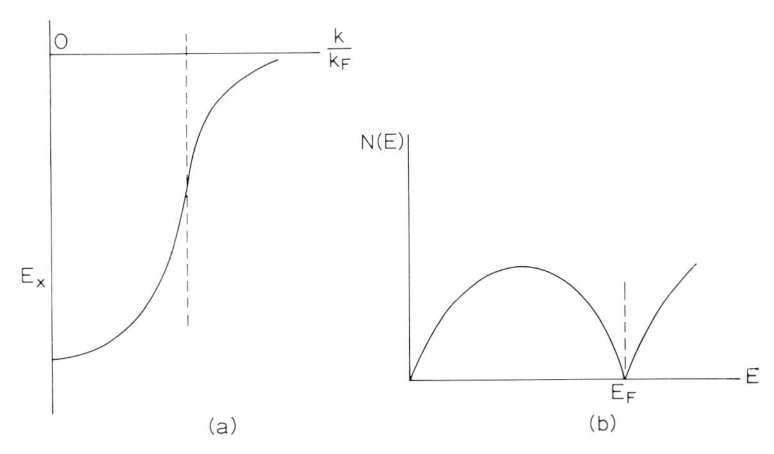

Fig. 3.1 (a) Exchange energy E_x and (b) density of states $N(E)$ according to the Hartree–Fock theory

to several anomalies. For example, the *density of states*, i.e. the number of states per unit volume in a given range of energy (*see* Appendix 3), which is

$$N(E) = \mathrm{d}N/\mathrm{d}E = \frac{\mathrm{d}N}{\mathrm{d}k} \Big/ \frac{\mathrm{d}E}{\mathrm{d}k} \tag{3.4}$$

goes to zero when $k \rightarrow k_F$ since $\mathrm{d}E/\mathrm{d}k \rightarrow \infty$ (as shown in Fig. 3.1), in contradiction with experimental measurements of energy bands in metals and also with the many physical properties, e.g. electronic specific heat, that depend on there being a large number of states at the Fermi surface.

The physical reason for this wrong result for the density of states is as follows. The exchange interaction, by providing each electron with a Fermi hole in which there is no electron with the same spin, produces a reduction in some of the e^2/r_{12} terms and thus a lowering of the energy levels of the quantum states, as indicated in equation (3.3). However, this applies only to quantum states that are *occupied*, i.e. those with $k \leqslant k_F$. The states above k_F are empty, so their energy levels are not depressed. As a result a kind of band splitting occurs at E_F, leading to a shortage of quantum states with energy levels in this region.

This difficulty, which stems from the k-dependence of E_x, has led to improved approximations, introduced by Slater, which replace the k-dependent exchange energy function by its average value,

$$\langle E_x \rangle = -\frac{3e^2 k_F}{4\pi} \tag{3.5}$$

i.e. one that is independent of k. This simplifying approximation restores the correct form of the density of states and has given good results when used in calculations of energy bands.

The trouble with the Hartree–Fock theory is that it does not allow for *Coulomb correlation*. The long-range character of the inverse square law of electrostatic force is the source of many difficulties which would disappear if the *effective* interaction between the metallic electrons were of shorter range. In fact, this is just what happens as a result of Coulomb correlation because then *all* electrons, not merely those of parallel spin, keep out of one another's way. A complete *correlation hole* is then formed around each electron, i.e. a 'sphere of influence' inside which no other electron is likely to be found. We are back to *screening* again, this time not the screening of ionic charge by surrounding electrons (*see* Chapter 2) but the screening of each electron by its companions. The correlation hole around each electron is a spherical region of positive charge from the jellium exposed by the absence of other electrons. This positive sphere, together with the negative electron at its centre, forms an electrically neutral unit known as a *quasiparticle*. In this guise the electrons behave at long range like neutral particles. A simple example of a screened-charge interaction potential was given by equation (1.29).

When there is Coulomb correlation and screened-charge interaction, the interaction integrals such as equation (3.1) do not effectively extend throughout the metal (even though the wavefunctions do) because the screened replacement of the e^2/r_{12} factor practically vanishes when r_{12} exceeds the screening radius, i.e. a few angstroms. The exchange interaction is then much weaker, and the difficulties in the Hartree–Fock model with the density of states and the related physical properties disappear.

3.3 EFFECTS OF CORRELATION

There are many other examples of the importance of Coulomb correlation. We have already seen the most important of all, the existence of the metal–non-metal transition, in Chapter 1.

Wigner and Seitz made great use of correlation in their cellular method (*see* Chapter 2). In applying their theory to monovalent metals such as sodium they assumed that correlation would make each Wigner–Seitz cell an exact, one-electron correlation hole, i.e. that each cell would contain one and only one metallic electron. Using the wavefunction derived for this cell (Fig. 2.3), they were then able to calculate the cohesion of the metal rather easily, since each electron in such a correlation hole experiences to a good approximation only the field of the ionic core in its *own* cell. All the cores and metallic electrons outside this cell behave almost like electrically neutral cells which make no contribution to the V field of this electron.

Another example of a correlative effect is *plasma oscillation*. As a result of their electrical interactions and consequent correlations in position and motion, the metallic electrons behave in some respects like a compressible, electrified fluid in a neutralizing background medium (i.e. a *plasma*), the macroscopic properties of which can be described by the methods of continuum mechanics and classical electrodynamics. This approach thus accepts that the long-range Coulomb interaction e^2/r_{12} is capable of organizing large numbers of electrons, over macroscopic regions of the material, into common modes of behaviour which are describable by classical continuum theory.

When this charged fluid is distributed uniformly over the background field of the positive ions or jellium, there is, on this macroscopic scale, electrical neutrality everywhere. However, if for some reason there is a region of low density in the fluid, the uncompensated part of the jellium field in this region attracts all the surrounding fluid inwards. Rushing in, this fluid acquires kinetic energy and overshoots, causing an upsurge of negative charge there. This is then electrically repelled outwards again and the process is repeated, so that a long-range oscillation of the fluid, i.e. a *plasma oscillation*, is set up, as first discussed by Bohm and Pines (1951).

Suppose that in such a plasma, with a density of n_0 electrons per unit volume, a *slab* of the fluid is moved wholesale through a distance x. This displaced slab acts as an electrical condenser and experiences a field F, opposing the displacement, of magnitude $4\pi n_0 ex$. An electron in the slab thus experiences a restoring force $-eF$ and is subjected to motion $m\ddot{x} = -4\pi n_0 e^2 x$, which gives a harmonic oscillation of frequency

$$\omega_p = (4\pi n_0 e^2/m)^{1/2} \tag{3.6}$$

This is the characteristic frequency of the plasma oscillations, typically about $10^{16}\ s^{-1}$. The energy quantum or *plasmon* $\hbar\omega_p$ of these oscillations, several electron-volts, is too large to be provided thermally at room temperature, but the oscillations can be excited externally by firing fast electrons through the metal.

The velocity of a free electron at the Fermi surface is about $(E_F/m)^{1/2}$. Dividing this by the plasma frequency, we obtain a characteristic length $(E_F/4\pi n_0 e^2)^{1/2}$. The commonly used value of this *Thomas–Fermi screening radius* is

$$\lambda_{TF} = \left(\frac{E_F}{6\pi n_0 e^2}\right)^{1/2} \tag{3.7}$$

which is about 1 Å in metals, and indicates the order of magnitude of the wavelength of a plasma oscillation below which it is meaningless to think of the system as a continuum rather than as separate particles.

The Coulomb interaction e^2/r_{12} between two electrons at a distance r_{12} thus falls into two parts. The first is composed of long-wave (Fourier) components, i.e. $\lambda > \lambda_{TF}$, and is completely accounted for by the plasma oscillations; the second, residual part is composed of short-wave components, $\lambda < \lambda_{TF}$. Beyond a distance $r_{12} \simeq \lambda_{TF}$ these short-wave components interfere destructively and cancel one another out, so that the residual part is a short-range interaction, very weak when $r_{12} > \lambda_{TF}$ (*see* Appendix 6). The detailed theory shows that it is roughly like the Thomas–Fermi screened-charge interaction of equation (1.29), with $\gamma^{-1} = \lambda_{TF}$, but more accurately like the Friedel oscillation screening described in Chapter 2 (*see* Fig. 2.7).

3.4 THE DIELECTRIC FUNCTION OF FREE ELECTRONS

The above electron–electron screening effects which result from Coulomb correlation are a form of *polarization* (Chapter 1) and so are describable in terms of a *dielectric constant* (*see* equations (1.21) and (1.26)). We recall that in a polarizing medium of dielectric constant κ the potential between two electrons at a distance r_{12} is $e^2/\kappa r_{12}$.

For a free-electron gas κ is not a constant but a *dielectric function*, $\kappa(\omega, q)$ of both the frequency ω and wavenumber q of an oscillatory disturbance. The general theory of this is usually simplified into two limiting approximate cases. The first is that of long-wave oscillations, $q \to 0$. This limit is the *plasma theory*, outlined above, in which the dielectric function $\kappa(\omega)$ at frequency ω takes the form

$$\kappa(\omega) = 1 - \omega_p^2/\omega^2 \tag{3.8}$$

and so is positive when $\omega > \omega_p$ and negative when $\omega < \omega_p$. This is interesting in connection with the *optical properties* of metals, i.e. the response to an applied electromagnetic wave of frequency ω. As its name implies, a dielectric function is normally associated with *dielectric materials*, e.g. transparent insulators. A positive value of κ indicates optical transparency. A negative value indicates the characteristic metallic optical properties of *opacity* and *high reflectivity*. At visible wavelengths, $\omega < \omega_p$ and this condition is satisfied, but for metals such as the alkalis ultraviolet radiation falls in the range $\omega > \omega_p$ where they become transparent. The effect is the same as that which causes radio waves to be reflected from the free electrons in the ionosphere at low frequencies. Such frequencies allow enough time for the free electrons to move completely, in response to the external oscillating radiation, and so to absorb it fully and emit a reflected beam.

At the low-frequency limit, $\omega \to 0$, the ability of the electron gas to give *Coulomb screening* is indicated by $\kappa(q)$. Suppose that the initially uniform jellium is perturbed by a spatially periodic distribution of background charge of density $\rho_0 = A \sin qx$. This gives a 'bare' electrostatic potential $\phi_0 \propto \sin qx$. The sinusoidal variation allows us to replace ∇^2 by $-q^2$ in Poisson's equation (equation (A 2.12), Appendix 2, with $\nabla^2 = d^2/dx^2$), so giving

$$q^2\phi_0 = 4\pi\rho_0 \tag{3.9}$$

The electron gas polarizes in this field, so that its initial (macroscopic) uniformity is modified by a periodic distribution ρ_1 of electrical charge, which again varies as sin qx and gives an *induced* potential ϕ_1. The total potential and charge distributions are thus

$$\left.\begin{aligned} \phi_t &= \phi_0 + \phi_1 = \phi_0/\kappa \\ \rho_t &= \rho_0 + \rho_1 \end{aligned}\right\} \tag{3.10}$$

and Poisson's equation again gives

$$q^2\phi_t = 4\pi\rho_t \tag{3.11}$$

In the macroscopic approximation, where the perturbation has a long wavelength compared with that of an electron at the Fermi surface, i.e. where

$q \ll k_F$, the Thomas–Fermi method can be used to estimate the electronic screening. Here we can think initially of a Fermi distribution of fixed energy 'thickness' E_F on top of the ground state, the energy of which undulates as $\sin qx$ in accordance with the variation of ϕ_t. However, this would give an undulating Fermi surface, which is as impossible an equilibrium state as an undulation of the surface of a still pond. The electrons spill down off the crests, fill up the troughs and so produce a level surface everywhere. When $\phi_t = 0$ the Fermi level (equation (1.3)) is

$$E_F = \frac{\hbar^2}{2m}(3\pi^2 n_0)^{2/3} \tag{3.12}$$

where n_0 is the number of electrons per unit volume. When $\phi_t \neq 0$ the local value $n(x)$ of n has to be adjusted to make the ensuing change ΔE_F balance the local deviation $e\phi_t$ in the ground energy level, i.e.

$$e\phi_t = \Delta E_F \approx (n - n_0)\frac{dE_F}{dn_0} \tag{3.13}$$

Taking dE_F/dn_0 from equation (3.12), we obtain

$$n - n_0 = \frac{3n_0 e\phi_t}{2E_F} \tag{3.14}$$

and thus

$$\rho_1 = -e(n - n_0) = -\frac{3n_0 e^2\phi_t}{2E_F} = -\frac{\rho_t}{q^2\lambda^2} \tag{3.15}$$

with $\lambda = \lambda_{TF}$ from equation (3.7). Finally,

$$\kappa(q) = \frac{\phi_0}{\phi_t} = \frac{\rho_0}{\rho_t} = 1 - \frac{\rho_1}{\rho_t} = 1 + \frac{1}{q^2\lambda^2} \tag{3.16}$$

We see that this static dielectric function depends on the wavelength of the variation in the field, as expected. A long-range field ($q \to 0$) gives $\kappa \to \infty$ and $\phi_t \to 0$, i.e. in this limit total screening results from the redistribution of the electrons.

The above result has been deduced for a simple sinusoidal field, $A \sin qx$. More general fields can now be analysed by superposing a number of such sinusoidal waves, with various q values and appropriate weighting factors A, to construct *Fourier sums* or integrals of the required form (*see* Appendix 6). As an example, consider the case where ρ_0 is the field of a fixed unit point charge, i.e. $\phi_0(r) = 1/r$. We have, from Appendix 6,

$$\phi_0(q) = \frac{1}{2\pi^2 q^2} \tag{3.17}$$

and hence for each component

$$\phi_t(q) = \frac{\phi_0(q)}{\kappa(q)} = \frac{1}{2\pi^2(q^2 + \lambda^{-2})} \tag{3.18}$$

The transform of this (*see* Appendix 6) gives

$$\phi_t(r) = \frac{1}{r}\exp(-r/\lambda) \tag{3.19}$$

which is the screened-charge Coulomb potential of the Thomas–Fermi approximation (equation (1.29)). In applying this approximation to electron–electron interaction, we simply regard each electron as a centre of a screened-charge distribution, with charge $-e$. Inside the radius λ_{TF} there is little screening, but outside this hole the screening reduces the effective field of the electron rapidly towards zero.

The Thomas–Fermi theory is too 'macroscopic' to be satisfactory. A better theory can be obtained by dropping the assumption that the potential varies slowly with distance and assuming instead that it is a *small* variation. A solution can then be found by applying perturbation theory to Schrödinger's equation. This is the Lindhard theory, which leads to the result

$$\kappa(q) = 1 + \frac{3\pi e^2 n_0}{2q^2 E_F}\left[2 + \frac{(1-\alpha^2)}{\alpha}\ln\left|\frac{1+\alpha}{1-\alpha}\right|\right] \tag{3.20}$$

where $\alpha = q/2k_F$. We recognize in this the same logarithmic expression as in equation (3.3); in both cases this occurs because, as α is increased, the region of the Fermi sphere over which the integral is taken, leading to the expression, increases up to the point where $\alpha = 1$, but not beyond. There is a singularity at $\alpha = 1$ which, when analysed, shows that for a point charge the screening is no longer exponential, as in equation (3.19), but is of an oscillating form, as in equation (2.27). The Lindhard theory thus leads back to the halo screening of the Friedel theory, discussed in Chapter 2 (*see* Fig. 2.7).

3.5 ELECTRON–ELECTRON COLLISIONS

The basic question posed in the title of this chapter still remains to be answered. If the electron gas were classical, its densely packed and strongly interacting particles could move individually only by a slow diffusion process, zigzagging with a mean free path of about 3 Å. However, the mean free path for electron–electron collisions in a metal is typically about 1 μm at room temperature and increases to about 0.1 m at 1 K. So why are electron–electron collisions so rare?

Perhaps it is because of Coulomb screening? As we have seen, this shortens the effective range of the electrostatic field of an electron from $1/r$ to about λ_{TF}, i.e. to an angstrom or so. Screening plays two roles in electron–electron interaction: first, this shortening of range; second, a more subtle effect we will discuss at the end of this section. The shortening of range does not solve the problem, however, because it leaves the electron field unscreened at close range. We recall from the kinetic theory of gases that collision probabilities are calculated in terms of the 'target area' or *collision cross-section* σ which one particle presents to another. In this case we expect the unscreened domain of an electron to present a target area $\sigma_0 \simeq 10^{-19}$ m^2. We also recall from kinetic theory that the mean free path l is about $1/n\sigma$, where n is the number of particles per unit volume. Thus $l \approx$ *atomic spacing* for the screened electron gas.

As we shall see, electron–electron collisions are scarce because of the Pauli principle and the laws of conservation of energy and momentum. The Pauli principle organizes a free-electron gas into a Fermi sphere in k-space, completely filled (at 0 K) up to the Fermi level E_F, completely empty above. An electron deep inside this sphere cannot collide effectively with any other electron because there are no empty quantum states *of similar energies* into which it could be deflected by such a collision. Only if it acquired an extra energy of order E_F could it find empty states in which to go, and in a metal in equilibrium at ordinary temperatures there are no sources with such large spare energy.

The Pauli principle thus limits the opportunity for electron–electron collisions to those electrons in states at or near the Fermi surface. Consider now the effect of conservation. Suppose that the Fermi surface is sharply defined (i.e. the temperature T is such that $k_B T \ll E_F$, a condition which is satisfied at room temperature) and, for convenience, choose the energy origin to lie at E_F. Consider an electron in a state A with an energy ε_A (i.e. its total kinetic energy is $\varepsilon_A + E_F$) which is *slightly* above the Fermi surface, i.e. $\varepsilon_A \ll E_F$, and which may collide with another electron. This other electron is overwhelmingly likely to be in a state B *inside* the Fermi sphere simply because that is where practically all the *filled* states are. It thus has an energy level ε_B, which on this energy scale is *negative*. The collision deflects the electrons into two other states, C and D, with energies ε_C and ε_D. The Pauli principle requires ε_C and ε_D to be *positive* since otherwise C and D would not be empty receptor states; energy conservation requires that $\varepsilon_A + \varepsilon_B = \varepsilon_C + \varepsilon_D$. It thus follows that $\varepsilon_A + \varepsilon_B$ is positive, i.e. that $|\varepsilon_B| < \varepsilon_A$. The states B must then lie in a thin shell of thickness $\leqslant \varepsilon_A$, just inside the Fermi surface. Since the Fermi sphere is uniformly filled with states, the number in this shell is only about a fraction ε_A/E_F of the total number of states. Similarly, the receptor states must also lie in a thin shell, of thickness about ε_A, just outside the Fermi surface. Such states are again about a fraction ε_A/E_F of the total. The collision probability is thereby reduced to about a fraction $(\varepsilon_A/E_F)^2$ of its classical value by this scarcity of both collision and receptor states; and the collision cross-section is reduced to about $(\varepsilon_A/E_F)^2\sigma_0$. Thermal excitation can provide electrons in states A, where $\varepsilon_A \simeq k_B T$. Hence the collision cross-section is about $(k_B T/E_F)^2\sigma_0$. At room temperature, where $k_B T/E_F \simeq 10^{-2}$, this gives a collision cross-section of about 10^{-23} m^2 and a mean free path of 1 μm. At 1 K the mean free path is about 0.1 m.

We thus have a good explanation of the independence of the electrons, *provided the sharp Fermi surface exists*. It is, however, reasonable to question this, since the concept of the Fermi energy distribution, with its sharp surface (at 0 K), although confirmed experimentally, is a deduction from free-electron theory in which the independence of the electrons is taken for granted. This question was elegantly tackled by Landau (1957) in his *Fermi liquid* theory, in which the key consideration is *continuity*. Suppose that we really do have a free-electron gas in which the particles are truly independent, i.e. the Coulomb interaction e^2/r_{12} is zero. Then all the results of free-electron theory are valid for this system, and we have a Fermi energy structure of the above type and with the above properties. Now 'switch on' the Coulomb interaction e^2/r_{12}, gradually and continuously. The energy levels of the quantum states begin to change gradually in consequence, but, because each stage in this process differs only infinitesimally from its predecessor, there is a continuous connection between each present quantum state and its immediate predecessor. In other words, the quantum states of the interacting electron gas are in one-to-one correspondence with those of the original free-electron gas. As a result, since the original states formed a Fermi distribution (i.e. a filled sphere with a well-defined Fermi surface), then so also do the states of the interacting system which, by this process, have evolved continuously from them.

There thus exists a Fermi sphere, a Fermi surface, *and the consequential independence property* as discussed above, for the occupants of these states. However, these occupants cannot be 'bare' electrons since electrons are no longer independent particles once the Coulomb interaction is switched on. Landau showed that the independent occupants of these states in fact are *quasiparticles*, i.e. 'clothed' electrons. As we have seen in the discussion of the correlation hole, around any given electron in a metal the other electrons are not distributed randomly. There is a 'sphere of influence' which tends to be avoided by the others. As this electron moves about, its sphere of influence moves with it. Landau's quasiparticle is the electron clothed in its sphere of influence. The detailed working of this theory shows that, because other electrons are involved in clothing the electron in question, each clothed electron appears to be a little heavier than a bare free electron. Typically, this effective mass is about 25% greater than the intrinsic mass. There is experimental evidence for such an effect.

3.6 THE DENSITY FUNCTIONAL METHOD

The plasma theory outlined in Section 3.3 focuses on the density of the electron gas and the oscillations of this density. A different approach, also based on electron density, was taken by Hohenberg and Kohn (1964). This *density functional* method provides a way of representing the difficult exchange and correlation effects as an *effective potential*, rather as the pseudopotential theory provided similarly for the effects of the ionic cores. In this way it enables the interacting electron gas to be described in terms of a one-electron Schrödinger equation and so justifies the independent electron approximation, provided an appropriately modified potential is used. The method is related to the Thomas–Fermi approach, which of course is based on the density of electronic charge (*see* equation (A2.12) in Appendix 2).

The energy E of the electron gas can be expressed in terms of the electron density $n(r)$:

$$E[n] = \int V(r)n(r)\,\mathrm{d}r + \frac{1}{2}\int\int \frac{e^2}{r_{12}} n(r_1)n(r_2)\,\mathrm{d}r_1\,\mathrm{d}r_2 + T[n] + E_{\mathrm{xc}}[n] \qquad (3.21)$$

where the first term is the potential energy in the (pseudopotential) field of the ionic cores, the second term is the classical Coulomb interaction energy as in the Hartree theory (the terms I_1 in equation (A2.9)), $T[n]$ is the kinetic energy (i.e. the Fermi energy) and $E_{xc}[n]$ represents the contribution of the exchange and correlation terms. The integrals are of course taken over the whole volume of the metal. The square brackets indicate that the quantity to which they refer, e.g. E_{xc}, is a function not of r directly but of a function, $n(r)$ of r, i.e. $E_{xc} = E_{xc}[n(r)]$. The term *functional* is used for such functions of functions. As a result, and as a generalization of the two-electron exchange interaction we considered in Section 3.2, E_{xc} is a *non-local* quantity, i.e. its value at any point r depends not only on the local condition at r but also on the condition at *every* point in the metal.

Hohenberg and Kohn's method is to find the $n(r)$ which minimizes $E[n]$ and then to take this $n(r)$ as the density distribution of the electrons in the ground state. The non-local $E_{xc}[n]$ makes such a procedure difficult. However, they showed that when the change of $n(r)$ with r is small, i.e. when the electron gas is *nearly uniform* in density, it is possible to simplify $E_{xc}[n]$ into a *local functional density*, i.e. one which depends only on the local condition at the point in question. This is done by writing

$$E_{xc}[n] = \int n(r)\varepsilon_{xc}\,dr \tag{3.22}$$

where $\varepsilon_{xc} = \varepsilon_{xc}(n(r))$, with the meaning that ε_{xc} is the exchange and correlation energy of an electron gas with a *uniform* density equal to the local value $n(r)$.

When equation (3.22) is substituted into equation (3.21), no non-local terms remain and it then becomes possible to express the variation of E with respect to $n(r)$ and so find the minimum energy from the condition

$$\delta E/\delta n(r) = 0 \tag{3.23}$$

subject to the requirement that all the local changes $\delta n(r)$ in $n(r)$ add up to a zero total change, so as to maintain a constant total number N of electrons in the gas. The application of this condition leads to the equation

$$\int \delta n(r)\left[\frac{\delta T}{\delta n(r)} + V_{eff}\right] dr = 0 \tag{3.24}$$

where

$$V_{eff} = V(r) + \int \frac{e^2}{r_{12}} n(r_2)\,dr_2 + V_{xc} \tag{3.25}$$

and

$$V_{xc} = \delta E_{xc}/\delta n(r) \tag{3.26}$$

The important result is that the difficult exchange and correlation terms have now become represented by a potential $V_{xc} = V_{xc}(r)$ which has a local value at any point r and which enters, along with the ionic potential and the Hartree–Coulomb term (equation (3.1)), into the Schrödinger equation. There are no non-local terms in this equation so that it is, like equations (1.8) and (2.16), a one-electron equation for an electron in the field V_{eff}. Provided we use a V_{eff} which includes V_{xc}, we can thus describe the interacting gas by the methods of the non-interacting, one-electron, theories presented in Chapters 1 and 2 and so confirm all characteristic results, e.g. the energy bands, Fermi surfaces and Brillouin zones, of these theories. The answer to our question, by this analysis, is then that the electrons *appear* to be independent because the effects of their interdependence can be largely absorbed into a local potential, just as they appear to be nearly free because their interaction with the cores can be largely absorbed in a pseudopotential.

The local density theory can be used, like the Hartree theory, for detailed calculations. The first step is to guess a V_{eff} and use this to calculate $n(r)$, which is then used in turn to calculate a better V_{eff}, and so on. The calculation leads to the simple result

$$V_{xc} = -\tfrac{1}{3}e^2[3\pi^2 n(r)]^{1/3} \qquad (3.27)$$

which, apart from a numerical factor, is the same as the simple averaged approximation proposed by Slater for the Hartree–Fock exchange term (equations (3.5) and (A 3.7) in Appendix 3).

REFERENCES

Bohm, D., and Pines, D., *Phys. Rev.*, **82**, 625 (1951).

Hohenberg, P., and Kohn, W., *Phys. Rev.*, **136**, B864 (1964).

Landau, L. D., *Soviet Phys. JETP*, **3**, 920 (1957).

Ziman, J. M., *Principles of the Theory of Solids*, Cambridge University Press (1964).

4

The pseudopotential

4.1 SIMPLIFICATIONS

As well as explaining why metallic electrons are so nearly free, the pseudopotential theory provides two practical benefits. First, the *weakness* of the pseudopotential enables a high-quality theory of metals to be built fairly simply by starting from the free-electron theory and then modifying it *slightly* by bringing in the effects of the ion cores through standard perturbation theory. This is an example of a general method for applying quantum mechanics to complicated systems, i.e. first find the *right* simple system (free electrons) – right in the sense that it is very similar to the complicated one in its significant aspects – and then go from the simple to the complicated by a short and safe extrapolation (perturbation theory).

The second benefit is that it brings the theory of metallic electrons, at least for simple metals, into the same form as the theory of X-ray diffraction, with its well-established mathematical methods and concepts, e.g. the *form* and *structure factors*. We have already seen in the NFE theory (Chapter 1 and Appendix 4) that Bragg's law links the idea of Brillouin zones to that of the reflection of waves from crystal planes. Pseudopotential theory takes us further in the same direction, because many of the characteristic features of X-ray analysis similarly depend on the *weakness* of the interaction of X-rays with individual atoms.

The form and structure factors in X-ray analysis arise from regarding a diffracted beam as the superposition of many component beams, each of which radiates from its own atomic scattering centre. The form factor gives the amplitude of an individual component beam radiating at an angle θ to the incident (unit) beam. The *structure factor*, adds these component beams together to make the total diffracted beam, taking into account their phase differences which determine the extent to which they reinforce or cancel one another (*see* Appendices 4 and 6). If the incident and diffracted beams have the wave-vectors $\boldsymbol{k}$ and $\boldsymbol{k}'$, respectively, and we define

$$\boldsymbol{q} = \boldsymbol{k}' - \boldsymbol{k} \tag{4.1}$$

where $q = |\boldsymbol{q}|$ indicates the angle θ between these beams, i.e.

$$q = 2k \sin \tfrac{1}{2}\theta \tag{4.2}$$

for elastic scattering (i.e. $|\boldsymbol{k}'| = |\boldsymbol{k}| = k$), then the structure factor of a system of N atoms, located at positions given by the vector $\boldsymbol{l}$, is

$$S(q) = N^{-1} \sum_{l} \exp(-\mathrm{i}\boldsymbol{q} \cdot \boldsymbol{l}) \tag{4.3}$$

(*see* equation (A4.35), Appendix 4). This expression sums the phase differences of the component beams in terms of their differences in path lengths as given by the positions of the scattering centres $\boldsymbol{l}$ and the beam angles θ, which enter through the product

$$\boldsymbol{q} \cdot \boldsymbol{l} = q_x l_x + q_y l_y + q_z l_z \tag{4.4}$$

The amplitude $W(q)$ of the total scattered beam is then given by the product of $S(q)$ with the form factor, which is also a function of angle, i.e. of q. For X-rays the form factor is $f(q)$, the *atomic form factor*. For metallic electrons it is $w(q)$, the *pseudopotential form factor*.

4.2 THE PSEUDOPOTENTIAL FORM FACTOR

As shown in Appendix 4, the key quantity in perturbation theory is the *matrix element* of the perturbation. If $V(r)$ is the perturbing potential, i.e. the pseudopotential in our case, and ψ_k and $\psi_{k'}$ are any two wavefunctions of the unperturbed system, i.e. free-electron wavefunctions in our case, then the matrix element $V_{kk'}$ is defined as

$$V_{kk'} = \int \psi_{k'}^* V \psi_k \, \mathrm{d}v \equiv \langle k' | V | k \rangle \tag{4.5}$$

where the integral is over the whole volume of the system and where the $\langle || \rangle$ notation represents the matrix element in Dirac's notation (*see* Appendix 1). The matrix element, which is an energy, measures the extent to which there is a threefold overlap of the two wavefunctions and the perturbing field in various regions of the system. Obviously, the perturbation can affect the system only if it is non-zero in those regions where both wavefunctions are also non-zero; and the strength of its effect, as indicated by the matrix element, depends on how much the high-amplitude regions of V, ψ_k and $\psi_{k'}$ all overlap.

In the particular case $k' = k$, the matrix element measures the first-order change in the energy level of the state ψ_k brought about by the perturbation. This is obvious since $\psi_k^* \psi_k$ measures the electron density in a volume element, and thus the matrix element V_{kk}, i.e. $\langle k | V | k \rangle$ is simply a weighted average of $V(r)$, the weighting factor in each volume element being equal to the electron density there. When $k' \neq k$, the matrix element measures the extent to which two states ψ_k and $\psi_{k'}$, which were quite distinct (i.e. *orthogonal, see* Appendix 1) in the unperturbed system, become linked by the perturbation. This coupling enables an electron which was in state ψ_k in the unperturbed system to transfer or *hop* into the state $\psi_{k'}$, and vice versa. For this reason the integral in equation (4.5) is sometimes known as a *hopping integral* (or *hopping matrix element*) and is important in problems of the scattering of electrons from one state to another. There is a 'golden rule' in quantum mechanics that the probability per unit time, $p_{kk'}$, for an electron in the state $\boldsymbol{k}$ to make a transition to a state $\boldsymbol{k}'$ is given by

$$p_{kk'} = \frac{2\pi}{\hbar} |\langle k' | V | k \rangle|^2 n(E) \tag{4.6}$$

where $n(E)$ is the density of states in the range of k'.

In the pseudopotential theory the unperturbed states are free-electron wavefunctions such as $\exp(-\mathrm{i}\boldsymbol{k} \cdot \boldsymbol{r})$. When these are substituted for ψ_k and $\psi_{k'}$ in equation (4.5), the matrix element becomes (*see* Appendix 6, equation (A 6.8))

$$W(q) = \int V(r) \exp(\mathrm{i}\boldsymbol{q} \cdot \boldsymbol{r})\, \mathrm{d}v \tag{4.7}$$

There is a mathematical serendipity here because this integral is also that *Fourier component* of $V(r)$ which has wave-vector $\boldsymbol{q}$ (*see* Appendix 6). In other words, if we express $V(r)$ in the form of a Fourier integral, i.e. as a continuous sum of sinusoidal waves with wave-vectors $\boldsymbol{q}$, then the weighting factors in this sum, which are by definition the Fourier coefficients $W(q)$, are also the matrix elements required in the perturbation theory:

$$V(r) = \frac{1}{(2\pi)^3} \int W(q) \exp(-\mathrm{i}\boldsymbol{q} \cdot \boldsymbol{r})\, \mathrm{d}q \tag{4.8}$$

It is thus convenient to describe pseudopotentials $V(r)$ in terms of their Fourier components $W(q)$. As shown in Appendix 6, $W(q)$ can then be expressed in the form

$$W(q) = w(q) S(q) \tag{4.9}$$

where $S(q)$ is the structure factor (equation (4.3)) and $w(q)$ is the *pseudopotential form factor*, given by

$$w(q) = \frac{1}{\Omega_{\mathrm{a}}} \int v(r) \exp(\mathrm{i}\boldsymbol{q} \cdot \boldsymbol{r})\, \mathrm{d}r \tag{4.10}$$

where Ω_{a} is the volume per atom and $v(r)$ is the pseudopotential of one atom.

This form factor has been derived for several metals (Harrison 1966). Figure 4.1 shows some examples in which $w(q)$ is expressed as a fraction of the Fermi energy E_{F}, and q as a fraction of the Fermi wavenumber k_{F}, i.e. the wavenumber

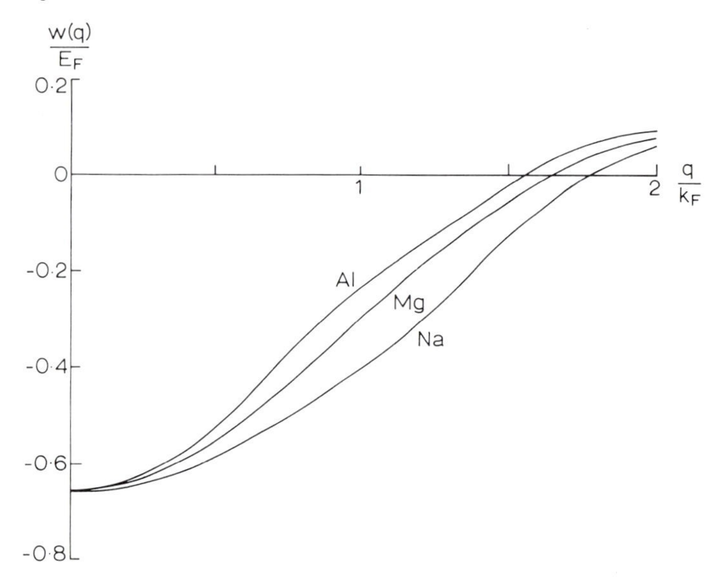

Fig. 4.1 Pseudopotential form factors, calculated by the method of Heine and Abarenkov (Harrison 1966)

of the surface of a sphere in k-space with the same volume as that of the actual Fermi distribution,

$$k_F = \left(\frac{3\pi^2 Z}{\Omega_a}\right)^{1/3} \tag{4.11}$$

where Z is the *metallic valence*, i.e. the number of metallic electrons per atom.

To understand Fig. 4.1 we begin by noting that $v(r)$ is made up of three factors. First is the energy of the Coulomb attraction $-Ze^2/r$ of a metallic electron to the net positive charge $+Ze$ of the ionic core. Second is the Pauli repulsion, according to which an electron cannot exist at a point already occupied by another electron (*see* Appendix 2); this is responsible for the additional nodes in the wavefunction of a metallic electron in the ionic core, from which the repulsive part of the pseudopotential derives. Added together, these give the *unscreened* potential, $v^0(r)$. Third is the electron screening discussed in Section 2.7 (Fig. 2.7) and Chapter 3. Combined with $v^0(r)$ it gives the *screened* potential $v(r)$, with the form factor $w(q)$ as shown in Fig. 4.1.

The simplest approach to $v^0(r)$ is to imagine all the electrons of the ionic core as being 'shrunk' down into the nucleus. This approximate *point-ion potential* leaves the Coulomb term $-Ze^2/r$ unchanged and converts the Pauli repulsion into a simple point repulsion which can be represented by a delta function $\delta(r)$ and which vanishes outside the central point $r = 0$. We then have

$$v^0(r) = -\frac{Ze^2}{r} + \beta\delta(r) \tag{4.12}$$

where β measures the strength of the Pauli repulsion. We now consider its Fourier components, $w^0(q)$. Since the Coulomb term is long-range compared with the delta function, it is the main contributor to the longer-wave components, i.e. those of small q; as a result, $w^0(q)$ is negative in this range. Conversely, the delta function predominates in the short-wave components and thus makes $w^0(q)$ positive when q is large. These effects are reflected in the general shape of the $w(q)$ curves in Fig. 4.1.

To obtain $w^0(q)$ we substitute from equation (4.12) into equation (4.10), which gives

$$w^0(q) = \frac{1}{\Omega_a}\left(-\frac{4\pi Ze^2}{q^2} + \beta\right) \tag{4.13}$$

using standard Fourier transforms (*see* Appendix 6). The effect of screening can then be dealt with to various degrees of approximation, as indicated in Chapter 3. The Thomas–Fermi approximation (*see* Section 3.4) leads, when q is small, to

$$w(q) = \frac{w^0(q)}{\kappa(q)} \tag{4.14}$$

and

$$\kappa(q) \simeq \frac{4me^2k_F}{\pi\hbar^2q^2} \tag{4.15}$$

where $\kappa(q)$ is the dielectric function (*see* equations (3.16), (3.7) and (1.13)). Thus, from equation (4.15) and its predecessors, we obtain for $q \rightarrow 0$

$$w(0) \simeq -\frac{k_F^2\hbar^2}{3m} \simeq -\frac{2}{3}E_F \tag{4.16}$$

which fixes the common starting point in Fig. 4.1. From equation (3.16) we have $\kappa(q) \rightarrow 1$ when q is large, and hence

$$w(q) \simeq w^0(q) \simeq \beta/\Omega_a \simeq \text{constant} \tag{4.17}$$

As a result of its shape, $w(q)$ passes through zero at a critical wavenumber q_0. One consequence of this is as follows. If, in a metal crystal, the Fermi surface happens to run close to a Brillouin zone boundary where the electrons could be strongly affected by the lattice field, i.e. in the region where $k_F \simeq \pi/a$ (*see* equation (1.17)), then Bragg reflection could change the wavenumber of such electrons from $+\pi/a$ to $-\pi/a$ (or vice versa), in which case $q = k' - k \simeq +\pi/a - (-\pi/a) \simeq 2\pi/a \simeq 2k_F$. Since $w(q)$ is nearly zero when $q \simeq 2k_F$, the scattering power of the pseudopotential is particularly weak for these electrons. This is an additional reason why such electrons behave very nearly as if they were free.

Other less extreme simplifications of the pseudopotential form factor have

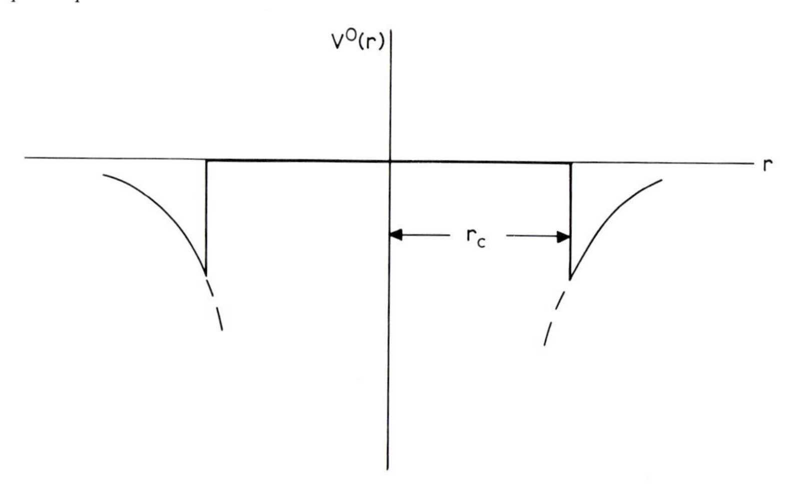

Fig. 4.2 The empty-core pseudopotential

been proposed. One of the most successful is the *empty core model* suggested by Ashcroft (1966). He argued that, since the Coulomb attraction and Pauli repulsion very nearly cancel each other inside the ionic core of radius r_c, then

$$v^0(r) = \begin{cases} 0 & (r < r_c) \\ -Ze^2/r & (r > r_c) \end{cases} \tag{4.18}$$

approximately, as shown in Fig. 4.2. The ensuing screened form factor is

$$w(q) = -\frac{4\pi Z e^2}{\Omega_a q^2 \kappa(q)} \cos q r_c \tag{4.19}$$

Closely related to it is the *model potential* proposed by Abarenkov and Heine (1965) in which the core is represented by a flat, non-zero potential that gives the same electron scattering (*see* Section 2.6) as the real core.

Despite its unconvincing shape (Fig. 4.2), the empty-core model is quite accurate when the core radius r_c is suitably chosen. Table 4.1 gives some values of this radius, which are roughly equal to the ionic radii in ionic crystals. The model has proved useful for calculating properties of metals from the pseudopotential theory.

More precise methods of calculating the form factor have been developed, but it is difficult, from theory alone, to achieve the high accuracy needed for some purposes. The greatest precision has been gained by using experimental measurements of electron band structures to fix the positions of key points on the theoretical $w(q)$ curve.

4.3 THE STRUCTURE FACTOR

The advantage of the splitting of the total pseudopotential $W(q)$ into $w(q)$ and $S(q)$ is that, once the form factor $w(q)$ of a given atom has been derived, it can

Table 4.1 Core Radii (Å) for the Empty Core Model (data from Harrison 1980)

Li	Be	B	C	Na	Mg	Al	Si	K
0·92	0·58	0·44	0·37	0·96	0·74	0·61	0·56	1·20
Ca	Zn	Ga	Ge	Rb	Sr	Cd	In	Sn
0·90	0·59	0·59	0·54	1·38	1·14	0·65	0·63	0·59
Sb	Te	Cs	Ba	Hg	Tl	Pb	Bi	
0·56	0·54	1·55	1·60	0·66	0·60	0·57	0·57	

then be used whenever that atom appears – in a perfect crystal, in an imperfect crystal, in a liquid, in an alloy, on a surface and so on – and whatever the feature being investigated – such as the cohesion of the crystal or liquid, the energy of a vacancy or a surface, or the electrical resistivity due to alloy atoms or thermal oscillations. All effects that depend upon the *positions* of the atoms relative to one another, which are the basis of these structure-dependent features, are represented by the structure factor $S(q)$ and for this the well-established methods of X-ray analysis are available.

The *intensity* $I(q)$ of scattering of an X-ray beam, at various angles θ, related to q as in equation (4.2), is given by

$$I(q) = [f(q)S(q)]^2 \tag{4.20}$$

per atom, where $f(q)$ is the (known) *atomic form factor* for X-ray scattering. Thus, if we are considering a structure for which $S(q)$ may not be known, e.g. a liquid, we can deduce $S(q)$ from the X-ray measurement of $I(q)$ and a knowledge of $f(q)$. This $S(q)$, combined with $w(q)$ in equation (4.9), then gives $W(q)$ and so opens the way to the calculation of the electronic properties of the structure.

One successful application has been to the theory of the *electrical resistivity* of simple liquid metals such as aluminium. Just as $W^2(q)$, with $f(q)$ in place of $w(q)$, measures the intensity of scattering of an X-ray beam through an angle θ, as given by equation (4.2), so $W^2(q)$ with the pseudopotential form factor $w(q)$ measures the intensity of scattering of a nearly free electron, from an initial state $\boldsymbol{k}$ into a final state $\boldsymbol{k}'$ of the same energy, at an angle θ to $\boldsymbol{k}$.

This scattering of conduction electrons is the basis of electrical resistivity. It does not exist in perfect crystals (because of the form of $S(q)$ for these, as discussed below) but results from irregularities which produce non-zero values of $S(q)$ at various q. There are many kinds of these, e.g. thermal vibrations of lattice atoms, impurity and alloy atoms, vacancies and other imperfections, and the irregular structure of the liquid. For the liquid the method has been to take $S(q)$ from X-ray measurements, $w(q)$ from pseudopotential curves such as those in Fig. 4.1, and then to use the ensuing $W(q)$ to calculate an $I(q)$ and from this, by standard methods, the resistivity of the liquid. Excellent agreement with measured resistivities has been obtained, e.g. for molten aluminium (Ashcroft and Guild 1965).

4.4 THE PERFECT CRYSTAL

The dominating feature for the perfect crystal is *Bragg reflection*. If a beam of waves, X-rays or electrons, travelling through the crystal fails to satisfy Bragg's law, then the crystal is perfectly transparent; there is no scattering. When the Bragg condition is satisfied the crystal scatters the waves strongly, but in such an organized, coherent way that a reflected wave similar to the incident one is formed. These two waves have equal status in the wavefunction of an electron at the Brillouin zone boundary (*see* equation (A 4.25) in Appendix 4). As regards $S(q)$ and $w(q)$, the sum of phase factors in equation (4.3) is zero for all $\boldsymbol{q}$ which do not give Bragg reflection, and each term in the sum is unity (*see* equation (A 4.29)) for those $\boldsymbol{q}$ that do give it.

It follows that, for the perfect crystal, we need consider the pseudopotential form factor $w(q)$ only for those wave-vectors $\boldsymbol{q}$ which are the *critical vectors* $\boldsymbol{g}$ at which Bragg reflection occurs. These $\boldsymbol{g}$ are the characteristic vectors of the crystallographer's *reciprocal lattice* (*see* equation (A 4.28)). For the b.c.c. lattice the reflecting planes which produce the first Brillouin zone boundary are the twelve planes of (110) type (*see* Fig. A 4.6) and the $\boldsymbol{g}$ for these is the reciprocal lattice vector in the $\langle 110 \rangle$ direction of length

$$\frac{2\pi}{a}(110) = \frac{2\pi}{a}(1 + 1 + 0)^{1/2} = \frac{2\sqrt{2}\pi}{a} \tag{4.21}$$

We write $w(\boldsymbol{q})$ for this particular $\boldsymbol{g}$ as $w(110)$. For the FCC lattice the reflecting planes of the first zone are the eight planes of (111) type and the six of (200) type. The corresponding reciprocal lattice vectors $\langle 111 \rangle$ and $\langle 200 \rangle$ have lengths $2\sqrt{3}\pi/a$ and $4\pi/a$. Again, the form factors are written as $w(111)$ and $w(200)$.

At points of high symmetry, such as N, L and X in Fig. A 4.6, it is often possible to calculate the form factor fairly accurately. For example, the values for aluminium are $w(111) = +0{\cdot}243$ eV and $w(200) = +0{\cdot}764$ eV (Ashcroft 1963). In the development of the NFE theory in Appendix 4, it is shown that the *band gap* ΔE (e.g. AB in Fig. 1.5) at such a point on a zone boundary is equal to twice the magnitude of the Fourier component of the potential there (equation (A 4.61)), i.e.

$$\Delta E(hkl) = 2\,|w(hkl)| \tag{4.22}$$

Thus, for aluminium, $\Delta E(111) = 0{\cdot}486$ eV and $\Delta E(200) = 1{\cdot}528$ eV. These values are small compared with the Fermi energy of this metal, about 10 eV, so this is a good NFE metal.

4.5 THE IMPERFECT CRYSTAL

For another example consider a simple form of *imperfect* crystal, i.e. one containing a single vacant atomic site. Removing one atom from an initially perfect crystal of N atoms has two effects. First, it slightly increases the atomic volume to $N\Omega_a/(N-1)$, which slightly decreases all the matrix elements. This is important for the energy of the crystal (*see* Chapter 5) but we shall ignore it here. Second, it changes $S(q)$ and hence $W(q)$. Starting from equation (4.3), let us remove the atom from the site $\boldsymbol{l}_0$, which we will make the origin of coordinates, $\boldsymbol{l}_0 = 0$. The structure factor is then

$$S(q) = \frac{1}{N}[\sum \exp(-\mathrm{i}\boldsymbol{q}\cdot\boldsymbol{l}) - \exp(-\mathrm{i}\boldsymbol{q}\cdot\boldsymbol{l}_0)] = \frac{1}{N}[\sum \exp(-\mathrm{i}\boldsymbol{q}\cdot\boldsymbol{l}) - 1] \quad (4.23)$$

where the sum is over the original perfect lattice. As before, when $\boldsymbol{q}$ gives Bragg reflection, every term in the sum is unity. Of greater interest, however, is the case where $\boldsymbol{q}$ does not give Bragg reflection. We no longer have $S(q) = 0$. Instead,

$$S(q) = -1/N \quad (4.24)$$

and hence

$$W(q) = -w(q)/N \quad (4.25)$$

For every q at which $w(q) \neq 0$, there is now a *non-zero* $W(q)$. Since $W(\boldsymbol{q})$ indicates a transition between any two states $\boldsymbol{k}$ and $\boldsymbol{k}'$ (on the Fermi surface) that differ by the vector $\boldsymbol{q}$, the vacancy is thus a centre of diffuse scattering of electrons in the crystal, i.e. a source of electrical resistance. By using the golden rule, equation (4.6), with the matrix element $W(\boldsymbol{q})$, it is possible to calculate this resistance. Satisfactory results have been obtained for simple metals (Harrison 1966).

This theory of the effect of a vacancy is a simple example of deviation from a perfect lattice. The theory can be extended directly to the theory of resistance due to alloy atoms, if the vacancy is replaced by such an atom, using a form factor for this which is different from that of the parent lattice. A further extension is to the theory of resistance due to thermal vibrations. In this case the lattice 'imperfections' are the *phonons*, i.e. the quanta of the thermal oscillations. The resistivity of the metal increases as a rising temperature brings more phonons into the structure.

REFERENCES

Abarenkov, I. V., and Heine, V., *Phil. Mag.*, **12**, 529 (1965).

Ashcroft, N. W., *Phil. Mag.*, **8**, 2055 (1963); *Phys. Lett.*, **23**, 48 (1966).

Ashcroft, N. W., and Guild, L. J., *Phys. Lett.*, **14**, 23 (1965).

Harrison, W. A., *Pseudopotentials in the Theory of Metals*, Benjamin, New York (1966); *Electronic Structure and the Properties of Solids*, Freeman, San Francisco (1980).

5

Cohesion of simple metals

5.1 COHESIVE ENERGIES

The strength, ductility, crystal structures and alloy behaviour of metals all emerge from the general cohesive properties, the forces and energies that hold atoms together in metals. The theory explains this cohesion and thus provides, in principle at least, a basis for understanding these metallurgical characteristics. It is of course a long step from the plane waves of free-electron theory to, say, the yield point of mild steel, and many intermediate concepts have been introduced to bridge the gap. When seeking to explain the strength of metals, we start, not from Schrödinger's equation, but usually from such things as grain boundaries, dislocations, elastic constants, impurity atoms and foreign inclusions. Nevertheless, for a deeper level of understanding - e.g. of why iron and aluminium are so different in their properties - we must turn to electron theory.

The method is to take a model of a metal (free electron, NFE, Wigner-Seitz or whatever) and calculate from it the dependence of the metallic electron and ionic core energies on the atomic volume. The existence of a minimum in this total energy function proves the mechanical stability of the metallic state at the equilibrium volume; and the curvature of the function about this minimum gives the elastic *bulk modulus of compression*.

The change in energy, when the metal is expanded from equilibrium to infinite spacing, gives the *metallic energy* relative to this infinitely expanded state. Depending on the model used, there are two general examples of this. In a free-electron or NFE model the metallic electrons remain as an electron *gas*, as the metal is expanded, and in the limit of infinite expansion the metal consists of a sparse array of positive ions in a dilute gas of free electrons. The metallic energy in this case is the *binding energy*, typically about 10 eV per atom. In other models, e.g. those put forward in Chapter 1 where a metal→non-metal transition occurs as the system is expanded, the limiting state is that of a dilute gas of *neutral atoms*. The metallic energy in this case is the *cohesive energy*, measured in practice as the *heat of vaporization* (more exactly, heat of sublimation extrapolated to zero temperature). Since the ionization of these atoms creates the gas of ions and electrons, there is the obvious relation *binding energy* = *cohesive energy* + *ionization energy*. In alkali metals the cohesive energy is only of the order of 1 eV per atom, and thus a more accurate theory is required to calculate it closely. Some of the other cohesive properties make still greater demands on the theory. For example, the *alloying energy* between high miscible metals is typically of the order of 0·1 eV per atom and the energy of transformation between (close-packed) crystal structures in good metals is even smaller. It is impossible to give sound *qualitative* explanations of many features of alloys and crystal structure when the distinctions rest on such minute energy differences.

5.2 SIMPLE METALS

In this chapter we consider only 'open metals', ones in which the ionic cores are 'not in contact' with each other and can be regarded as immutable objects in both the atomic and the metallic states. We thus assume that the energy levels of the core electrons are so far below the energy bands of the metallic electrons that there is no significant mixing of the core and metallic quantum states (*see* Appendix 5). This excludes the transition metals and those near them (copper, silver, gold and, to a lesser extent, zinc, cadmium and mercury), in which the outermost d electrons of the cores are strongly involved in the metallic state. We shall discuss these later.

We limit ourselves here to *simple metals*, those for which the NFE model is a good approximation. These are mainly alkalis, alkaline earths, aluminium, gallium, indium and lead. Beryllium and to some extent lithium are marginal cases because they have somewhat large pseudopotentials, and hence E, k curves which deviate appreciably from the free-electron parabola. This is because there are no p electrons in their ionic cores, so that the Pauli repulsion of the metallic electrons is rather small, which leaves the Coulomb attraction of the positive ionic charge less fully compensated than in the 'good' NFE metals.

5.3 PRINCIPLES OF METALLIC COHESION

There is one simple general cause of metallic cohesion: the *electrostatic attraction* between the ionic cores and the metallic electrons. Quantization complicates this in various ways, and often disguises it heavily. The amplitudes of the wavefunctions dictate the spatial distribution of the electrons and so limit the amount of potential energy gained from the ion–electron attraction, and the curvatures of these functions produce important kinetic energy effects. As a result it is possible to regard metallic cohesion in at least four different ways, as being due to:

(i) electrostatic ion–electron attraction,
(ii) reduced curvature of valency wavefunctions in the metal, compared with the free atom,
(iii) filling of only the lower levels of energy bands,
(iv) resonating covalent bonds between the ions.

As regards (ii), we recall that curvature of the wavefunction implies kinetic energy (*see* for example equation (A 1.8)). In the free atom there is an outermost curving of the wavefunction downwards, representing the dropping of electron density asymptotically towards zero at large distances from the atom. In the metal this outermost downward curving disappears because here the outer regions of one atom overlap the ionic cores of the neighbouring ones. The potential energy remains low at large distances from any one atom, which removes the need for the wavefunction to drop towards zero (*see* Figs. 2.2 and A 1.4). This reduction of curvature gives a corresponding drop in electronic kinetic energy, relative to the free atom, but the prime cause is still (i), since it is the electrostatic attraction of a valency electron to the surrounding regions of low potential energy provided by the neighbouring ions which changes the shape of the wavefunction. There is also, of course, the offsetting kinetic energy associated with the band structure of the Fermi distribution, in the metallic state. In the simple metals this Fermi energy is responsible for preventing inward collapse under the influence of (i).

The splitting of atomic levels into energy bands, (iii) above, provides a view of metallic cohesion similar to that of molecular orbital theory (*see* Figs. A 2.2 and A 4.9), but the effect is nevertheless essentially the same as (i) and (ii) above. The bottom of the band owes its low energy to the favourable positioning of electrons when neighbouring atomic orbitals combine in a bonding orientation (Fig. A 2.2), and the filling of part of the band, above this ground state level, is just the Fermi distribution again. For monovalent metals such as sodium the cohesion results from the filling of only the lower half of the band, so that the average electron energy is less than that of the atomic level from which the band derives. A simple rough estimate of the cohesive energy can be made from this. To a first approximation the splitting into the energy band is spread symmetrically about the atomic level (*see* equation (A 4.22) and Fig. A 4.9). Since a band derived from an atomic state is half-filled in a monatomic metal, the Fermi level E_F should be halfway up the band, i.e. at about the atomic level. The average Fermi energy is $\frac{3}{5}E_F$. Thus the average energy of the electrons in the band should be about $-\frac{2}{5}E_F$ per atom, below the level of the free atom. This gives a surprisingly good first estimate of the cohesive energy in alkali metals. In sodium, for example, $\frac{2}{5}E_F = 1{\cdot}28$ eV, which compares with the observed evaporation energy of 1·13 eV per atom. Such an explanation of cohesion would fail for a divalent metal such as magnesium were it not for the fact that the p orbitals have energy levels only slightly above the corresponding s valency levels, which broaden into overlapping bands in the solid, thus increasing the density of states and enabling the metallic electrons all to be fitted into low energy states.

Although it does not offer fundamentally new insights, the energy band method will prove very useful when we come to discuss the cohesion of the transition metals, where the NFE model is inappropriate and a tight binding approach has to be used (*see* Chapter 6).

The move towards a more chemical explanation of metallic cohesion is completed in method (iv), which envisages a definite *metallic bond* between neighbouring atoms, much like a covalent bond. This approach, pioneered by Pauling (1938, 1949, 1984), pictures a metal as held together by a network of nearest-neighbour atomic bonds. These bonding orbitals are only partly occupied by electrons, so that each such electron is able to move, via holes in neighbouring ones, from orbital to orbital through the metal. To a chemist this is an example of the *electron-deficient covalent bond* in which the valence is unsaturated because there are more bonding orbitals than electrons to fill them. The simplest example of an electron-deficient bond is the molecular hydrogen ion H_2^+ discussed in Appendix 1; molecules such as boron hydride provide the usual examples. A physicist might reinterpret this picture along the lines of the other approaches mentioned above. For example, in a metal crystal the lines of linked bonds, considered as lines of orbitals, could be represented as one-dimensional Bloch functions, as in Appendix 4 (equation (A 4.9)). All the concepts of band theory then reappear (March 1974), and lead the theory back, once again, to (i).

The two primary effects, then, in metallic cohesion are the *electrostatic attraction* between ions and metallic electrons, which on account of its $-e^2/r$ form gives a negative energy term $-B/r$, where r is the atomic radius in the metal; and the *Fermi repulsion*, which gives a positive energy term $+A/r^2$ (*see* equation (A 3.8)). The total energy of the metal is thus, to a first approximation,

$$E = \frac{A}{r^2} - \frac{B}{r} \tag{5.1}$$

Although a rough estimate, this expression correctly represents the main effects in the cohesion of the simple metals.

With precise calculations there are numerous complications and difficulties. Outstanding among these are the effects of electron–electron interaction, the exchange and correlation aspects discussed in Chapter 3. There have been two main ways through these problems. First was the brilliant early theory of Wigner and Seitz (Chapter 2) which elegantly bypassed many of the problems to give a sophisticated account of cohesion under some rather limiting restrictions. Second was the (nearly) free-electron approach, which starts off much more naively but, with the help of the concepts of the pseudopotential and the density functional, can now go a long way forward.

5.4 COHESION BY THE WIGNER–SEITZ METHOD

The Wigner–Seitz cellular method, introduced in sections 2.2 and 3.3, has three great advantages for the theory of the cohesion of a simple monovalent metal such as sodium:-

(i) It provides a realistic ground state wavefunction of a metallic electron, the energy E_{WS} of which can be determined as a function of the radius r_s (*see* equation (2.1)) of the WS spherical cell.

(ii) As noted in Section 3.3, the assumption that in a monovalent metal each WS cell contains a *single* electron disposes of electron correlation in a simple but effective way. Each WS cell simply becomes a correlation hole.

(iii) This assumption also disposes of the long-range electrostatic interactions. Each cell is electrically neutral and so, being also nearly spherical in crystals such as f.c.c. and b.c.c., has hardly any electrical interaction with any other cell. The whole theory of cohesion is thus reduced simply to evaluating the energy of one electron from its interaction with the ion in its own cell. This problem was already solved in the calculation of E_{WS} from the wavefunction and core potential in

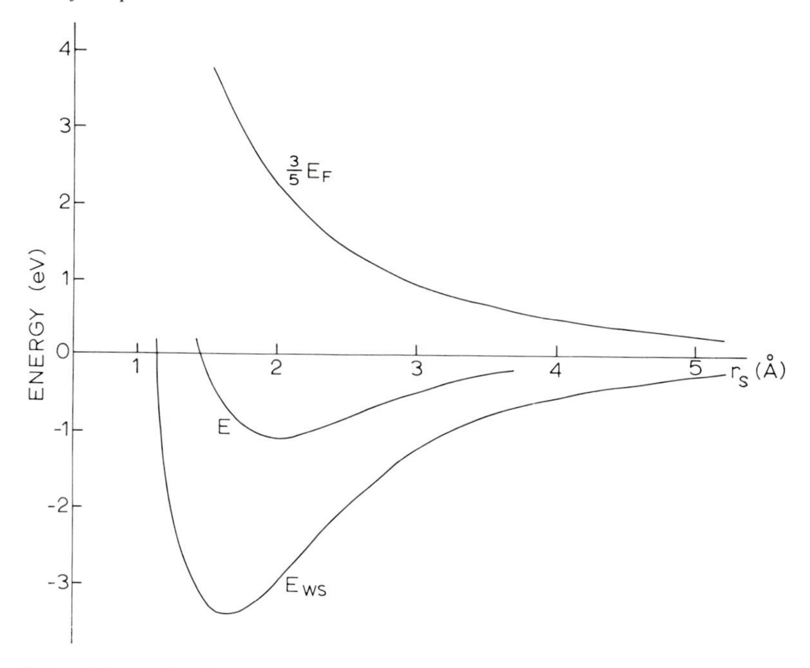

Fig. 5.1 **The calculated ground state energy E_{ws}, average Fermi energy $\frac{3}{5}E_F$ and total metallic electron energy E of sodium, as functions of the Wigner–Seitz radius r_s**

the cell. As noted in the discussion of Fig. 2.2, this E_{WS} varies at large r_s approximately as r_s^{-1}, in accordance with the electrostatic nature of the cohesion and the nearly uniform distribution of the electronic charge in the cell.

To obtain the average total electron energy we have to add to E_{WS} the Fermi energy for which, as a first approximation, we can use the free electron value $\frac{3}{5}E_F$. Thus the simplest form of the WS theory gives the electronic energy per atom of the metal as

$$E = E_{WS} + \tfrac{3}{5}E_F \tag{5.2}$$

As shown in Fig. 5.1, the minimum in E, which represents the equilibrium state of the metal, occurs at a slightly greater radius than that of the minimum in E_{WS}. The depth, 1·1 eV, of this minimum in E (relative to the value at $r_s = \infty$) is the calculated *cohesive* energy. This is because in Fig. 2.2 the asymptotic value E_0 of E_{WS}, as $r_s \rightarrow \infty$, is the energy level of the free neutral atom. The WS wavefunction is not a free-electron function and, in the limit $r_s \rightarrow \infty$, it becomes, not the function of a free electron in a dilute gas, but that of the valency electron, with energy E_0, in the neutral atom.

There have been many refinements of such calculations, for example taking better account of exchange and correlation effects, introducing an effective mass m^* in place of m in the calculation of E_F, and estimating the effect on the potential energy of replacing the polyhedral WS cell by the equivalent sphere. An accurate method of deriving the potential of the core from atomic spectroscopy data, known as the *quantum defect method*, has led to precise calculations of the cohesive properties of the alkali metals (Kuhn and van Vleck 1950, Ham 1955, Brooks 1956).

The main limitations of the WS approach are that it provides a wavefunction only for the symmetrical case $k = 0$, that it is not easily generalized to multivalent metals where the correlation effects cannot be so simply disposed of, and that the replacement of the WS polyhedron by the equivalent sphere rules out all effects of crystal structure. Not only are the Brillouin zone effects eliminated (these of course make only minor contributions to cohesion in 'good' NFE metals); the WS sphere approximation removes the means of explaining the choice of specific crystal structures, or of understanding the difference between crystalline and liquid metals.

5.5 CLOSE PACKING IN METALS

Why are metals close-packed? A convenient measure of close packing is the ratio

$$\eta = b/2r_s \tag{5.3}$$

where b is the spacing of nearest neighbours in the structure and r_s is the radius of a sphere of atomic volume, i.e. the WS sphere. Some values are given in Table 5.1. The more complex crystal structures of metals such as mercury and indium are also close-packed, being slight distortions of f.c.c. The small volume increase (about 3%) which occurs on melting shows that metals remain fairly close-packed in the liquid state.

To explain close packing we must look for a feature which, at constant atomic volume, favours a high value of η. The Fermi energy is of no use, for it depends *only* on volume, not structure. The feature we need is in fact the *total electrostatic energy* of the whole system of ions and metallic electrons. To evaluate this we abandon the WS assumption of electrically neutral spherical cells, and sum or integrate the various e^2/r Coulomb energy terms over the entire system. The simplest model we can use for this is a lattice of *point ions*, each of charge $+Ze$, embedded in a *uniform distribution of free electrons*, Z per ion, N per unit volume:

$$N = Z/\Omega_a \tag{5.4}$$

Table 5.1 Close-packing parameters

Parameter	f.c.c.	c.p.h. (ideal)	b.c.c.	s.c.	Diamond	WS sphere
η	0·905	0·905	0·879	0·806	0·698	...
α	1·79	1·79	1·79	1·76	1·67	1·80

This is rather like the Hartree theory described in Section 3.2, except that there the ions were 'spread out' uniformly as jellium, the continuous charge of which everywhere neutralized the uniform electronic charge distribution so that the electrostatic energy (in the absence of exchange and correlation effects) was zero. We now condense the jellium into a lattice of point ions and consider the electrostatic energy of this modified Hartree model.

Since the charge distributions are already fixed by the above assumptions, the electrostatic interaction energies are given precisely as integrals of e^2/r in which the quantum-mechanical factors, $\psi^*\psi$, are replaced by the simple constant N. Nevertheless, the long-range nature of the e^2/r interactions creates mathematical difficulties, but these are similar to those encountered in the electrostatic theory of ionic crystals which have been solved by the calculation of the *Madelung constant*. A similar method of solution applicable to the present problem, developed by Ewald (1921) and Fuchs (1935, 1936a, b), gives the total electrostatic energy E_{el} of the point ions and the uniform electron gas as

$$E_{el} = -\frac{1}{2}\frac{Z^2e^2}{r_s}\alpha \tag{5.5}$$

per ion, where α is a *structure constant*, equivalent to the Madelung constant, values of which for various structures are given in Table 5.1. The value given there for the WS sphere relates to the energy of a point ion in a *single* WS sphere containing a uniform distribution of electronic charge totalling $-Ze$.

We see from Table 5.1 that the electrostatic energy clearly favours the close-packed structures; also that, to this level of accuracy, there is no perceptible difference between the energies of the f.c.c., b.c.c. and c.p.h. structures at the same atomic volume; and further that the α factor for these three structures is practically the same as for the WS sphere, which justifies the WS assumption that the electrostatic interaction between these spheres is negligibly small.

5.6 COHESION BY THE PSEUDOPOTENTIAL METHOD

As we saw in the discussion of the Hartree theory (Section 3.2), such an energy calculation inadequately assesses the electrostatic binding because it neglects the contribution of electron–electron correlation. This keeps the electrons out of one another's way and so reduces the repulsive component of the total electrostatic interactions in the metal. It is clear, by analogy with the WS sphere (the electrostatic energy of which is represented by $\alpha = 1{\cdot}8$ in Table 5.1), that the correlation energy must be of comparable magnitude to E_{el}. From equation (3.5), we have

$$E_x = -12{\cdot}5/r_s \tag{5.6}$$

in electron-volts per atom, for alkali metals, where r_s is measured in units of the Bohr radius a_0 ($= \hbar^2/me^2 = 0{\cdot}53$ Å). For the same metals

$$E_{el} = -24{\cdot}35/r_s \tag{5.7}$$

which is only twice as great.

By adding these values to the average free-electron Fermi energy $\frac{3}{5}E_F$, which is $+30{\cdot}1/r_s^2$ in the same units, we obtain a first estimate from free-electron theory of the total metallic energy:

$$E_t = \frac{30{\cdot}1}{r_s^2} - \frac{36{\cdot}8}{r_s} \tag{5.8}$$

in electron-volts per atom (compare this expression with equation (5.1)). The minimum occurs at $r_s = 1{\cdot}6a_0$, which compares badly with the observed range, from about $2a_0$ to $6a_0$; moreover, it gives the same r_s for every alkali metal.

These faults stem of course from the assumption of point ions. The next step is to replace the point ions by finite ion cores, which increase roughly in proportion to the ionic radii of the alkali metals. The simplest (unscreened) pseudopotential we can use is the *Ashcroft empty core model* (equation (4.18)),

$$v^0(r) = \begin{cases} 0 & (r < r_c) \\ -Ze^2/r & (r > r_c) \end{cases} \tag{5.9}$$

The values of r_c can be determined by fitting this expression to free atoms, choosing r_c so that the valence s state has an ionization energy equal to the observed value. Values obtained by a slight improvement of the method (Harrison 1980) are given in Table 5.2. The values of r_i, also given, are standard ionic radii deduced from the observed lattice constants of ionic crystals. We see that their magnitudes are fairly similar for the NFE metals.

The best way of using equation (5.9) is to improve the E_{el} of equation (5.5). Since this was calculated on the assumption of a uniform electron density, the improvement consists simply in deleting, from this E_{el}, that energy which is due to ion–electron Coulomb interaction inside a sphere of radius r_c; this corresponds to the change from the point-ion to the empty-core potential inside this region. In a volume element dv the amount of electronic charge is $-eN\,dv$. If this element is at a distance r ($\leqslant r_c$) from the point-ion of charge $+Ze$, this element of Coulomb interaction is $-(Ze^2N/r)\,dv$. Thus we have to add to the E_{el} of equation (5.5) the *positive* energy

$$E_{ec} = \int \frac{Ze^2N}{r}\,dv = 2\pi Ze^2 r_c^2 N \tag{5.10}$$

Table 5.2 Core radii r_c and ionic radii r_i (Å) (data from Harrison 1980)

Radius	Li	Na	K	Rb	Cs	Be	Mg	Ca	Sr	Ba
r_c	0·92	0·96	1·20	1·38	1·55	0·58	0·74	0·90	1·14	1·60
r_i	0·7	0·98	1·33	1·49	1·7	0·34	0·75	1·05	1·18	1·38

Radius	Al	Ga	In	Tl	Zn	Cd	Hg	Si	Ge	Sn	Pb
r_c	0·61	0·59	0·63	0·60	0·59	0·65	0·66	0·56	0·54	0·59	0·57
r_i	0·55	0·62	0·92	1·05	0·83	0·99	1·27	...	...	...	...

Table 5.3 Comparison of theoretical and observed cohesive properties for Na, Mg and Al (data from Harrison 1980)

Parameter		Na	Mg	Al
r_s (Å)	theoretical	2·26	1·87	1·65
	observed	2·11	1·76	1·58
K (10^{11} J m^{-3})	theoretical	0·074	0.54	1·63
	observed	0·064	0·35	0·76
E_t, binding energy (eV per ion)	theoretical	5·3	21·6	52·2
	observed	6·3	24·4	56·3
Cohesive energy (eV per ion)	observed	1·1	1·5	3·4

where the integral is over all volume elements for which $r \leqslant r_c$.

The total *binding energy* per ion of the metal is thus

$$E_t = \tfrac{3}{5}E_F + E_{ec} + E_{el} + E_x \tag{5.11}$$

the binding being provided by the negative terms E_{el} and E_x, the resistance to collapse by the positive terms $\frac{3}{5}E_F$ and E_{ec}. Predictions from this are compared with observation, for three 'good' NFE metals, sodium, magnesium and aluminium, in Table 5.3. Here r_s is of course obtained by minimizing E_t with respect to the atomic radius, and K is the bulk modulus, given by

$$K = \Omega_a \frac{\partial^2 E_t}{\partial \Omega_a^2} \tag{5.12}$$

where Ω_a is the atomic volume. The agreement is fairly good, considering the drastic simplifications made. The E_t comparison is of course with the *binding energy*, which contains the large contribution from ionization of the free atom. By contrast, the observed *cohesive energy* is much smaller and comparable to the errors in the theoretical estimates. There have been many more accurate calculations, based on better pseudopotentials, more precise exchange and correlation corrections, and quantum defect methods, culminating in some massive and impressively accurate computations (Moruzzi *et al.* 1978) based on the *local functional density* method (*see* Chapter 3).

5.7 BAND-STRUCTURE ENERGY

We have seen that the Ewald electrostatic energy E_{el} does not differentiate between close-packed structures (except to an insignificant degree, not apparent from Table 5.1). We must look elsewhere for structural differences. It will be convenient to divide E_{el} into a major, structure-independent part E_S, represented by the WS sphere ($\alpha = 1{\cdot}8$) in Table 5.1, and a minor, structure-dependent residue E_E:

$$E_{el} = E_S + E_E = -\frac{1}{2}\frac{Z^2e^2}{r_s}[1{\cdot}8 + (\alpha - 1{\cdot}8)] \tag{5.13}$$

and then to combine E_S with $\frac{3}{5}E_F$, E_{ec} and E_x into a total term E_V, dependent only on volume, not on structure:

$$E_V = \tfrac{3}{5}E_F + E_{ec} + E_S + E_x \tag{5.14}$$

The term E_E is insignificant for highly symmetrical close-packed structures, because their WS polyhedra are so nearly spherical, but it is much larger for less symmetrical structures and is thus an important source of resistance to shear distortion, from which the shear coefficients of elasticity derive.

When we go beyond the free-electron theory to take account of Bragg reflections, a second structure-dependent term appears. This is the *band-structure energy*, E_{BS}. The total energy of the metal is then

$$E_t = E_V + E_E + E_{BS} \tag{5.15}$$

Figure 5.2 shows the general effect. The presence in k-space of a Brillouin zone boundary, due to Bragg reflecting planes in the crystal which have a reciprocal

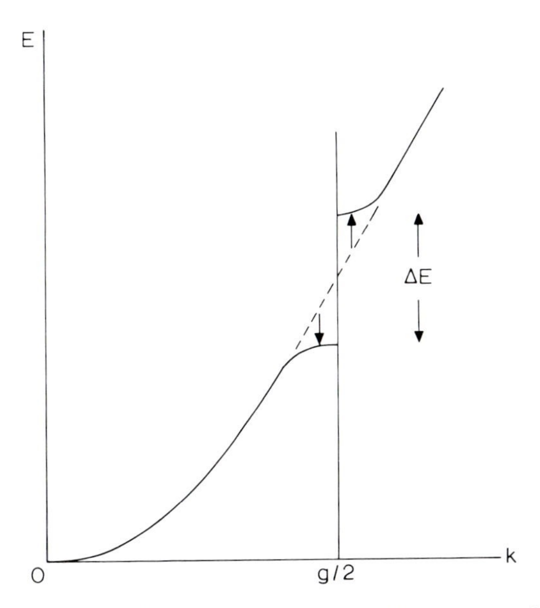

Fig. 5.2 Change of band energy due to a Brillouin zone boundary

lattice vector $\boldsymbol{g}$ (*see* Appendix 4), distorts the E, k relation from the free-electron parabola to the form shown. The effect of this deviation is *always* to lower the electron energy below the free electron value, so that E_{BS} is always negative and the cohesion is increased. This is obvious when the wavenumber at the Fermi surface lies just below or just reaches $\frac{1}{2}g$; but even when the electron distribution overlaps into the second zone, where the deviation is positive, the overall effect is still negative because most of the electrons reside in the lower zone where the deviation is negative.

The magnitude of E_{BS}, which is small compared with E_V, depends on the band gap ΔE. This, as shown in Appendix 4 and Chapter 4 (equations (A 4.61) and (4.22)), goes as $\Delta E = 2|w(g)|$, where $w(g)$ is the pseudopotential form factor with the reciprocal lattice vector $\boldsymbol{g}$, corresponding to a Bragg reflection from $\boldsymbol{k}$ to $\boldsymbol{k}'$, where $\boldsymbol{k}' = \boldsymbol{k} + \boldsymbol{g}$ and $|\boldsymbol{k}| = |\boldsymbol{k}'|$, off the crystal planes responsible for this zone boundary. It is shown in Appendix 7 (equations (A 7.4) and (A 7.5)) that E_{BS} is proportional to the square of the band gap:

$$E_{BS} \propto -|w(g)|^2 \tag{5.16}$$

Physically, the reason for this is that an electron, in the course of its movement through the structure, has to interact with at least two ions in order to sense the structure; and each of these two interactions is proportional to the form factor of the ion involved. Heine and Weaire (1970) have related this result to the general form of the pseudopotential function, as shown in Fig. 4.1, and extracted from it explanations of several features of the crystal structures of metals, which we shall now summarize.

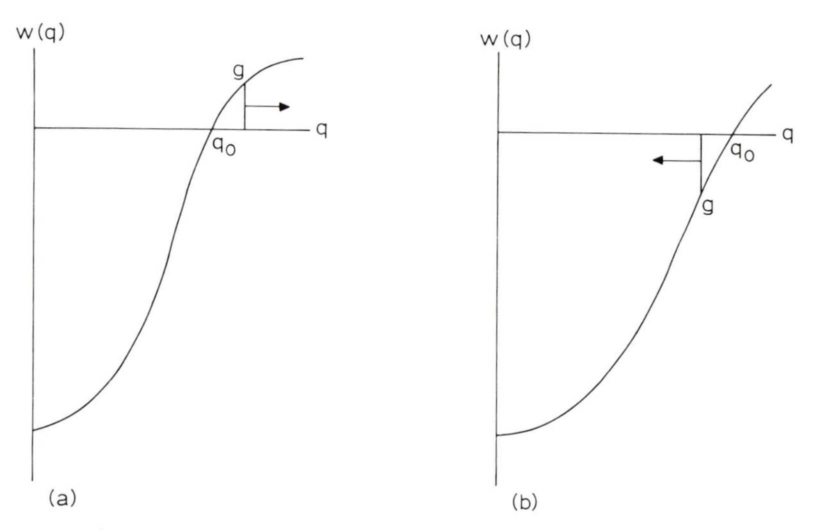

Fig. 5.3 **The pseudopotential form factor, showing typical positions of the first reciprocal lattice vectors; the band structure energy may cause lattice contraction (a) or expansion (b) as indicated by the direction of the arrow**

The key feature is that, as q is increased from zero towards $2k_F$, the form factor $w(q)$ passes through zero, at the point q_0 in Fig. 5.3, which is where the Coulomb attraction and Pauli repulsion in the pseudopotential balance. Figure 5.3 also shows two typical positions of the reciprocal lattice vector g for the most widely spaced Bragg reflection planes of the crystal. If g were to coincide with q_0, the band-structure energy would be zero. Although the Bragg planes would be geometrically capable of reflecting, the pseudopotential would then be zero in that component which matches the spacing of these planes. The further g is from q_0, in *either direction*, the greater is $|w(q)|$ and hence the greater is the (negative) E_{BS}. The cohesion of the crystal thus increases as g moves away from q_0, and the E_{BS} term in the cohesive energy thereby provides a 'force' driving g away from q_0. Since $g = 2\pi/d$, where d is the spacing of the lattice planes giving this Bragg reflection, then crystals with $d > 2\pi/q_0$ are *expanded* by E_{BS} to greater spacings, while those with $d < 2\pi/q_0$ are *contracted* to smaller spacings. Heine and Weaire showed that this could account for the directions in which the observed lattice constants of a large group of non-transition metals deviated from the values predicted by a free-electron theory of E_V similar to that described in Section 5.6 above.

As an example, we give in Table 5.4 Heine and Weaire's comparison of observed and calculated r_s values for aluminium, gallium, indium and thallium (from the same group of the Periodic Table). Also given are the ratios g/q_0, where g is the reciprocal lattice vector of the main reflecting planes which determine the first Brillouin zone. We see that for aluminium the situation is as shown in Fig. 5.3(a), so that the band-structure energy is expected to *contract* the crystal structure, in agreement with the r_s values given. For thallium the band-structure energy is expected to expand the structure, again in agreement with the figures.

The crystal structures of gallium and indium are less symmetrical, but can be regarded as distortions of f.c.c. If they were f.c.c., then the g/q_0 values would be as given, i.e. the 'average' g of the first zone would fall rather close to q_0, in which case the band-structure energy would be very small. Heine and Weaire emphasized that there is a further possibility in such cases. Suppose (*see* Fig. 5.4) that a simple close-packed crystal structure has its main g value very near to the q_0 of the metal concerned. Then if the crystal cell distorts, e.g. from cubic to tetragonal as shown in Fig. 5.4, so bringing some of the reflecting planes closer together and moving others further apart, the g splits into higher and lower

Table 5.4 Values of r_s and g/q_0 for Al, Ga, In and Tl (data from Heine and Weaire 1970)

Parameter		Al	Ga	In	Tl
r_s (Å)	observed	1·58	1·67	1·84	1·90
	theoretical	1·73	1·64	1·79	1·64
g/q_0	g_{111} (f.c.c.)	1·04	0·94	0·93	. . .
	g_{200} (f.c.c.)	1·2	1·09	1·08	. . .
	g (c.p.h.)	. . .	. . .	. . .	0·92

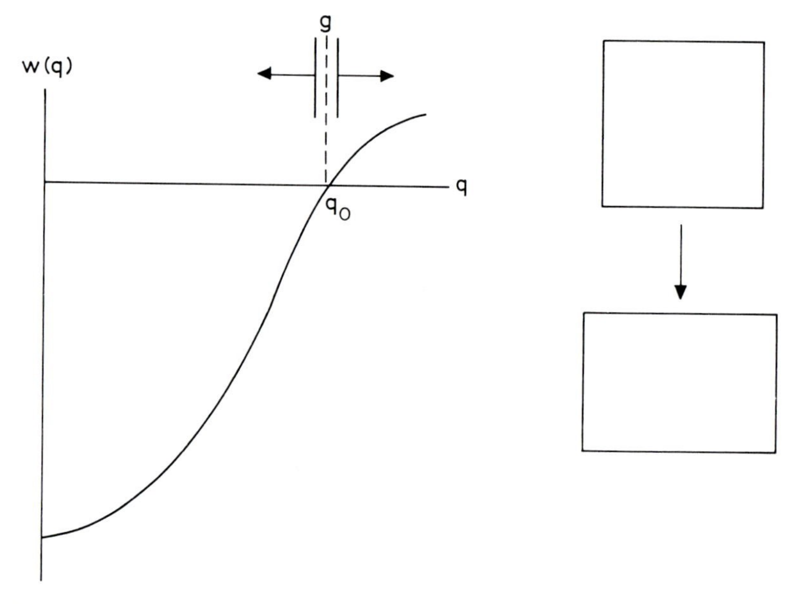

Fig. 5.4 When g coincides with q_0 the crystal structure may transform to one of lower symmetry in order to split g into values above and below q_0

values for these two sets of planes, both of which the crystal finds energetically advantageous. The E_{BS} effect thus favours a slightly asymmetrical variant of the close-packed structure in such cases.

Indium has a face-centred tetragonal structure with an axial ratio $c/a = 1{\cdot}08$, while gallium has a more complicated structure with seven nearest neighbours. By studying the positions of the main reflecting planes for this structure in relation to the function which determines E_{BS}, Heine and Weaire concluded that seven neighbours is the most that can be placed at the optimum distance without reducing the atomic volume of gallium below the best value for the volume-dependent energy, E_V in equation (5.15).

By similar arguments they explained the complexity of the crystal structure of mercury and the large axial ratios of the hexagonal structures of zinc and cadmium ($c/a = 1{\cdot}86$ and $1{\cdot}89$, respectively), compared with the nearly ideal $c/a = 1{\cdot}63$ of magnesium. Zinc and cadmium have large axial ratios for much the same reason that gallium has a small ratio: it enables them each to have six close neighbours, the most that can satisfy the band-structure requirement if the volume is to remain at that which minimizes E_V.

The question of why some of those NFE metals with simple structures are f.c.c., some b.c.c. and some c.p.h. presents a considerable challenge to the theory. An early calculation by Harrison (1965) of E_{BS} (at constant volume) for these three structures correctly predicted c.p.h. for sodium and magnesium, and f.c.c. for aluminium. (Sodium becomes c.p.h. at low temperatures, and its preference for b.c.c. at higher temperatures is an entropy effect, due to a low-frequency vibrational mode in b.c.c., which reduces the free energy.) Subsequent calculations for other metals proved less successful, however, and it is likely that second-order perturbation theory, which is the basis of the calculations outlined in this Section and Appendix 7, is too drastic an approximation to identify with certainty the small energy differences between these rival close-packed structures. Calculations to the third order of refinement are probably necessary.

In view of this it might seem impossible to understand qualitatively why a particular simple metal has a particular crystal structure. Heine and Weaire (1970) have, however, succeeded in this. Their key point was that one of the factors in E_{BS}, i.e. $\chi(q)$ (the *perturbation characteristic*, as given by equations (A 7.1)–(A 7.5)) varies with q in the same manner as E_x varies with k/k_F in Fig. 3.1: it increases very rapidly from a negative lower limit to a zero upper limit as q, increasing, passes through the value $2k_F$. They therefore approximated it drastically by the step function

$$\chi(q) = \begin{cases} C & (q < 2k_F) \\ 0 & (q > 2k_F) \end{cases} \tag{5.17}$$

where C is a negative constant, and assumed that its effect on E_{BS} is so dominant that E_{BS} also behaves in the same way. The value of E_{BS}, on this basis, is simply proportional to the number of reflecting planes with $g \leqslant 2k_F$ in the crystal structure.

The alkali metals, with $Z = 1$, have the smallest k_F. For them the c.p.h. structure has the advantage because its widely spaced reflecting planes, $\{1\bar{1}00\}$, give $g \simeq 2k_F$ for $Z = 1$ and so can contribute a strong E_{BS}, whereas the first reflecting planes in f.c.c. and b.c.c. are too closely spaced and give only a weak E_{BS} at this k_F. However, when $Z = 1{\cdot}5$, b.c.c. is best favoured, with its twelve $\{110\}$ reflecting planes that have $g < 2k_F$ for this Z; which will be of interest when we consider alloys such as the b.c.c. β-brass. For $Z = 2$, c.p.h. is strongly favoured again because all the planes of its first Brillouin zone, i.e. $\{1\bar{1}00\}$, $\{1\bar{1}01\}$ and $\{0002\}$, have $g < 2k_F$ for this Z, which is consistent with the predominance of this structure among the divalent metals. When $Z = 3$, as in aluminium, the eight $\{111\}$ and six $\{200\}$ planes give f.c.c. the biggest E_{BS}.

5.8 CENTRAL FORCES

The structure terms E_E and E_{BS} in equation (5.15) can be interpreted as if they represented *central forces*, acting between ion and ion along the line between their centres. This is obvious for the electrostatic E_E but is perhaps surprising for E_{BS}, which comes from the perturbation of the electron gas by the ionic fields. However, this perturbation leads to the screening of the ionic fields by the electrons and, by means of this screening, it reproduces (with opposite sign) some of the main features of the electrostatic fields responsible for it. The two terms can in fact be conveniently taken together as one composite *effective ion–ion interaction,* $\Phi(r)$ (*see* Appendix 7 for details).

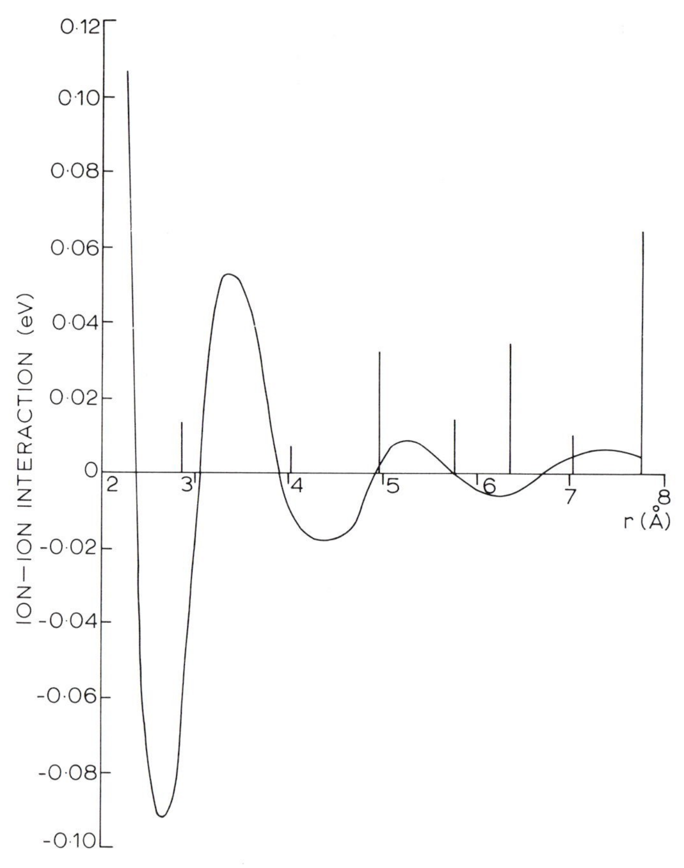

Fig. 5.5 The effective ion–ion interaction for aluminium

Figure 5.5 shows an example, for aluminium, from Harrison (1964). In interpreting this curve we must remember that it does not include the much larger E_V component of the total energy and so, when it is used to discuss alternative structures, we should think of it as describing interactions between ions swimming about in an electron gas *of fixed volume*. We see that when two such ions approach closely there is a sharp repulsion – almost as if they were hard spheres – as the electron screening is inoperative at such distances and there is then a bare E_E repulsion between the ions. The quantum-mechanical basis of the electronic perturbation leads to successive *haloes* of screening which, at large distances, become the Friedel oscillations (*see* Fig. 2.7). This is to be expected since the ion–ion interaction consists of the electrostatic energy of one ion sitting in the haloes of the screened field of the other ion.

Also shown in Fig. 5.5 are vertical bars at the positions of neighbouring ions in f.c.c. aluminium, the height of each indicating the number of neighbours at that distance. The twelve nearest neighbours obviously have the strongest

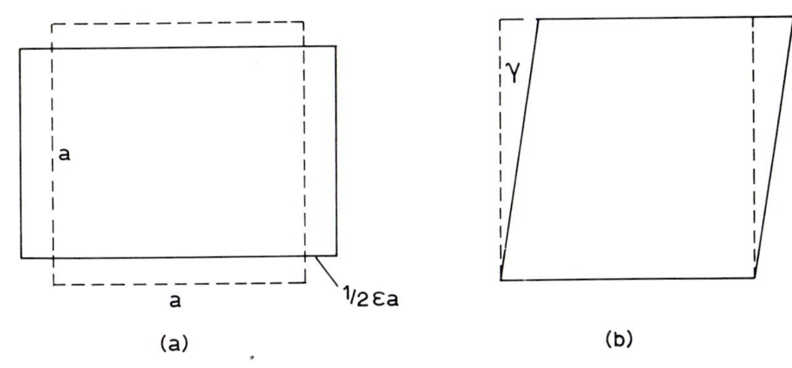

Fig. 5.6 **Two modes of shear distortion for a cubic crystal (*see* the text)**

interaction, but they are not set at quite the ideal point on the curve. The E_V term and the geometry of the f.c.c. structure thus exert a constraint on the approach of nearest neighbours. If this constraint were lifted, perhaps as in the liquid state or at a lattice defect, nearest neighbours might move a little closer, provided the volume were kept constant through a compensating decrease in the coordination number.

Curves such as that of Fig. 5.5 can be useful when considering alternative structures, always at the same volume, for metals. We shall see examples of this later.

The ion–ion central force does not of course mean that the total cohesion of a metal can be represented in terms of 'bonds' between the ions, for it involves only the minor contribution of $\Phi(r)$ to E_t. The major E_V term is indifferent to where the ions are, so long as the volume stays the same, and so cannot possibly be represented in terms of bonds.

5.9 SHEAR ELASTIC CONSTANTS AND ATOMIC OSCILLATIONS

The cubic → tetragonal distortion shown in Fig. 5.4 is an example of *shear* at constant volume. The effects which govern it are obviously related to those which govern the resistance of a crystal to elastic shear produced by an external force. Elastic *shear constants* are defined in terms of infinitesimal distortion at constant volume. For a cubic crystal this takes two basic forms, shown in Fig. 5.6. The first is closely related to the distortion shown in Fig. 5.4 and consists of an extension of one cube edge, i.e. $a \rightarrow a(1+\varepsilon)$, and a corresponding contraction, $a \rightarrow a(1-\varepsilon)$, along one other cube edge. The second consists of shear through an angle γ, along a cube face in the direction of a cube edge. In terms of the standard cubic crystal elastic constants, c_{11}, c_{12} and c_{44}, these distortions give

$$\frac{1}{2\Omega_a}\frac{\partial^2 E_t}{\partial \varepsilon^2} = c_{11} - c_{12} \qquad (5.18)$$

$$\frac{1}{\Omega_a}\frac{\partial^2 E_t}{\partial \gamma^2} = c_{44} \qquad (5.19)$$

where Ω_a is the atomic volume. We also have

$$3K = c_{11} + 2c_{12} \tag{5.20}$$

where the bulk modulus, K, is defined by equation (5.12).

In equation (5.15) for E_t, the structure terms E_E and E_{BS} give the shear elasticity, whereas the volume term E_V determines the resistance to compression and expansion. We thus do not expect close correlation between the bulk modulus and the two shear moduli. One indication of this is that, particularly in the alkali metals, the *Cauchy relations* (i.e. $c_{12} = c_{44}$ in cubic crystals) are not even approximately valid. These relations follow from the assumption that the forces in the crystal are central forces, acting between ions along the lines between their centres; but the major E_V term is unaffected by the relative positions of the ions, so long as the optimum atomic volume is preserved.

Although the electrostatic term E_E contributes almost negligibly to the cohesive energy of a highly symmetrical crystal structure because the WS polyhedron is nearly spherical in this case, it does make a strong positive contribution to elastic shear strength because the effect of shear deformation is to distort the polyhedron away from its near-spherical shape. If we think of the crystal as a lattice of positive point ions embedded in a uniform, neutralizing electronic fluid, then we see that an elastic shear at constant volume reduces some ion–ion distances, and the consequent increase in their repulsive energy more than compensates (in second order, for infinitesimal deformation) for the drop in this energy that results from the increase in other ion–ion distances. The contribution of E_E was calculated by Fuchs (1935, 1936a, b) using Ewald's method, as in equation (5.5), to give the values in Table 5.5. These values have to be multiplied by a^{-4}, where a is the lattice constant, and are then in units of 10^{11} J m^{-3}. We see that the electrostatic energy is more sensitive to the c_{44} type of distortion, especially in f.c.c.. For metals of higher valency the same values apply, but multiplied by Z^2.

For sodium the calculated values (in 10^{11} J m^{-3}) are $\frac{1}{2}(c_{11} - c_{12}) = 0{\cdot}0068$ and $c_{44} = 0{\cdot}05$, which compare well with the observed values of 0·0045 and 0·045 respectively, showing that in this monovalent metal the E_{BS} contribution is very small, as would be expected since the first Brillouin zone is only half-full, up to a near-spherical Fermi surface. For aluminium the calculated values are

Table 5.5 Contribution of electrostatic energy to the shear constants of monovalent metals

	f.c.c.	b.c.c.
$\frac{1}{2}(c_{11} - c_{12})$	4·52	2·26
c_{44}	40·5	16·8

$\frac{1}{2}(c_{11} - c_{12}) = 0{\cdot}153$ and $c_{44} = 1{\cdot}37$, which are very different from the observed ones, $0{\cdot}23$ and $0{\cdot}28$ respectively, showing that here E_{BS} plays a big part, as would be expected since the Fermi surface in this $Z = 3$ metal overlaps the Brillouin zone boundaries. The c_{44} discrepancy is particularly striking and indicates a large *negative* E_{BS} contribution to this elastic constant. Heine and Weaire (1970) attribute this to the fact that g_{111} lies close to q_0 for this metal (*see* Table 5.4), so that there is a tendency for g_{111} to split, i.e. for some (111) planes to move further apart and others to move closer together. A splitting of just this type is produced by the c_{44} type of deformation, since this stretches some [111] directions and compresses others. The positive E_E contribution and negative E_{BS} thus almost balance, for c_{44}, so that aluminium is *almost mechanically unstable* with respect to this mode of deformation. Heine (1969) has argued that in mercury ($Z = 2$) the balance tips the other way, so that a deformed variant of the f.c.c. structure is preferred.

The theory of elastic constants leads directly on to that of long-wave *elastic vibrations*. For the more general theory of *atomic oscillations* in solids, which extends also to the short-wave components where the discreteness of the atomic structure has to be taken into account, it has been usual to assume that the atoms are bound together by pairwise central forces acting between nearest neighbours. When differences have appeared between such theories and observation, for example in the number of vibrational modes in various ranges of wavelength, attempts have sometimes been made to improve the theories by assuming that the atomic forces are *non-central* in their action. However, as we have seen from the theory of E_{BS}, for shear oscillations in simple metals the assumption of central forces is a good one. The right way to improve the theory of atomic vibrations in this case is, as indicated by Fig. 5.5, to take account of central-force interactions between more distant neighbours.

An approximate method for finding the *velocity of sound* in a metal is to regard the oscillating ions as performing plasma oscillations (Bardeen and Pines, 1955). For the frequency ω_p of these oscillations we can rewrite equation (3.6) in the form

$$\omega_p = (4\pi\rho_0 Z^2 e^2/M)^{1/2} \tag{5.21}$$

where ρ_0 is the number of ions per unit volume, and Ze and M are the charge and mass of an ion. Since n_0 in equation (3.6) is $Z\rho_0$, we have

$$\omega_p = (4\pi n_0 Z e^2/M)^{1/2} \tag{5.22}$$

We have not yet, however, taken account of the screening of the ionic charges by the electrons. For this we can use the dielectric function from equation (3.16) in the long-wavelength limit (i.e. small wavenumber q) to modify Ze to $Zeq^2\lambda^2$, where λ is the screening radius (equation (3.7)). We obtain a ω_p proportional to q and hence a velocity of sound v_s given by

$$v_s = \omega_p/q = (\tfrac{2}{3}ZE_F/M)^{1/2} \tag{5.23}$$

from equation (3.7). Typically, $v_s \simeq 10^3$ m s^{-1}.

From standard statistical mechanical arguments, the *vibrational specific heat* of the metal can then be deduced. For example, the *Debye temperature* Θ_D, which characterizes this specific heat, is given by (Mott and Jones, 1936)

$$\Theta_D = (3/4\pi\Omega_a)^{1/3} h v_s / k_B \tag{5.24}$$

where Ω_a is the atomic volume and k_B is Boltzmann's constant.

5.10 STACKING FAULTS

Because of their crystallographic simplicity and regularity, *stacking faults* and *coherent twin interfaces* in f.c.c. and c.p.h. structures are particularly amenable to analysis by the above methods. The f.c.c. and c.p.h. structures can be regarded as stacks of hexagonal close-packed sheets, the atoms of each sheet 'sitting' in positions centred above triangles of nearest-neighbour atoms in the sheet below. For each sheet there are three sets of such positions: those occupied by its own atoms, labelled A, and two sets of positions, B and C, in its triangles. The perfect f.c.c. and c.p.h. structures are represented by stacking sequences such as . . . ABCABC . . . and . . . ABABAB . . . , respectively. The simpler kinds of twins and faults that can exist in these sequences are given in Table 5.6. In each structure the underlined faulted region represents a sandwiched layer of the other crystal structure.

Since the numbers and distances of nearest neighbours are preserved in these sequences, the fault energies are in all cases small compared with, for example, large-angle grain boundaries and free surfaces. To a first approximation each fault energy can be regarded as being provided by the excess energy of the 'crystal structure' of the sandwich, relative to that of the main crystal. It follows from this that the fault energies are lowest when the unfaulted structures, f.c.c. and c.p.h., have almost equal energies which, from the kind of argument outlined in Section 5.7, is expected for certain values of Z ($\simeq$ 1·3, 1·7, 2·2). The absence of nearest-neighbour differences means that only the outlying region of the $\Phi(r)$ curve (e.g. as in Fig. 5.5), where the interaction produces Friedel oscillations, is important. A convenient approximation to use for $\Phi(r)$ in this case is thus the Friedel form, as given by equation (A 7.11).

On this basis Blandin has calculated various fault energies (Blandin 1966,

Table 5.6 Simple twins and stacking faults

f.c.c. twin	ABCABACBA
c.p.h. twin	ABABACACA
f.c.c. intrinsic fault	ABCABABCAB
c.p.h. intrinsic fault	ABABACBCBC
f.c.c. extrinsic fault	ABCABACABCA
c.p.h. extrinsic fault	ABABACABABA

Blandin *et al.* 1966). The method was to sum $\mathbf{\Phi}(r)$ between the atoms in two close-packed planes P and P′ a distance d apart, and to take the difference, $I_{PP'}$, of this sum for these two cases in which (i) P′ is and (ii) P′ is not displaced by the vector $\boldsymbol{b}$ that translates the plane from the unfaulted to the faulted position. The fault energy, γ (per unit area), is then the sum of $I_{PP'}$ over all pairs of such planes:

$$I_{PP'} = N_P \sum_j [\mathbf{\Phi}(\boldsymbol{r}_j + \boldsymbol{b}) - \mathbf{\Phi}(\boldsymbol{r}_j)] \tag{5.25}$$

$$\gamma = \sum I_{PP'} \tag{5.26}$$

where $\boldsymbol{r}_j$ is the position vector of the jth atom in plane P′, relative to a reference atom in P, and N_P is the number of atoms per unit area of plane. The $\mathbf{\Phi}$ terms are evaluated through their Fourier transforms, i.e.

$$\mathbf{\Phi}(\boldsymbol{r}_j) = \int \mathbf{\Phi}(\boldsymbol{q}) \exp(-\mathrm{i}\boldsymbol{q}\cdot\boldsymbol{r}_j)\,\mathrm{d}\boldsymbol{q} \tag{5.27}$$

(*see* equation (A 6.3)), in which form equation (5.25) becomes

$$I_{PP'} = N_P \int \mathbf{\Phi}(\boldsymbol{q}) \sum_j \exp(-\mathrm{i}\boldsymbol{q}\cdot\boldsymbol{r}_j)[\exp(-\mathrm{i}\boldsymbol{q}\cdot\boldsymbol{b}) - 1]\,\mathrm{d}\boldsymbol{q} \tag{5.28}$$

The wave-vector $\boldsymbol{q}$ of $\mathbf{\Phi}(\boldsymbol{q})$ is then resolved into components q_n and $\boldsymbol{q}_p$, normal and parallel to the plane of a sheet. We recognize in the sum over the atoms j of the sheet P′ a two-dimensional version of the *structure factor* (*see* e.g. equation (4.3)). Since these atoms are arranged in a perfect two-dimensional lattice, in this sheet, of which the two-dimensional reciprocal lattice vectors may be denoted by $\boldsymbol{g}$, the argument used in Section 4.4 also applies here, i.e. $\boldsymbol{q}_p = \boldsymbol{g}$. This leads to

$$I_{PP'} = N_P \sum \int \mathbf{\Phi}(\boldsymbol{q}) \exp(-\mathrm{i}\boldsymbol{q}\cdot\boldsymbol{r}_0)[\exp(-\mathrm{i}\boldsymbol{q}_p\cdot\boldsymbol{b}) - 1]\,\mathrm{d}q_n \tag{5.29}$$

where the sum is over all the allowed values $\boldsymbol{g}$ of $\boldsymbol{q}_p$; $\boldsymbol{r}_0$ is the vector joining an atom in P to the equivalent atom in P′ in the unfaulted crystal, and $q_p = (q^2 - q_n^2)^{1/2}$. The integration over q_n, from zero upwards, means that q varies from the smallest value, g_{min}, of q_p up to infinity (the value $g = 0$ is excluded by the term in brackets in equation (5.29)).

In a monovalent (alkali) metal the low electron density gives a small wavenumber k_F at the Fermi surface, so that $g_{min} > 2k_F$. In these circumstances $\mathbf{\Phi}(q)$, with $q > 2k_F$, approaches zero with increasing q (*see* Fig. A 7.1), and Blandin showed that then

$$I_{PP'} \propto \exp(-qd) \tag{5.30}$$

Table 5.7 Calculated fault energies, mJ m^{-2}

	Li	Na	K	Be	Mg	Al
Twin	2	0·03	0·4	390	60	60
Intrinsic fault	4	0·06	0·8	760	120	150
Extrinsic fault	6	0·10	1·2	1240	195	125

where $q = (g_{min}^2 - 4k_F^2)^{1/2}$ and d is the distance between P and P′. The larger d values in this case make negligible contributions to the total fault energy γ, and when q is large (as in monovalent metals) all contributions to γ are small. By contrast, in multivalent metals $2k_F > g_{min}$ and q in equation (5.30) becomes imaginary, so that the exponential term is converted into a sinusoidal function. We are then in the region of the long-range oscillations of the $\Phi(r)$ function (*see* Fig. 5.5), and many pairs of P, P′ sheets make significant contributions to a fairly large total γ.

In summary, the fault energy in f.c.c. and c.p.h. structures depends strongly upon the number Z of electrons per atom, for two reasons. At small Z the Fermi surface of a simple metal lies too far inside the Brillouin zone for there to be any appreciable deviation from the free-electron sphere; structure effects are very small in this case. Sodium is the best example of this, and it is significant that the c.p.h. structure sodium adopts at low temperatures is always heavily faulted. As Z increases, the Fermi surface expands into the region of the Brillouin zone boundaries where structure effects make themselves felt, giving the possibility of a larger fault energy. Whether the fault energy does become large then depends on the second effect of Z: on the extent to which the spacings of the reflecting planes favour f.c.c. or c.p.h., as outlined in Section 5.7. The fault energy is of course small when these structures are equally favoured. Table 5.7 gives some values of fault energies calculated by Blandin (loc.cit) and others. Similar theoretical and experimental values have been obtained for aluminium (Hodges 1967, Edington and Smallman 1965).

5.11 VACANCIES AND DISLOCATIONS

Stacking faults are exceptional in that the packing of nearest-neighbour atoms is everywhere the same as in 'good' crystal. The defects to which we now turn are more irregular; the immediate neighbourhood of some of their atoms differs grossly from that in the good crystal. This presents the electron theory with much more difficult problems, and drastic approximations usually have to be made. Nevertheless, it is possible to obtain a qualitative understanding of their structures and the magnitude of their energies.

When a *vacant atomic site* is created in an otherwise perfect crystal, there are two energy changes. The potential energy increases because the electron gas in this vacancy loses the benefit of the positive ionic charge which previously existed there. The kinetic energy decreases because the crystal, of N atoms, is enlarged to the volume of $N + 1$ sites, so that all the Bloch waves are slightly stretched. Fumi (1955) was the first to show how to estimate these for a monovalent metal.

For the potential energy we start with a perfect crystal and annul the positive monovalent ionic charge at a central site by placing a negative charge there. Following the method outlined in Appendix 4 (equation (A 4.48)) and Section 4.2 (equation (4.5)), we express the resultant change in the energy of an electron in a (plane wave) state exp $(\mathrm{i}\boldsymbol{k}\cdot\boldsymbol{r})$, to first order, as

$$\delta E = \frac{1}{N\Omega_\mathrm{a}}\int \psi_k^* V \psi_k \,\mathrm{d}\boldsymbol{r} = \frac{1}{N\Omega_\mathrm{a}}\int \exp(-\mathrm{i}\boldsymbol{k}\cdot\boldsymbol{r})V(r)\exp(\mathrm{i}\boldsymbol{k}\cdot\boldsymbol{r})\,\mathrm{d}\boldsymbol{r} \qquad (5.31)$$

where $N\Omega_\mathrm{a}$ is the volume of the (perfect) crystal. With $V(r) = +e^2/r$, the problem is similar to that of the point-ion potential $w(q)$ discussed in Section 4.2 (equation (4.10)) in the limit $q \to 0$ (since $\boldsymbol{q} = \boldsymbol{k}' - \boldsymbol{k}$ and $\boldsymbol{k}' = \boldsymbol{k}$ in equation (5.31)). Equation (4.16) gave this limit as $w(0) = -\frac{2}{3}E_\mathrm{F}$, where E_F is the Fermi energy. Thus, $\delta E = \frac{2}{3}E_\mathrm{F}/N$. Summing this over all the N electrons, we obtain for the total potential energy of formation of the vacancy

$$N\,\delta E = +\tfrac{2}{3}E_\mathrm{F} \qquad (5.32)$$

This vacancy increases the volume of the crystal by the fraction N^{-1}. Since the Fermi energy varies as (volume)$^{-2/3}$, the kinetic energy per electron is reduced by the fraction $\frac{2}{3}N^{-1}$. Since the average Fermi energy is $\frac{3}{5}E_\mathrm{F}$, the change in kinetic energy is

$$N\,\delta(\tfrac{3}{5}E_\mathrm{F}) = -\tfrac{2}{3}\times\tfrac{3}{5}E_\mathrm{F} = -\tfrac{2}{5}E_\mathrm{F} \qquad (5.33)$$

The energy of formation of the vacancy E^f is thus

$$E^\mathrm{f} = (\tfrac{2}{3}-\tfrac{2}{5})E_\mathrm{F} = \tfrac{4}{15}E_\mathrm{F} \qquad (5.34)$$

which is Fumi's result. Fumi argued that refinements of this simple argument could bring the value of E^f down to about $\frac{1}{6}E_\mathrm{F}$, which compares well with the measured value for monovalent metals, as summarized in Table 5.8.

This monovalent theory is easily generalized to higher valencies, and the corresponding figure then becomes $\frac{1}{6}ZE_\mathrm{F}$, but for metals such as magnesium, aluminium and lead this gives values which are much too high compared with

Table 5.8 Vacancy formation energies (eV) of the monovalent metals (Evans 1977) compared with $\frac{1}{6}E_\mathrm{f}$

	Li	Na	K	Rb	Cs	Cu	Ag	Au
$\frac{1}{6}E_\mathrm{F}$	0·79	0·54	0·35	0·31	0·26	1·17	0·92	0·92
Measured E^f	0·34	0·42	0·39	0·27	0·28	1·17	1·01	1·01

measured ones such as 0·81 (Mg), 0·66 (Al) and 0·53 (Pb). Evans (1977) has critically reviewed this and later theories and concluded that none gives a quantitatively satisfactory treatment of the electron distribution near the vacancy. As mentioned above, this is a general problem for any kind of structural irregularity in which the local environment deviates grossly from the structure of a good crystal.

The theory of *dislocations*, their structures and their energies, can be developed largely without referring explicitly to the electron theory. Most of the energy of a dislocation resides in its long-range elastic field and is calculable from the elastic properties. Even the structures and properties of dislocation cores can be described largely in terms of those of stacking faults. The relatively small energy belonging to structurally 'bad' crystal in the core raises the same kind of problems, for electron theory, as a vacancy does. This energy is often roughly estimated by analogy with the disordered structures of liquids or large-angle grain boundaries, with known energies, or by the use of simple, assumed atomic-bond force laws. These dislocation core structures play a major role in determining cross-slip behaviour and Peierls–Nabarro stresses, particularly in the b.c.c. metals (Vitek 1985), and also in pipe diffusion.

5.12 LIQUIDS AND GRAIN BOUNDARIES

The energy band gaps in crystals of simple metals are small because the pseudopotentials are shallow. In the disordered liquid state of such metals these weak Brillouin zone effects virtually vanish. Another way to look at this is to note that, in simple monovalent metals, the mean free path of the electrons between successive ionic collisions is of the order of 50 atomic spacings at the melting point. The non-crystallinity of the liquid is fully established on this scale, so that the correlations in ionic positions, sensed by such pairs of collisions, conform to a liquid distribution, not a crystalline one. It follows that the electron theory of liquid metals can be based on the perturbation of simple plane-wave functions by the ionic pseudopotentials, and that the effects of this perturbation can be summarized by means of an isotropic effective mass (*see* equation (A 4.17)) with no Brillouin zone complications.

A calculation of the energy of a simple liquid metal from the pseudopotential theory can thus be made by the methods presented in Sections 4.3 and 5.6. The structure of the liquid has to be taken as a given input, using a structure factor (equation 4.3)) derived from X-ray or neutron diffraction analysis, or from an empirical theoretical model of a liquid such as that proposed by Percus and Yevick (*see* e.g. Faber 1972, Ziman 1979). A calculation of this type, which contrasted such a structure factor for the liquid with a harmonic oscillator one to represent the hot crystal, has been made by Hartmann (1971). For sodium it gave a latent heat of melting of 0·02 eV per atom, compared with the observed value of 0·03 eV per atom.

In order to understand the physical aspects of the liquid metallic state, we can imagine melting to occur in two stages. First, at the melting point, the *crystal* is expanded to the volume of the liquid. Second, at constant volume and temperature, it is disordered into the structure of the liquid. The work required for the first stage can be estimated from the bulk modulus and the volume change. For aluminium we take $K \approx 7 \times 10^{10}\,\mathrm{J\,m^{-3}}$ and $\Delta\Omega/\Omega = 0{\cdot}04$. The work done is then $\frac{1}{2}(7 \times 10^{10}) \times (0{\cdot}04)^2 \simeq 6 \times 10^7\,\mathrm{J\,m^{-3}} \simeq 160\,\mathrm{cal\,mol^{-1}}$, whereas the latent heat of melting is $2500\,\mathrm{cal\,mol^{-1}}$. The disparity is even greater for sodium.

Thus, almost all of the latent heat is absorbed by the process of disordering at constant volume. How is this to be understood in terms of electronic energies? We return to the effective ion-ion interaction, as shown in Fig. 5.5, from which we see that the nearest neighbours to the central ion play the dominating role. The interactions of the more distant neighbours are very small in comparison. The twelve nearest neighbours in the f.c.c. crystal fall near the minimum in the interaction curve, where the depth is about 0·1 eV per neighbouring ion. The curve rises so steeply about this minimum that a 5% deviation from the optimal spacing raises the energy by the order of 0·1 eV per ion–ion interaction. A similar increase would be obtained if, as appears to happen in melting, the structural rearrangement (at constant volume, starting from the expanded crystal) were to reduce the number of nearest neighbours by about one per ion.

Thus we see that the nearest-neighbour interactions provide sufficient opportunity for the observed increase in energy ($2500\,\mathrm{cal\,mol^{-1}} \approx 0{\cdot}11$ eV per atom) when the structure disorders. The sources of this increase are the electrostatic energy E_{E}, which increases as a structure deviates from symmetry, and the band-structure energy E_{BS} which falls when the Brillouin zone effects disappear through the loss of crystallinity. Countering these energy increases, of course, is the gain ΔS in entropy due to the disordering, which at the melting point T_{m} contributes a favourable term $-T_{\mathrm{m}}\,\Delta S$ to the free energy of the liquid.

The entropy of melting ΔS of all simple-crystal substances is about the same, of the order of k_{B} (Boltzmann's constant) per atom or molecule. It is thus a thermodynamic characteristic of the liquid state generally, independent of the particular electronic basis of cohesion. Melting occurs when $T_{\mathrm{m}}\Delta S$ equals the latent heat of melting L_{m}. Metals have fairly low L_{m} and T_{m} compared with their cohesion against vaporization. Since L_{m} is determined mainly by change of structure at constant volume, it correlates better with the elastic shear constants than with the bulk modulus (which correlates better with the energy of vaporization). We saw in Section 5.9 that the electronic structure of simple metals leads to relatively small resistance to elastic shear, and much the same applies for melting. The high resistance of metals to volume change, compared with their small resistance to structural rearrangement at constant volume, also explains why they show only a small increase in volume (2–3%) on melting (Faber 1972).

Because of its incoherent structure, a *large-angle grain boundary* is expected to resemble a monolayer of the liquid, and its energy can be roughly estimated on this basis without need to attempt explicit electron theory calculations, which would be difficult. In fact, if such a boundary is regarded as a layer of liquid, two atoms thick, the latent heat of melting provides a value of the grain boundary energy in fair agreement with observation (*see* e.g. Cottrell 1975).

REFERENCES

Bardeen, J., and Pines, D., *Phys. Rev.*, **99**, 1140 (1955).

Blandin, A., in *Phase Stability in Metals and Alloys* (ed. P. S. Rudman *et al.*), McGraw-Hill, New York (1966) p. 115.
Blandin, A., Friedel, J., and Saada, G., *J. Phys. (Paris),* Suppl. C **3**, 128 (1966).
Brooks, H., in *Theory of Alloy Phases*, American Society for Metals, Cleveland, Ohio (1956), p. 199.
Cottrell, A. H., *An Introduction to Metallurgy*, Edward Arnold, London (1975).
Edington, J. W., and Smallman, R. E., *Phil. Mag.*, **11**, 1109 (1965).
Evans, R., in *Vacancies '76* (ed. R. E. Smallman and J. E. Harris), The Metals Society, London (1977).
Ewald, P. P., *Ann. Phys.,* **64**, 253 (1921).
Faber, T. E., *An Introduction to the Theory of Liquid Metals*, Cambridge University Press (1972).
Fuchs, K., *Proc. R. Soc. A.* **151**, 585 (1935); *Ibid.*, 153, 622 (1936a); *Ibid.*, **157**, 444 (1936b).
Fumi, F. G., *Phil. Mag.*, **46**, 1007 (1955).
Ham, F. S., *Solid State Phys.*, **1**, 127 (1955).
Harrison, W. A., *Phys. Rev.,* **136**, 343, 388, 393 (1964); *Pseudopotentials in the Theory of Metals*, Benjamin, New York (1966); *Electronic Structure and the Properties of Solids*, Freeman, San Francisco (1980).
Hartmann, W. M., *Phys. Rev. Lett.*, **26**, 1640 (1971).
Heine, V., in *The Physics of Metals, – Electrons* (ed. J. M. Ziman), Cambridge University Press (1969), Chap. 1.
Heine, V., and Weaire, D., *Solid State Phys.*, **24**, 249 (1970).
Hodges, C. H., *Phil. Mag.*, **15**, 371 (1967).
Kuhn, T. S., and van Vleck, J. H., *Phys. Rev.*, **79**, 382 (1950).
March, N. H., *Orbital Theories of Molecules and Solids*, Clarendon Press, Oxford (1974).
Moruzzi, V. L., Janak, J. F., and Williams, A. R., *Calculated Electronic Properties of Metals*, Pergamon Press, Oxford (1978).
Mott, N. F., and Jones, H., *The Theory of the Properties of Metals and Alloys*, Oxford University Press (1936).
Pauling, L., *Phys. Rev.*, **54**, 899 (1938); *Proc. R. Soc. A,* **196**, 343 (1949); *J. Solid State Chem.*, **54**, 297 (1984).
Vitek, V., in *Dislocations and the Properties of Real Materials* (ed. M. H. Loretto), The Metals Society, London (1985).
Ziman, J. M., *Models of Disorder*, Cambridge University Press (1979).

6

The d band in transition metals

6.1 A FRESH START

In Appendix 4 it is pointed out that there is an alternative to the NFE theory for describing the metallic state. This is the *tight-binding method*. Its starting point is not free electrons, but electrons bound in the atomic orbitals of separate atoms; and its key idea is that when these atoms are brought sufficiently close together to feel slightly the influence of one another's fields of core potential, the electrons are no longer permanently confined to their parent atomic orbitals but can occasionally hop from atom to atom, into vacant atomic orbitals in these neighbouring atoms.

Despite its success with the simple metals, the NFE method has to make way for the tight-binding method when it comes to describing the transition metals, and a fresh start must be made. This is because the properties of the transition metals - atomic, chemical and metallic - are determined largely by the presence of partly filled d shells in their atoms, and because at the usual interatomic spacings of these metals the atomic d orbitals overlap only slightly, just enough to admit the above hopping process but not enough to destroy the basically atomic character of these states (*see* Fig. A 1.4). It is therefore natural to describe these d states by the tight-binding approximation.

In this chapter we shall consider only the partly filled d orbitals and the energy bands formed from them, ignoring both the metallic sp electrons (which were our entire concern in the previous chapter) and the mixing of the d and sp states. Despite these simplifications, many properties of the transition metals can be explained in this way.

6.2 A SIMPLE THEORY OF d BANDS

We consider the structure of atomic d orbitals (Fig. A 1.3). The dominating feature is the pair of nodal planes through the nucleus (or nodal cones, in the d_{z^2} representation, shown in Fig. A 1.3), which corresponds to the value $l = 2$, characteristic of d states, of the angular momentum quantum number (*see* Appendix 1). Because of this the amplitude of the d orbitals is zero at the centre of the atom, and the orbitals themselves are divided by the nodes into two pairs of lobes extending outward from the origin. Each pair of lobes has one ψ sign, $+$ or $-$, opposite to that of the other pair.

In the metal it is expected that only the tips of the lobes of neighbouring

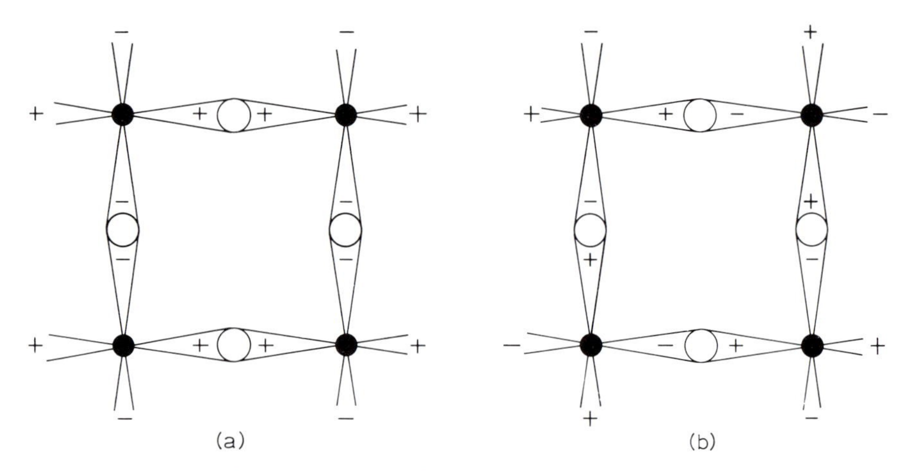

Fig. 6.1 **(a) Bonding and (b) antibonding orbitals in the d band; (after Friedel 1969)**

atoms overlap. Two such overlapping tips may belong to lobes of the same sign, in which case there is a doubling of the amplitude of the wavefunction ψ and hence a quadrupling of electron density there. Alternatively, their lobes may be of opposite sign, in which case the amplitude and local density go to zero there. This is reminiscent of the *bonding* and *antibonding* orbital overlaps of molecular chemistry (Fig. A 2.2), in which the same-sign overlaps concentrate the electron distribution along the axes between the nuclei, thus gaining (negative) electrostatic energy. As a result the energy level is lowered in this bonding configuration; and the antibonding level is raised correspondingly. Thus in a molecule the original atomic energy level is split into lower and upper levels (Fig. A 2.2).

Friedel (1969) has extended these ideas to the d state structure of transition metals. Figure 6.1 shows this schematically for the two extreme cases: where all the overlaps are bonding, and where they are all antibonding. We expect the lower and upper energy levels of these two extreme cases to form the lower E_l and upper E_u limits of the d band in the metal. Between these two extremes there are many intermediate possibilities in which some overlaps are bonding and others are antibonding. These have intermediate energy levels, of course, and so a quasi-continuous *band* of allowed energy levels extends between E_l and E_u. (This is the description, in the tight-binding scheme, of the formation of energy bands in the metal, as explained in Appendix 4.)

The patterns of bonding and antibonding orbitals shown in Fig. 6.1 suggest a method which is often used to estimate the approximate width of an energy band. Imagine the crystal structure to be divided up into Wigner–Seitz polyhedra, as described in Section 2.2. The pattern of bonding orbitals then corresponds to the case, e.g. as in Fig. 2.3, where the Bloch wavefunction has zero slope, but finite amplitude, across a polyhedron boundary. This defines the ground state wavefunction, the energy of which is at the bottom of the band.

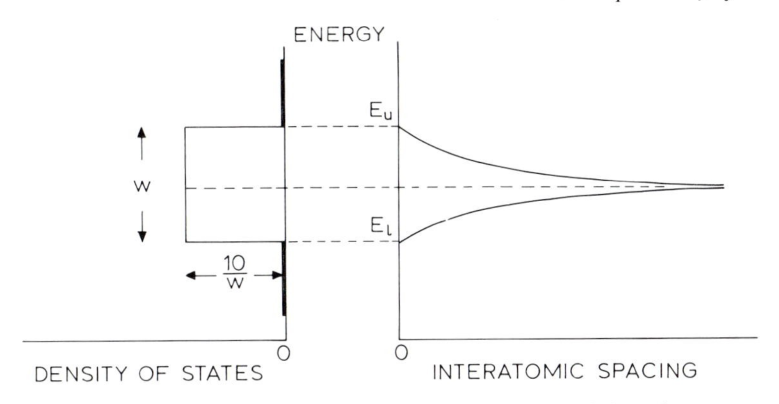

Fig. 6.2 Assumed density of states per atom in the d band

Correspondingly, the completely antibonding pattern, the energy of which is at the top of the band, is equivalent to the opposite case where the amplitude of the wavefunction is zero, but the slope is finite, across the polyhedron boundary. By applying this alternative boundary condition, the Wigner–Seitz method, in which the polyhedron is replaced by the equivalent WS sphere, can be generalized to estimate the energy of the top of the band. The procedure in this application is to solve the single-atom Schrödinger equation to obtain the states and energy levels for each of the two conditions $\partial\psi/\partial r = 0$ and $\psi = 0$ across the boundary, $r = r_s$, of the WS sphere.

For a simple calculation of cohesion, Friedel assumed that the d states are evenly distributed over their band, so that the density of d states is constant within the band and zero outside, as shown in Fig. 6.2. Let z_d be the number of electrons per atom in the d band ($0 < z_d < 10$), and let w be the width of the band, i.e. $w = E_u - E_l$. The electrons fill the band up to the energy level $w(z_d/10)$, so the *average* Fermi energy in the band is $wz_d/20$. Guided by the molecular theory (*see* e.g. Fig. A 2.2) and also by the calculation in Appendix 8, we assume as a first approximation that the splitting of the atomic d states into the d band is distributed symmetrically, in energy, about the energy level of these d states. The extreme bonding state then has an energy level $E_l \simeq -w/2$, relative to the free atom. The average energy of a d electron, relative to the free atom, is thus $(wz_d/20) - w/2$. The (positive) *cohesive energy* per atom E_d, due to the d-band, follows:

$$E_d \simeq -z_d\left(\frac{wz_d}{20} - \frac{w}{2}\right) = \frac{w}{20}[z_d(10 - z_d)] \tag{6.1}$$

This varies parabolically with z_d, being zero for a completely empty ($z_d = 0$) or completely full ($z_d = 10$) band and a maximum, $5w/4$, when the band is half-full ($z_d = 5$), as shown in Fig. 6.3. This roughly reproduces the actual variation of cohesion along the transition metal series (*see* Fig. 6.4), although there are some complications in the middle, particularly in the first long period, and also

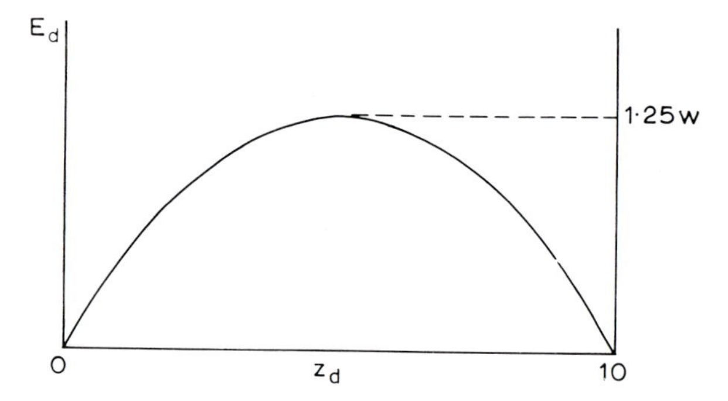

Fig. 6.3 Predicted variation of cohesive energy along a transition metal series

at the noble metal ends, which we shall discuss later. It is known, e.g. from X-ray and ultraviolet spectroscopy, that the d band width is typically $w \simeq 5$ to 10 eV. On this basis the maximum cohesive energy is about 6 eV per atom, which is of the same order as the observed peak values (Fig. 6.4).

This simple theory is developed in more detail in Appendix 8, where it is shown that two coefficients mainly determine the d band. These are, in Dirac's bra–ket notation (*see* equation (A 1.31)),

$$\alpha_{ip} = \langle ip|V|ip\rangle \tag{6.2}$$

$$\beta_{ip}^{jq} = \langle ip|V_i|jq\rangle \tag{6.3}$$

where $i \neq j$. In these formulae $|ip\rangle$ represents an *atomic* d state in atom i, i.e. one of the five atomic orbitals such as those shown in Fig. A 1.3, and

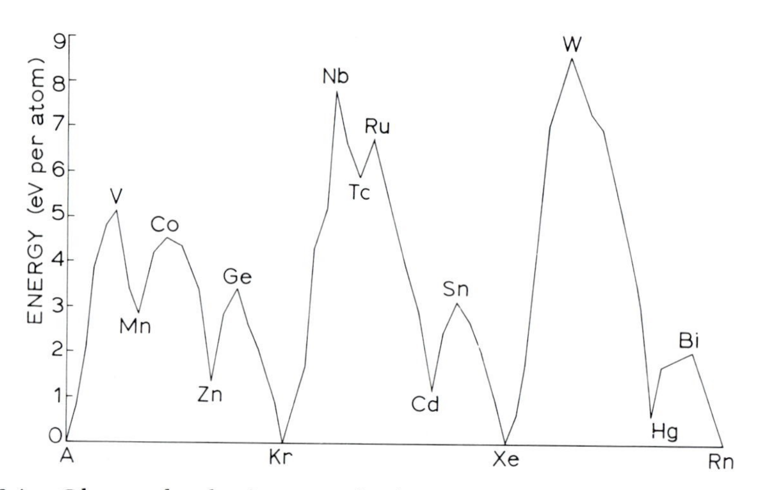

Fig. 6.4 Observed cohesive energies in the long periods

p (= 1, 2, 3, 4, 5) specifies which of these five it is; $\langle ip|$ represents the complex conjugate form of this state. The two coefficients are *matrix coefficients*, obtained by using perturbation theory to calculate the effects, on the free atom states and energy levels, of the (weak) potentials V and V_j of neighbouring atoms of the same species, when a large number of them are brought together to form the metal. Atoms i and j are nearest or next-nearest neighbours (e.g. in b.c.c. metals) and $V_i|jq\rangle$ represents the modification of the state $|jq\rangle$ in atom j by the field V_i from its neighbour, atom i. Similarly, $V|ip\rangle$ represents the total effect of the fields of all the neighbours (and next-nearest neighbours in b.c.c. metals) l on the state $|ip\rangle$ in atom i, i.e.

$$V = \sum_{l} V_l \quad (6.4)$$

The coefficient α measures the change in the energy level of the pth d state of the atom i in the metal, brought about by the fields of the neighbours; and coefficient β determines the extent to which different atomic d orbitals of these atoms combine to form 'molecular orbital' states extending throughout the metal. In these combined states the electrons can 'hop' from atom to atom; so in a partly filled d band they provide metallic conductivity, as was envisaged by Pauling (1938, 1984) in his *resonating valence bond* theory of the metallic state.

The wavefunction $\psi(E)$ of an electron of energy E in this band is given approximately, in the perturbation theory, as a weighted sum of the free-atom d orbitals:

$$\psi(E) = \sum_{i,p} a_{ip}|ip\rangle \quad (6.5)$$

and a standard procedure exists for determining the weighting coefficients a_{ip} (*see* Appendix 8). Similarly, the energy level E of this state in the d band is derived as

$$E = E_0 + \sum_{i,p} a_{ip}^* a_{ip} \alpha_{ip} + \sum_{j,q} \sum_{i,p} a_{jq}^* a_{ip} \beta_{ip}^{jq} \quad (6.6)$$

Here E_0 is the unperturbed energy level of the atomic d state; and the sum over j is, from the meaning of β, a sum over the atoms which are the neighbours (and next-nearest neighbours, in b.c.c. metals) of atom i. Since $\psi(E)$ is a sum of atomic wavefunctions each of which can be added in positive or negative orientation (*see* Fig. A 2.2), the coefficients a_{ip} can be either positive or negative. However, their sign does not affect the α sum since $a_{ip}^* a_{ip} = |a_{ip}|^2$, so this sum contributes a uniform (downward) shift of the atomic energy level when the atoms are brought together to form the metal, a shift that was ignored in the

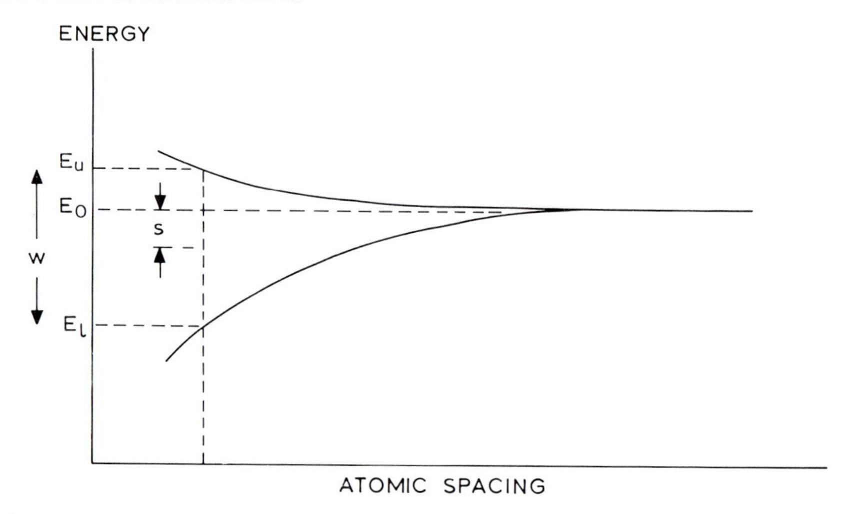

Fig. 6.5 Splitting of the d band according to equation (6.6)

simple theory as represented by equation (6.1) and Fig. 6.2. We show its effect in Fig. 6.5. The signs of the a_{ip} do, however, affect the β sum in equation (6.6) because the terms there are products of different coefficients, e.g. $a_{jq}^{*}a_{ip}$. Because of this, the β sum produces a *band* of values, ranging from a lower limit where a maximum number of terms is negative to an upper limit where a maximum number is positive. Between these limits is a spectrum of intermediate energy levels, corresponding to various sets of positive and negative products, from which the density-of-states distribution in the d band is built up. The β coefficients are thus responsible for splitting the atomic energy level E_0 into the d band, while the α coefficients determine the mean position s, of this band relative to the level E_0, as shown in Fig. 6.5. Friedel has shown that s ($\simeq$ 1 eV) is small compared with the bandwidth w, so that the approximation $s = 0$ implicit in Fig. 6.2 is qualitatively justified.

6.3 THE DENSITY OF STATES

The next step in the theory is to replace the crude rectilinear form assumed for the density of states in Fig. 6.2 by a more refined distribution. Since the position of each energy level in the d band is determined by a sum of various positive and negative terms in equation (6.6), the arrangement of these levels in the band – the density of states – bears some resemblance to a probability distribution. This has led to a simple extension of the theory, beyond the stage represented by Fig. 6.2, in which the density of states, $n(E)$ per atom, is expressed, like a probability distribution, in terms of the *moments* of the distribution, a standard method in probability theory (Cyrot-Lackmann 1968, Ducastelle and Cyrot-Lackmann 1970, Friedel 1969). Taking E_0 as the mean energy of the d band distribution, the mth moment about this mean is defined as

$$M_m = \int n(E)(E - E_0)^m \, \mathrm{d}E \tag{6.7}$$

where the integral is over the band. The hope is that $n(E)$ can be sufficiently well represented by its first few moments, which can be related to the coefficients β of the energy levels, as shown in Appendix 9. The following results are there derived. The first moment vanishes:

$$M_1 = \int n(E)(E - E_0)\,\mathrm{d}E = 0 \tag{6.8}$$

This simply expresses the fact that we have centred the distribution on E_0. The second term is, for N atoms,

$$M_2 = N^{-1} \sum \sum \beta_{ip}^{jq} \beta_{jq}^{ip} \tag{6.9}$$

where the sum is over i, p, j and q, with $j \neq i$. The third term is

$$M_3 = N^{-1} \sum \beta_{ip}^{jq} \beta_{jq}^{ko} \beta_{ko}^{ip} \tag{6.10}$$

again summed over all subscripts and superscripts, with $i \neq j \neq k$. Generalizing, the mth term is

$$M_m = N^{-1} \sum \beta_{ip}^{jq} \beta_{jq}^{ko} \ldots \beta_{rt}^{ip} \tag{6.11}$$

i.e. a sum of m products of the β coefficients, with the same summation rules as before.

To understand the meaning of these expressions we recall that β_{ip}^{jq}, defined by equation (6.3), measures the probability amplitude that an electron in the state $|ip\rangle$ on atom i will (in unit time) hop into the state $|jq\rangle$ on a neighbouring atom j. Thus M_2 measures the probability amplitude of processes in which the electron hops from $|ip\rangle$ into $|jq\rangle$, and then back to $|ip\rangle$ again. Similarly, M_4 measures the amplitude for closed circuits of four such hops, as shown in Fig. 6.6.

6.4 SOME APPLICATIONS

In the range of validity of the tight-binding model, these hopping processes are relatively rare since they correspond to a weak perturbation of the free-atom d orbitals. It thus follows that M_2 is the main term. On this basis, Friedel (1969, 1970) has made some simple applications, as follows. First, there is the rough approximation

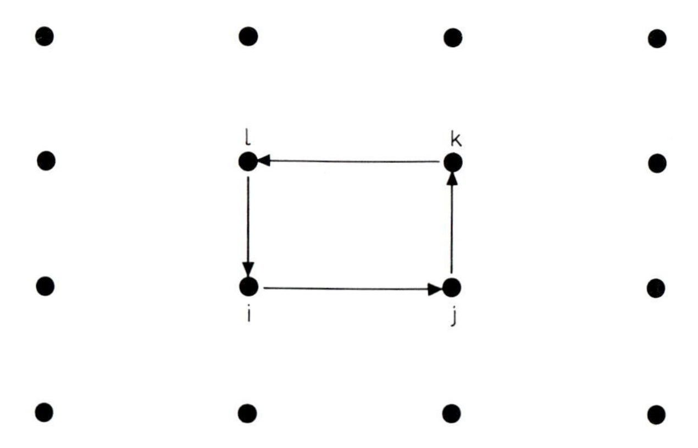

Fig. 6.6 A closed circuit of four jumps (M_4)

$$M_2\left(= N^{-1}\sum \beta_{ip}^{jq}\beta_{jq}^{ip}\right) \simeq z\overline{\beta^2} \tag{6.12}$$

where z is the number of nearest neighbours of atom i and $\overline{\beta^2}$ is the average value of $(\beta_{ip}^{jq})^2$. This approximation recognizes that there is one β^2 contribution to M_2 from each of the z neighbours. We note from equation (6.7) that M_2 measures the *standard deviation* or *dispersion* of the distribution $n(E)$. The root-mean-square deviation is of the order of half the bandwidth:

$$\tfrac{1}{2}w \simeq \sqrt{M_2} \simeq \sqrt{z}\sqrt{\overline{\beta^2}} \tag{6.13}$$

and hence, from equation (6.1),

$$E_{\mathrm{d}} \simeq \sqrt{(M_2)}\,[z_{\mathrm{d}}(1 - 0{\cdot}1\,z_{\mathrm{d}})] \tag{6.14}$$

When a close-packed crystal melts or forms a large-angle grain boundary, the coordination number z of the atoms drops from 12 to about 10 or 11. The resulting change in cohesive energy is thus determined approximately by the change in $\sqrt{M_2}$, i.e. by

$$\Delta E_{\mathrm{d}} \simeq E_{\mathrm{d}}(1 - \sqrt{X}) \tag{6.15}$$

where $X \simeq 10/12$ or $11/12$. It follows that the latent heat of melting and the grain boundary energy, per atom, are about 5–10% of the cohesive energy.

It should be noted that this value depends on the square root of the coordination number. Thus, despite the resemblance of Fig. 6.1 to a simple 'nearest-neighbour bond' structure, it would be wrong to attempt to estimate the energy of a structure by simply counting the number of nearest-neighbour bonds and allocating a standard unit 'bond' energy to each. This popular procedure is not validated by the d band theory. The recent trend has been towards the use of empirical *many-body* bonding interactions; a successful example of this, for calculating energies in b.c.c. transition metals, has been provided by Finnis and Sinclair (1985).

Since the distances and numbers of nearest neighbours are independent of the stacking sequence of close-packed planes in f.c.c. and ideal c.p.h. structures, M_2 is unaffected by the presence of stacking faults or coherent twin interfaces. To estimate the energies of these faults it is therefore necessary to go to higher moments than M_2. This also means that in transition metals these energies must be small compared with the cohesive energy, since they do not depend on M_2. Such faulting is therefore a fairly common feature of f.c.c. and c.p.h. transition metals.

In some lattices it is geometrically necessary that $M_3 = 0$. For example, in the square lattice of Fig. 6.6 it is impossible to return to the starting point in three steps (although this might of course be possible at the centre of an irregularity such as a dislocation). In f.c.c. and c.p.h. structures M_3 circuits are possible, but M_3 is not changed by the presence of stacking faults or coherent twin interfaces, so it is necessary to go to M_4 to find an energy of faulting. More elaborate calculations of these fault energies have now been made, up to M_{22}, which agree well with measured values (Papon *et al.* 1979). In summary, they give

$$2\gamma_t \simeq \gamma_i \simeq \gamma_e \simeq A\,|E_{\text{c.p.h.}} - E_{\text{f.c.c.}}| \tag{6.16}$$

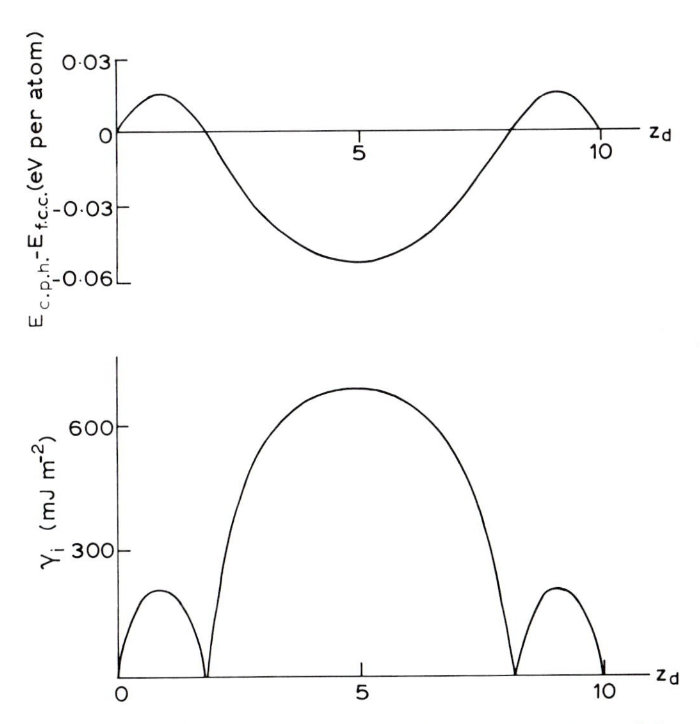

Fig. 6.7 **Calculated difference in energy between c.p.h. and f.c.c. structures, together with intrinsic stacking fault energy**

Table 6.1 Distribution of crystal structures across the transition metals

f.c.c.		c.p.h.		b.c.c.		c.p.h.		f.c.c.	
Ca	Sc	Ti	V	Cr	(Mn)	(Fe)	Co	Ni	Cu
Sr	Y	Zr	Nb	Mo	Tc	Ru	Rh	Pd	Ag
(Ba)	La	Hf	Ta	W	Re	Os	Ir	Pt	Au

for the (paramagnetic) 3d transition metals, where γ_t, γ_i and γ_e are the energies per unit area of the twin, intrinsic and extrinsic faults, $E_{c.p.h.}$ and $E_{f.c.c.}$ are the energies per atom of the c.p.h. and f.c.c. structures, and A converts energy per atom into energy per unit area. Figure 6.7 shows the calculated variation across the 3d series.

Although at this level of the theory there is no place for the b.c.c. structure which is found in the middle of the transition series, the occurrence of c.p.h. and f.c.c. near the ends of the series is fairly consistent with the pattern suggested by Fig. 6.7. The distribution of crystal structures is broadly as given in Table 6.1. The observed very low stacking fault energies of cobalt and high-cobalt alloys (e.g. 70Co–30Ni) are also explained by this theory.

6.5 MORE COMPLETE THEORIES

While the above theory gives a broad understanding of cohesion in the transition metals, it leaves many questions unanswered. For example, why does the middle group of transition metals prefer the b.c.c structure? What is the explanation of the cohesion of copper, silver and gold? Why is there a local minimum in the cohesion at and near manganese (Fig. 6.4)? Where does ferromagnetism fit into the picture? Beyond these questions there is a more basic limitation of the above theory: it has provided no calculated value of β. Many refinements and additions have been made which enable more complete versions of the theory to deal with these points.

The first step is to find the overlap energy of two d orbitals on two neighbouring atoms. The problem is similar to that of molecular orbital theory (*see* Appendix 2). Viewed along the line joining these two atoms, the atomic orbitals have *rotational symmetry*, i.e. along a circular path around this line the wavefunction ψ of a d orbital changes sign $m = 0$, 1 or 2 times, depending on the orientation of the orbital in relation to the line (*see* Fig. A 1.3). By the arguments developed in Appendix 2, the overlap energy of the two orbitals is zero unless they have the same symmetry about this line (*see* Fig. A 2.4). Figure 6.8 shows that there are three non-zero cases, labelled ddσ, ddπ and ddδ, using the notation of Appendix 2.

In these diagrams the two orbitals are in each case shown in identical orientation on the two atoms, which corresponds to a wavenumber $k = 0$ in the Bloch function. In this orientation the ddσ overlap is a bonding one since the overlapping lobes have the same sign, the ddπ is antibonding, and the ddδ is bonding. Obviously the σ bond is the strongest since the lobes point towards

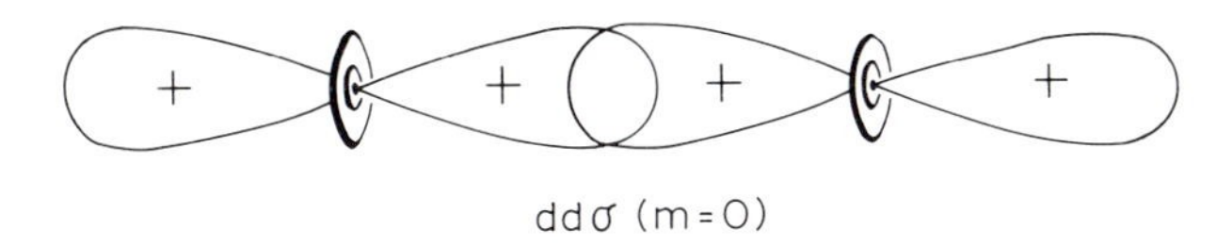

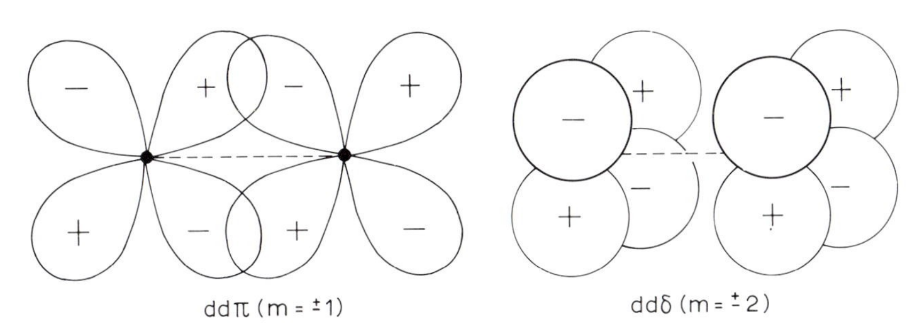

Fig. 6.8 Three orientations of overlapping d orbitals

each other, and the δ bond is the weakest since the lobes then point sideways. The values of the coefficient β or hopping integral (equation (6.3)) which give the overlap energy have been estimated for these three cases. *Andersen's method (Andersen 1973) leads to the values

$$\beta = cw(r_s/d)^5 \tag{6.17}$$

where r_s is the radius of the WS sphere, d is the interatomic distance, w is the width of the d band, and $c = -2{\cdot}4$, $+1{\cdot}6$ and $-0{\cdot}4$ for the σ, π and δ bonds respectively. We notice the rapid fall-off with increasing distance, as the inverse fifth power. This justifies our neglecting all but the nearest neighbours in most structures; in the b.c.c. structure the 15% additional distance between second nearest neighbours reduces their dd interaction energies by about 50%.

The next step is to express, in terms of the above β values, the coefficients β_{ip}^{jq} (equation (6.3)) for the interactions of d states of types p and q on neighbouring atoms i and j, positioned according to the crystal structure in question. The general method is that developed by Slater and Koster (1954). For cubic crystals p and q are taken to be one or other of the basic d states of equation (A 1.44), oriented as in Fig. A 1.3, and varying with direction like one of the functions xy, yz, zx, $x^2 - y^2$, $3z^2 - r^2$. The orientation of each neighbouring atom j relative to the central atom i is specified by the direction cosines hkl of the line from i to j. Each of the five basic d states then exhibits a certain symmetry about this line, the σ, π and δ proportions of which are determined by its character (xy, yz, . . .) and by the cosines hkl. For example, the $x^2 - y^2$ function shown in Fig. A 1.3 is

*See, for example, Fletcher and Wohlfarth (1951), Asdente and Friedel (1961), Heine (1969), Andersen (1973), Harrison (1980), Pettifor (1983).

of $m = 0$ symmetry about the x and y axes, $m = \pm 1$ about the $x = y$ directions, and $m = \pm 2$ about the z axis.

The values tabulated by Slater and Koster for the σ, π and δ proportions in the various d states, in various hkl directions, enabled them to express the dd overlap energies (in our notation, the β_{ip}^{jq} coefficients) in terms of the three β values and the direction cosines hkl. (They also tabulated values for overlaps of other states: ss, sp, pp, sd and pd.) For simplicity we shall ignore the small δ overlap terms. We denote the β overlap energies for the σ and π symmetries as (ddσ) and (ddπ), and write the coefficients β_{ip}^{jq} as β_{xy}^{xy}, β_{zx}^{x2-z2}, ... From Slater and Koster's list we note the following four expressions, of particular importance:

$$\left.\begin{aligned}
\beta_{xy}^{xy} &= 3h^2k^2(\mathrm{dd}\sigma) + (h^2 + k^2 - 4h^2k^2)(\mathrm{dd}\pi) \\
\beta_{zx}^{zx} &= 3h^2l^2(\mathrm{dd}\sigma) + (h^2 + l^2 - 4h^2l^2)(\mathrm{dd}\pi) \\
\beta_{x2-y2}^{x2-y2} &= \tfrac{3}{4}(h^2 - k^2)^2(\mathrm{dd}\sigma) + [h^2 + k^2 - (h^2 - k^2)^2](\mathrm{dd}\pi) \\
\beta_{3z2-r2}^{3z2-r2} &= [l^2 - \tfrac{1}{2}(h^2 + k^2)]^2(\mathrm{dd}\sigma) + 3l^2(h^2 + k^2)(\mathrm{dd}\pi)
\end{aligned}\right\} \tag{6.18}$$

At the centre of the Brillouin zone, where the wave-vector is zero and the d states are identically oriented on every atom, symmetry requires that all the $\beta_{ip}^{jq} = 0$ except those above (and of course β_{yz}^{yz}). Different types of d states on neighbouring atoms are approximately orthogonal and, in a more rigorous generalization of the theory, can be regrouped into orthogonal states.

6.6 THE f.c.c. AND b.c.c. STRUCTURES

To examine other points in the Brillouin zone it is necessary to express the wavefunction $\psi(E)$ in equation (6.5) as a Bloch wave by introducing the usual phase factor with a wave-vector $\boldsymbol{k}$ ($= k_x, k_y, k_z$, to be distinguished from the direction cosine k in equations (6.18)) into the coefficients a_{ip}:

$$a_{ip} = a_{ip}^0 \exp(\mathrm{i}\boldsymbol{k}\cdot\boldsymbol{r}) \tag{6.19}$$

If we move out into the zone along a cubic axis, e.g. the z axis so that $k_z \neq 0$ and $k_x = k_y = 0$, then the cancellation of terms from the neighbours in a cubic crystal leaves the terms of equations (6.18) as the only ones we have to consider. For this particular direction, [001], in the zone the above coefficient reduces to

$$a_{ip} = a_{ip}^0 \exp(\mathrm{i}k_z z) \tag{6.20}$$

where z is measured from the central atom.

Consider first f.c.c. crystals. The 12 nearest neighbours have direction cosines of the type [110], (i.e. $[hk0]$ where $h^2 = k^2 = \frac{1}{2}$, together with all its orientational variants. Four of them lie in the plane $z = 0$, giving $\exp(\mathrm{i}k_z z) = 1$, and four lie in each of the two planes $z = \pm a/2$, where a is the lattice constant, giving the phases $\exp(\pm \mathrm{i}k_z a/2)$. For each of equations (6.18) we sum the overlap energies over the 12 neighbours, using the appropriate values of h^2, k^2

and l^2, and introducing the appropriate phase factor for each group of four neighbours. Writing $k_z a/2 = \theta$ and recalling that $\exp(i\theta) + \exp(-i\theta) = 2\cos\theta$, we obtain for these sums

$$\left.\begin{aligned} E_{xy} &= 3(\mathrm{dd}\sigma) + 4(\mathrm{dd}\pi)\cos\theta \\ E_{zx}(=E_{yz}) &= 3(\mathrm{dd}\sigma)\cos\theta + 2(\mathrm{dd}\pi)(1+\cos\theta) \\ E_{x^2-y^2} &= \tfrac{3}{2}(\mathrm{dd}\sigma)\cos\theta + 2(\mathrm{dd}\pi)(2+\cos\theta) \\ E_{3z^2-r^2} &= (\mathrm{dd}\sigma)(1+\tfrac{1}{2}\cos\theta) + 6(\mathrm{dd}\pi)\cos\theta \end{aligned}\right\} \quad (6.21)$$

These give sections along k_z through five d zones formed from the five basic atomic d states. For less symmetrical directions through the zones the energy cannot be expressed as simply as this because there is some *hybridization* between them (*see* Appendix 4 and Fig. A 4.13). The E, k relation for each of these bands in the k_z direction is given by letting $\cos\theta$ range from $+1$ at the zone centre to -1 at the zone boundary (for (002) Bragg reflection). The qualitative form of these bands can be found by using the relative values $(\mathrm{dd}\sigma) = -2{\cdot}4$ and $(\mathrm{dd}\pi) = +1{\cdot}6$ from equation (6.17). This band structure is shown in Fig. 6.9. The energy scale is derived from the value of the width w of the d band given for nickel by Harrison (1980). Each of the two subclasses (the three xy, yz, zx states and the two x^2-y^2, $3z^2-r^2$ states) extends over practically the whole energy range of the total d band. There is a confluence of states near the upper range of energy, so

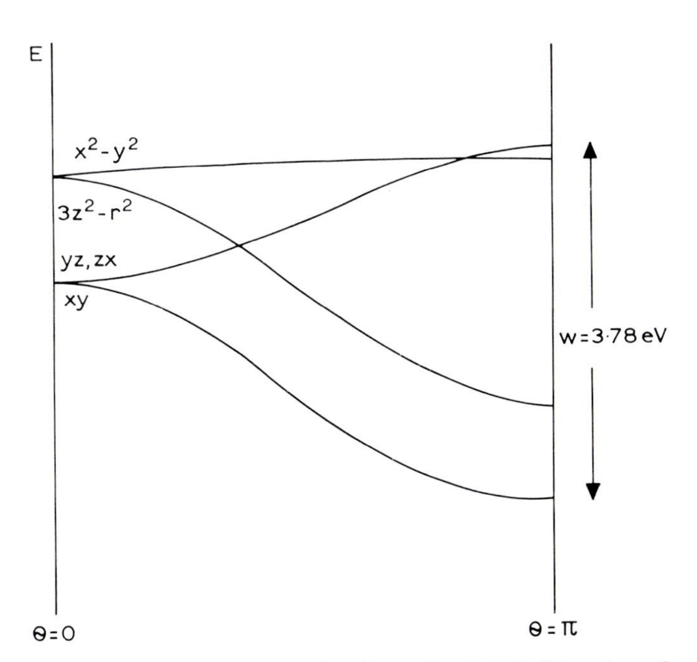

Fig. 6.9 Approximate energy bands along the [001] direction, for f.c.c. nickel

that the density of d states in the f.c.c. structure is particularly high near the top of the band. Friedel (1969) pointed out that this high density of states near the top of the f.c.c. band is expected to lead to a low Fermi energy in metals such as nickel, palladium and platinum, which have nearly full d bands; this is consistent with the stability of the f.c.c. structure in these metals.

A similar analysis can be made for the b.c.c. structure. The main difference here is that the six second-nearest neighbours have to be taken into account, as well as the eight nearest neighbours. Thus two values have to be used, i.e. $(\mathrm{dd}\sigma)_1$ and $(\mathrm{dd}\sigma)_2$, $(\mathrm{dd}\pi)_1$ and $(\mathrm{dd}\pi)_2$, the relative magnitudes of which can be obtained from equation (6.17) with $d_2 = 1{\cdot}15d_1$. The direction cosines of the nearest neighbours are of the type $[hkl]$ with $h^2 = k^2 = l^2 = \frac{1}{3}$, and those of the second neighbours are of the type [100] and its variants. As before, we consider a wave-vector $\boldsymbol{k} = 0, 0, k_z$. Thus, four of the nearest neighbours have a phase factor $\exp(\mathrm{i}ka/2)$ and the other four have a phase factor $\exp(-\mathrm{i}ka/2)$; four of the second-nearest neighbours are in the plane $z = 0$ and the other two have phase factors $\exp(\pm\,\mathrm{i}ka)$.

Using these values to sum the overlap energies of equations (6.18) over the nearest and second-nearest neighbours, we obtain the equivalent of equations (6.21) for the b.c.c. structure:

$$\left.\begin{aligned}
E_{xy} &= [\tfrac{8}{3}(\mathrm{dd}\sigma)_1 + \tfrac{16}{9}(\mathrm{dd}\pi)_1]\cos\theta + 4(\mathrm{dd}\pi)_2\\
E_{zx}(= E_{yz}) &= [\tfrac{8}{3}(\mathrm{dd}\sigma)_1 + \tfrac{16}{9}(\mathrm{dd}\pi)_1]\cos\theta + 2(\mathrm{dd}\pi)_2(1 + \cos 2\theta)\\
E_{x^2-y^2} &= \tfrac{16}{3}(\mathrm{dd}\pi)_1\cos\theta + 3(\mathrm{dd}\sigma)_2\\
E_{3z^2-r^2} &= \tfrac{16}{3}(\mathrm{dd}\pi)_1\cos\theta + (\mathrm{dd}\sigma)_2(1 + 2\cos 2\theta)
\end{aligned}\right\} \qquad (6.22)$$

where again $\theta = k_z a/2$.

The general form of these equations is shown in Fig. 6.10, the width of the d band again being taken from Harrison (1980), this time for chromium. There is a striking reversal of the bonding and antibonding roles of the two groups of states as k_z increases from the centre of the zone to the boundary. The pattern is quite different from that of the f.c.c. structure.

Both Fig. 6.9 and 6.10 represent special cases because of the high symmetry of the particular $\boldsymbol{k}$ direction, along the [001] axis, considered in these diagrams. To explore the more general structure of the bands we can examine, for the b.c.c. structure, another $\boldsymbol{k}$ line, running from the origin to the zone boundary, close to k_z but diverging slightly (i.e. $\boldsymbol{k} = k_x, k_y, k_z$ with, for example, $k_x < k_y \ll k_z$). This introduces additional cosine terms into equations (6.22), i.e. $\cos(k_x a/2)$, $\cos(k_x a)$, $\cos(k_y a/2)$ and $\cos(k_y a)$, which produce a splitting of the energy degeneracies of Fig. 6.10, as shown qualitatively in Fig. 6.11(a).

However, Fig. 6.11(a) is misleading as it stands because, now that we are no longer on a symmetry line, it is no longer true that $\beta^{jq}_{ip} = 0$ for the coefficients linking *different* types of d states on different atoms. It becomes necessary to consider the effects of all the cross-coefficients, such as, for example $\beta^{x^2-y^2}_{zx}$. The general method in all such cases where two or more quantum states ψ_1 and ψ_2 become linked by two such matrix elements is to consider combined

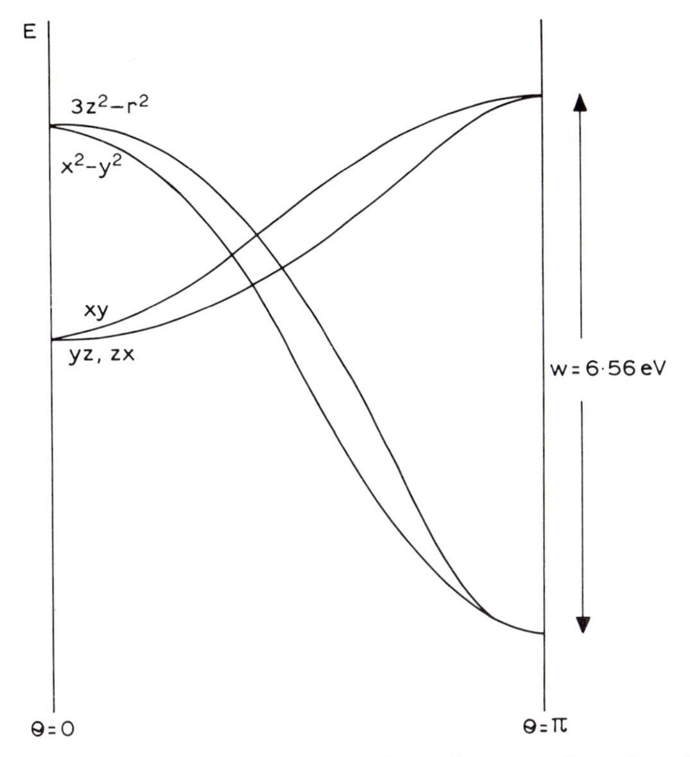

Fig. 6.10 Approximate energy bands along the [001] direction, for b.c.c. chromium

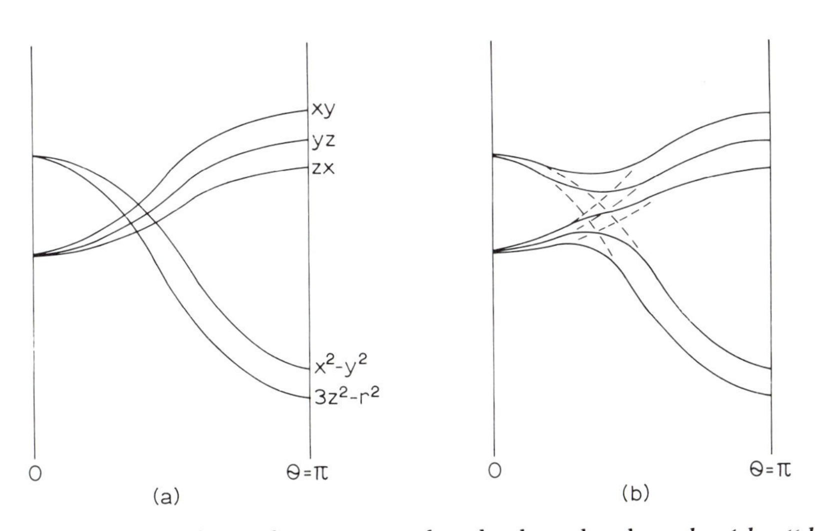

Fig. 6.11 Approximate b.c.c. energy bands along k, where $k_x < k_y \ll k_z$ (a) without and (b) with hybridization

wavefunctions $c_1\psi_1 + c_2\psi_2$, in which the weighting factors c_1 and c_2 are determined by minimizing the energy, using the method of variations and the secular equations, as outlined in Appendix 5. As shown there, when the energy levels of the original states are very different ψ_1 and ψ_2 do not combine well and the combined wavefunctions then approximate closely to either ψ_1 and ψ_2 in the two energy ranges. But where the original energy levels are close or degenerate there is strong combining, i.e. $|c_1| \simeq |c_2|$, and the combined wavefunctions become separated in energy by a gap, as in equation (A 5.9). The practical effect of this is that *hybridization* occurs, where the original energy levels cross, as in the example in Fig. A 4.13. In such a case, as the energy range is transversed, going upwards from below, there is at first little hybridization, and the primary wavefunction of lowest energy predominates in the combined wavefunction (e.g. $|c_1| \gg |c_2|$, with ψ_1 = the s state in Fig. A 4.13). Then, as the original energy cross-over point is approached, c_2 increases and eventually predominates over c_1, so that the wavefunction changes towards the pure ψ_2 form (e.g. the p state in Fig. A 4.13). There is then a gap until the upper energy range of the hybridized states is reached, in which the weighting coefficients again change from $|c_1| \gg |c_2|$ to $|c_2| \gg |c_1|$ as the energy is further increased.

Figure 6.11(b) shows qualitatively how the band structure of Fig. 6.11(a) becomes modified by this process of hybridization. The original d states can be recognized at the two extremes, $\theta = 0$ and $\theta = \pi$. We see that, for example, the lowest band starts out from the origin as the zx type of state but meets the zone boundary as the $3z^2 - r^2$ type. There is also a further hybridization, which we shall consider in Chapter 7, where the band of s states overlaps the d band.

We see from Fig. 6.11(b) that the total d band of the b.c.c. structure is made up of five overlapping hybridized bands, two which lie in the low energy range, two in the high range, and one which rises more steeply from the low to the high range. This is a very general characteristic which leads to some important features of the density of states distribution in b.c.c. transition metals. In

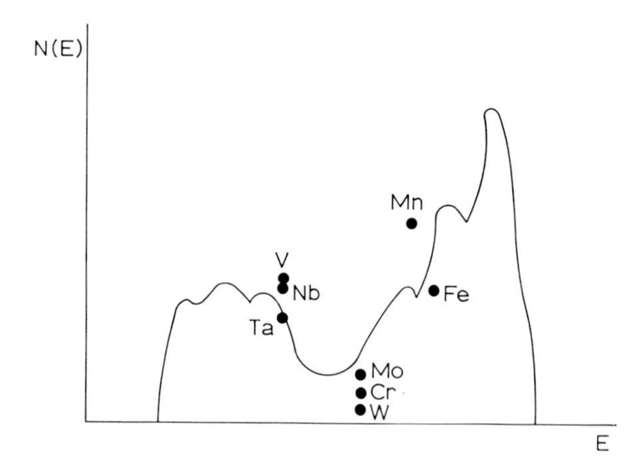

Fig. 6.12 Density of states $N(E)$ for the d band in b.c.c. chromium, the labelled points indicating values determined by specific heat measurements for various b.c.c. metals (Friedel 1969)

particular, the density of states is high in both the low and high energy ranges, but low in the intermediate region between them. Figure 6.12 (after Friedel 1969) compares experimentally deduced densities of states at the Fermi surface for several b.c.c. metals. An important consequence of this band structure, in the transition metals nearly halfway along their periods where the d band is not more than about half-filled with electrons, is that the Fermi energy is expected to be lower than it would be in other structures, such as f.c.c., because of the high density of states in the low energy range. This is a stabilizing feature for the b.c.c. structure, and it is significant that all the transition metals in this part of the Periodic Table (V, Cr, Nb, Mo, Ta, W) are b.c.c. Pettifor (1970) has shown that the distribution of the main crystal structures (b.c.c., f.c.c., c.p.h.) along the three transition metal rows can thus be satisfactorily explained.

6.7 CORRELATION EFFECTS: MAGNETISM

All of what we have so far considered in this chapter is of course a one-electron theory. As in all such theories the Coulomb repulsion and spin correlations (exchange effect) between the electrons are ignored, so allowing electrons to approach one another too closely. We recall (*see* Appendix 2, and Chapter 3) that in a partly filled shell of an atom the exchange effect favours the parallel alignment of electron spins (*Hund's rule*) because within this atom the aligned electrons then keep out of one another's way and so have less repulsive Coulomb energy. This alignment is opposed by the kinetic energy of the electrons because, when the spins are parallel, the Pauli principle allows only one electron to go into each state, and so some of the electrons may have to be promoted into otherwise empty states of higher energies.

Heisenberg and Dirac were the first to recognize that this exchange effect could provide the basis for *ferromagnetism*. The d bands of the transition metals are a particularly suitable system for its realization since, being only partly filled, they provide empty states into which electrons can be promoted, and, being narrow, they do not exact too large a kinetic energy penalty for the promotion.

We can thus imagine forming a solid by bringing together a large number of spontaneously magnetized atoms of, for example, iron, cobalt or nickel. With iron and cobalt, in which the 3d shell of the stable atom contains six and seven electrons, respectively, we certainly expect the Hund effect to provide some spontaneous alignment of spins within each atom when incorporated into the solid. With nickel, in which the d shell is more nearly full, the position is less clear, but at least we can expect many of the atoms (in the solid) to be in the $3d^9$ configuration and thus to have one uncompensated electron spin each. In forming the solid, then, we are bringing together atoms many of which act as elementary bar magnets of strength equivalent to one, two or, in iron, three electron spins.

The difficult problem in the ferromagnetism of the transition metals is to understand the *coupling* by which these individually magnetized atoms align their spins in parallel so as to produce the spontaneous, macroscopic magnetization which is ferromagnetism. This coupling is too large to be explained by the purely magnetic forces which the magnetized atoms exert on each other. It is also too large to be explained by the *direct* exchange interaction (as was envisaged in the original Heisenberg theory). To understand this we go back to the form of the *exchange integral* which gives the energy of this interaction, and is exemplified by equation (A 2.10). This is an integral of the repulsive electrostatic energy, e^2/r_{12}, over a product of four wavefunctions, $\psi_A(1)\psi_B(2)$ $\psi_A(2)\psi_B(1)$, where for example $\psi_A(1)$ refers to electron 1 in orbital A. The localized character of the d orbitals means that each ψ is small in the region between two neighbouring atoms, and so the fourfold ψ product is very small – much smaller, for example, than the twofold ψ product in the β integrals (equation (6.3)) which determine the width of the d band.

The most likely mechanism of coupling involves the *itineracy* of the d electrons (Hubbard 1963, Friedel 1969). Consider the possibility of such an itinerant d electron hopping onto a particular neighbouring atom where the magnetic spins are aligned 'upwards'. If its own spin is downwards, then such a jump could – and, unless the band is rather empty, would – bring it into an orbital already occupied by an electron of upward spin from which it would experience a large Coulomb repulsion. This repulsion generally causes the electron to avoid such an atom. There is thus a tendency for the itinerant electrons to avoid the neighbourhoods of atoms of opposite spin; and so for the predominance of a given spin direction to spread from each atom to its neighbours. At low temperatures this spin-ordering influence can spread cooperatively until one domain of common spin takes command over a macroscopic region of the metal.

The effective value U of the repulsive energy between two electrons of opposite spins in the same orbital is not easy to calculate accurately, but Friedel (1969) showed that $U \simeq 1$ eV. Suppose that we have a transition metal in the non-magnetic state, i.e. the numbers of its atoms with up and down spins are equal, $n\uparrow = n\downarrow = \frac{1}{2}N$. The total repulsive energy is then $Un\uparrow n\downarrow$, i.e. $\frac{1}{4}UN^2$. Now transfer a small number Δn from the down to the up spin state, so that the repulsive energy is diminished to $(\frac{1}{4}N^2 - \Delta_n^2)\ U$. This gain $-\Delta_n^2 U$ in potential energy is opposed by an increase in kinetic energy. The promotion of Δn electrons by an average energy ΔE increases the total kinetic energy by $\Delta n\,\Delta E$, i.e. $\simeq \Delta_n^2/n(E_F)$, where $n(E_F)$ is the density of states (per atom, per spin) at the Fermi level. We see that the potential energy decrease predominates over the kinetic energy increase when

$$Un(E_F) \geqslant 1 \tag{6.23}$$

This is the *Stoner criterion* for the occurrence of ferromagnetism. Guided by Fig. 6.2, we can assume as a first approximation that $n(E_F) \simeq 5/w$, in which case the criterion becomes

$$5U \geqslant w \tag{6.24}$$

Taking $U \simeq 1$ eV we see from Fig. 6.9 that nickel ($w \simeq 3{\cdot}8$ eV) satisfies the criterion easily. However, this is rather exceptional because the width of the d band increases considerably, both in the earlier members of each transition series (e.g. $w \simeq 6{\cdot}6$ eV for chromium in Fig. 6.10) and also in the 4d and 5d series (by about 40 and 70%, respectively, relative to the 3d series). Conditions for ferromagnetism in the Periodic Table are thus favourable only in the vicinity of nickel in the first long period.

Even iron is excluded on this criterion, which is consistent with the non-magnetism of f.c.c. iron. However, the double sub-band structure of the b.c.c. band, shown in Fig. 6.12, saves the day for b.c.c. iron because it gives a high density of states in the upper sub-band, in which the spin alignments take place. In fact the net energy, $n(E_F) - U$, is about $-0{\cdot}3$ ev per atom for b.c.c. iron (Janak and Williams 1976), which not only makes it ferromagnetic, below the Curie temperature, but also stabilizes the b.c.c. structure in this metal, at low temperature, relative to the non-magnetic f.c.c. structure. The return of the b.c.c. structure at high temperatures (> 1665 K) is thought to be an entropy effect, associated with a low frequency of some of the vibrational modes in this structure (Zener 1948).

The double sub-band structure of the d band in b.c.c. metals also explains the well-known observed curve of the *saturation magnetization* of ferromagnetic metals and alloys (Fig. 6.13). The upper sub-band, which is responsible for the ferromagnetism, lies so far above the lower band that, when the number of d electrons per atom is decreased, it empties totally while the lower one remains full. This upper sub-band, which provides the ferromagnetism, holds about five electrons per atom (with opposite spins) and so cannot provide more than about 2·5 spins of magnetism per atom. When the electron concentration in this sub-band drops below about 2·5 the aligned half of the band begins to empty, so reducing the saturation magnetization, as in the region Fe → Cr of Fig. 6.13. Although manganese and chromium are not themselves ferromagnetic, many ferromagnetic b.c.c. alloys have been examined which establish this part of the diagram. This explanation of Fig. 6.13 confirms Pauling's intuitive view that the d band consists of an upper half-band responsible for ferromagnetism and a lower one responsible for cohesion (Pauling 1938).

When a system becomes magnetized it also expands slightly, so as to reduce the size of the kinetic energy penalty which has to be paid for the promotion of

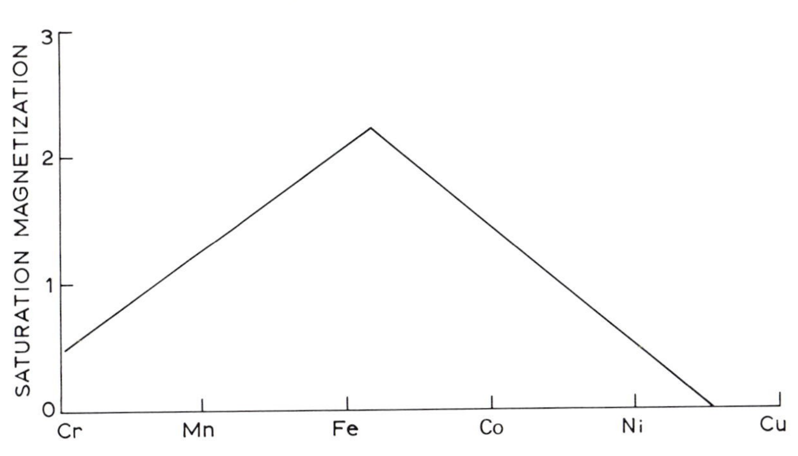

Fig. 6.13 **Saturation magnetization per atom of ferromagnetic metals and alloys**

the electrons. One consequence of this is that the bulk modulus of elasticity is also reduced. The effect can be large: for example, the bulk modulus of iron is only about two-thirds of that of ruthenium, the (non-magnetic) element immediately below it in the second long period.

In metals with a wide bandwidth, i.e. $w > 5U$, the Fermi energy suppresses ferromagnetic ordering and equal numbers of opposite spins have then to be accepted by the system. The spin correlation forces are still present, however, and the system will try to accommodate them while maintaining equality between the spins. As recognized by Slater (1930, 1951), this leads to a tendency for the opposite spins to keep apart as much as possible by becoming localized in neighbouring atoms, i.e. to form an *antiferromagnetic* superlattice in which alternate atomic sites carry alternate spins. The b.c.c. structure is ideally suited for this, since it consists of two interpenetrating simple cubic sublattices, one which provides the sites at the body centres, and the other which provides the sites at the b.c.c. cubic cell corners. The basic b.c.c. antiferromagnetic structure then has up spins on one sub-lattice and down spins on the other.

In metals (in contrast to antiferromagnetic non-metallic crystals, such as FeO) this effect is generally too weak to be manifested. The b.c.c. alkali metals, for example, are not antiferromagnetic. However, chromium is antiferromagnetic below 311 K owing to an unusual stabilizing effect first explained by Lomer (1962). Its magnetic structure is a form of the b.c.c. antiferromagnetic superlattice, described above, but with an important complication, known as a *spin density wave*. The spin polarization $\boldsymbol{P}(\boldsymbol{r})$ of each sublattice is not constant throughout the structure, but varies in amplitude as a sinusoidal function of position $\boldsymbol{r}$:

$$\boldsymbol{P}(\boldsymbol{r}) = \boldsymbol{P}(\boldsymbol{Q}) \exp(\mathrm{i}\boldsymbol{Q}\cdot\boldsymbol{r}) \qquad (6.25)$$

with a wave-vector $\boldsymbol{Q}$ that is generally rather less than half the length of the reciprocal lattice vector of the crystal structure in the [100] direction.

Lomer noticed that in chromium $\boldsymbol{Q}$ happens to be almost exactly equal to the vector $\boldsymbol{k}$ which can be drawn between various parts of the Fermi surface in the Brillouin zone. Recalling the b.c.c. d band structure in Fig. 6.10, we see that when this zone is about half-full, as in chromium, the Fermi level must cross several of the individual bands. Each individual band is empty on one side of a crossing (i.e. where it is higher than the Fermi level) and full on the other side. Each such crossing thus denotes a point on the Fermi surface and, since there are several such crossings, there are similarly several distinct pieces of Fermi surface in the zone. It turns out that some of these are *nested*, which means that parts of them are approximately flat and parallel to similar parts on other pieces, and so are separated from them by a single common vector which is small compared with a reciprocal lattice vector.

Lomer realized that this opens up the possibility for an energetically favourable coupling to develop between the conduction electrons, in the Fermi surface, and the antiferromagnetic spin structure of the d band, provided this structure has a wave-vector $\boldsymbol{Q}$ that is related simply to the $\boldsymbol{k}$ of the nested surfaces, e.g. $\boldsymbol{k} + \boldsymbol{Q} = -\boldsymbol{k}$. The periodic spin structure will then provide a weak periodic potential (with a period generally unrelated to that of the lattice potential) with which the Bloch waves of the conduction electrons can interact. This effect, although small, appears to be sufficient to make chromium antiferromagnetic (Windsor 1972, Skriver 1981).

A complicating factor which is often important in splitting degenerate energy levels, particularly in systems containing heavy atoms, is *spin–orbit coupling*. This is a relativistic effect and so is small unless an electron is moving at near-relativistic speed. When an electron moves slowly through an electric field, e.g. that of an atomic nucleus, it experiences this field essentially as an *electrostatic* one, as we have assumed throughout this book. But when it goes through at high speed the field also presents a magnetic aspect to the electron, which interacts with the magnetic moment of the electron spin. There is then a coupling between the spin and the orbital motions of an electron. In heavy atoms, where the strong Ze/r field of the nucleus can lead to high orbital speeds, this effect is important; for example, in lead there is a splitting of nearly 2 eV. While spin–orbit coupling is a general effect in all atomic systems, its particular significance in transition metals is that some of the degeneracies between overlapping d bands are removed. It is one of the factors which prevent the antiferromagnetism of chromium from being repeated in the corresponding elements, molybdenum and tungsten, in the two later transition series (Cracknell and Wong 1973).

REFERENCES

Andersen, O. K., *Solid State Commun.*, **13**, 133 (1973).

Asdente, M., and Friedel, J., *Phys. Rev.*, **124**, 384 (1961).

Cracknell, A. P., and Wong, K. C., *The Fermi Surface*, Oxford University Press (1973).

Cyrot-Lackmann, F., *J. Phys. Chem. Solids*, **29**, 1235 (1968).

Ducastelle, F., and Cyrot-Lackmann, F., *J. Phys. Chem. Solids*, **31**, 1295 (1970).

Finnis, M. W., and Sinclair, J. E., *Phil. Mag. A*, **50**, 45 (1985).

Fletcher, G. C., and Wohlfarth, E. P., *Phil. Mag.*, **42**, 106 (1951).

Friedel, J., in *The Physics of Metals, I – Electrons* (ed. J. M. Ziman), Cambridge University Press (1969), Chap. 8; in *Physics of Modern Materials*, Vol. I, International Atomic Energy Agency, Vienna (1980), p. 163.

Harrison, W. A., *Electronic Structure and the Properties of Solids*, Freeman, San Francisco (1980).

Heine, V., in *The Physics of Metals, I – Electrons* (ed. J. M. Ziman), Cambridge University Press (1969), Chap. 1.

Hubbard, J., *Proc. R. Soc. A*, **276**, 238 (1963).

Janak, J. F., and Williams, A. R., *Phys. Rev. B*, **14**, 4199 (1976).

Lomer, W. M., *Proc. Phys. Soc.*, **80**, 489 (1962).

Papon, A. M., Simon, J. P., Guyot, P., and Desjonquères, M. C., *Phil. Mag., B*, **40**, 159 (1979).

Pauling, L., *Phys. Rev.*, **54**, 899 (1938); *J. Solid State Chem.*, **54**, 297 (1984).

Pettifor, D. G., *J. Phys. (Paris)*, C **3**, 367 (1970); in *Physical Metallurgy* (ed. R. W. Cahn and P. Haasen), Elsevier, Amsterdam (1983), Chap.3.

Skriver, H. L., *J. Phys. F*, **11**, 97 (1981).

Slater, J. C., *Phys. Rev.*, **35**, 509 (1930); *Ibid.*, **82**, 538 (1951).

Slater, J. C., and Koster, G. F., *Phys. Rev.*, **94**, 1498 (1954).

Windsor, C. G., *J. Phys. F*, **2**, 742 (1972).

Zener, C., *Elasticity and Anelasticity of Metals*, Chicago University Press (1948).

7

The metals of the long periods

7.1 HYBRIDIZATION OF THE s AND d BANDS

According to Fig. 6.3, the contribution of d electrons to the cohesive energy reaches zero at the noble metal end of the transition series, where the number z_d of these electrons per atom is 10. The band is then full and, from the model of Fig. 6.2, makes no contribution to cohesion. Then, since the atoms of copper, silver and gold have one s electron, outside the filled d shell, we might expect the cohesive properties of these metals to be much like those of the corresponding alkalis, potassium, rubidium and caesium, at the beginning of the three long periods. Clearly an important part of the story, which explains why the cohesion remains strong immediately beyond the end of each transition series, is missing from Chapter 6.

What is also missing from that account – and which, as we shall see, provides the answer to the above problem – is the influence of the *conduction* s *band* (or, more accurately, the sp band, since the s states hybridize with p states; *see* Fig. A 4.13). This sp band, which spreads over a broad range of energy, overlaps the energy range of the d band in the transition metals and the overlap produces significant changes in the total band structure. The s band by itself has of course the usual NFE 'parabolic' form of a conduction band (*see* e.g. Fig. 1.5). It is a broad, low-density band, capable of holding only two electrons per atom with opposite spins. We can see the overlap effect qualitatively by superposing a steep NFE parabola on, for example, the d band structure of Fig. 6.11(b), as shown in Fig. 7.1.

The resulting total band structure after hybridization for a non-magnetic b.c.c. transition metal thus starts at the bottom with a band which has an s-like form, but which changes into a d-like form (1) at the zone boundary ($\theta = \pi$). There follow, in order of increasing energy, a narrow bonding band (2), a band which is at first bonding and then climbs steeply to a high-energy region (3), another similar band (4), a narrow antibonding band (5), and finally an upper band which starts out as d-like and then changes at very high energy into an s-like band (6). As explained for the simpler example of sp hybridization in Appendix 4 (Fig. A 4.13), the effect of this hybridization is to 'push' the energy levels of the overlapping bands apart, so reducing the density of states in the middle of the total band and increasing it on the two sides of this central region.

Similar effects occur in the f.c.c. structure. Thus Fig. 7.2 represents the result of a detailed spd band calculation in the [001] direction for copper. Friedel

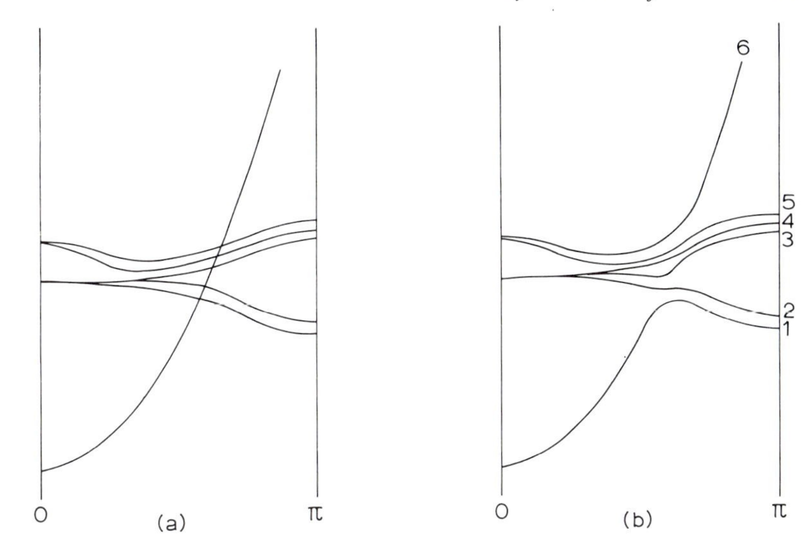

Fig. 7.1 Hybridization (b) of a b.c.c. d band (from Fig. 6.11(b)) and a parabola representing an NFE sp band (a)

(1969) has shown that the pushing apart of the energy levels by spd hybridization can explain the cohesion of copper, silver and gold, even though the d band is completely filled (as indicated by the position of the Fermi level shown in Fig. 7.2). Because the downward-shifted part of the hybridized spd

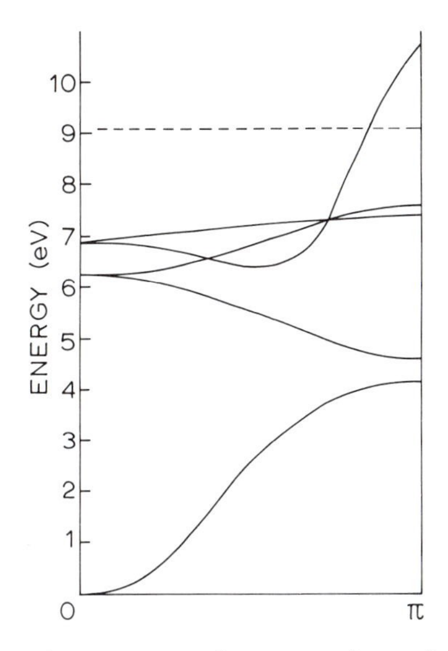

Fig. 7.2 The spd band structure of copper along the [001] direction; the horizontal dashed line indicates the Fermi level (after Mattheiss 1964)

band is filled, whereas the upward-shifted part contains a smaller number of electrons, there is on average a *lowering* of energy as a result of hybridization which Friedel estimates to be of the order of 1 eV per atom. This additional contribution is sufficient to account approximately for the greater cohesion of the noble metals, relative to the alkali metals.

7.2 THE RENORMALIZATION THEORY

We now examine the spd band theory in greater detail by outlining a powerful extension of the Wigner–Seitz method (*see* Chapters 2 and 3) to the transition metals and their neighbours (Gelatt *et al.* 1977). We recall that Wigner and Seitz introduced their method as a way of calculating a wavefunction for the $k = 0$ state of an energy band, by applying the condition $d\psi/dr = 0$ at the WS cell boundary to the solution of the single-atom Schrödinger equation. We also noted in Section 6.2 that if we similarly apply instead the boundary condition $\psi = 0$ the WS method can also give the approximate wavefunction and energy at the *top* of the band. Gelatt *et al.* used these methods to estimate the band limits, but supplemented them with the feature from which their method gets its name; and which is closely related to an *atomic sphere approximation* introduced by Andersen (1973). The took the well-established *free-atom* wavefunction, cut it off sharply at the WS cell boundary and then brought the electronic charge, in the external cut-off region, back into the cell by simply increasing the amplitude of the atomic wavefunction by a suitable amount (about one electron). This method of *renormalizing* the amplitude of the wavefunction inside the cell has proved to be a good approximation for deducing a potential energy which is more realistic, for the calculation of the band limits, than the free-atom Hartree-Fock version used in the original WS method (Section 2.2).

The two boundary conditions are conveniently taken together by examining the function $\psi^{-1}(d\psi/dr)$, which goes to zero when $d\psi/dr = 0$, at the bottom of the band, and to infinity when $\psi = 0$, at the top. Figure 7.3 shows this function at various energies for the renormalized copper atom at its Wigner–Seitz radius (1·41 Å). We see that the bottom of the s band and the bottom and top of the d band are at about −11, −8 and −4 eV, respectively.

The cohesion of the solid is calculated in a series of steps, represented by the successive terms in the equation

$$E_t = E_{prep} + E_{ren} + E_{sp} + E_d + E_{sd} \qquad (7.1)$$

each of which corresponds to a stage in the process of condensing a gas of free atoms into a solid. Here E_{prep} is the energy which must be supplied to promote a free atom from its normal electronic configuration, usually $d^{n-2}s^2$, into the $d^{n-1}s$ configuration which more nearly resembles that in the solid. Its value, known from atomic spectroscopy, is typically 1–3 eV, but is zero for copper. E_{ren} is the energy which must be supplied to renormalize this prepared free atom, and is typically of the order of 1 eV. In the next stage, renormalized free atoms of the gas, each already in a WS cell of equilibrium radius, are packed together to make the solid. E_{sp} is the change in energy brought about by the ensuing spreading of the atomic s level into the sp band. This is normally a small

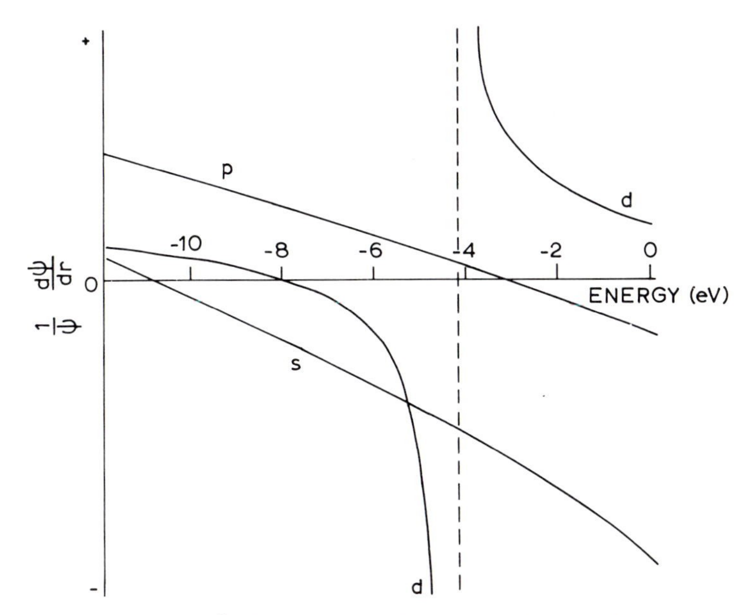

Fig. 7.3 Values of ψ^{-1} $(d\psi/dr)$ at various energies for the renormalized copper atom, evaluated at the Wigner–Seitz radius (after Gelatt *et al.* 1977)

contribution to cohesion, although it exacts a slight energy cost in chromium, because of the small equilibrium cell radius in this metal which compresses the conduction electrons and so forces up their energy. In copper E_{sp} contributes about half the total cohesive energy. E_d is the energy gain from the formation of the d band. As we saw in Chapter 6, this is the large term which is responsible for the strong cohesion of the transition metals, although it is taken as zero in copper. Finally, E_{sd} is the spd hybridization energy, discussed in Section 7.1, responsible for about half the cohesion of copper.

Figure 7.4 shows the values, joined in order from left to right, of these five components of the cohesive energy, as estimated by Gelatt *et al.* for each of the metals of the first transition series. For copper, of course, $E_{prep} = E_d = 0$. Note the large value of E_{prep} for manganese; this is mainly responsible for the relatively weak cohesion in manganese and is a consequence of its rather stable $3d^5\,4s^2$ structure. The E_{ren} term varies as an inverse function of the WS radius. Also shown in Fig. 7.4 is the sum of these contributions, E_t, which is seen to agree well with the observed cohesive energy. Gelatt *et al.* have obtained similar results for the second transition series.

7.3 THE CHANGE OF ENERGY WITH ATOMIC RADIUS

In calculating the results shown in Fig. 7.4 it was assumed that the metals are at their equilibrium radii, but Gelatt *et al.* also examined the effect of varying the WS radius. Figure 7.5 shows their results for copper.

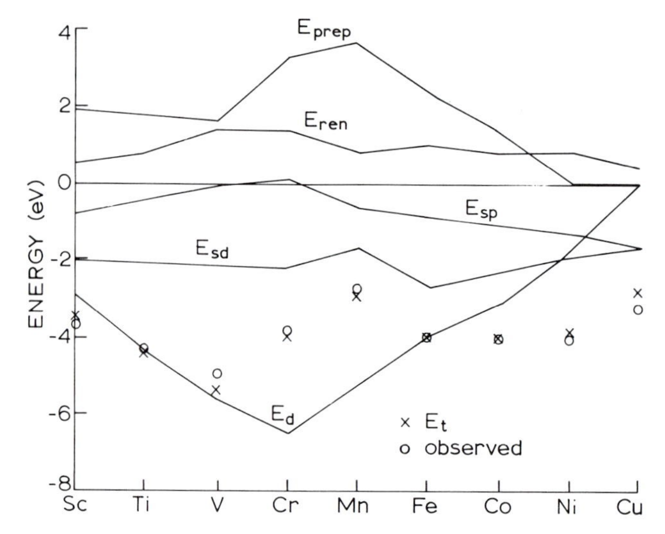

Fig. 7.4 Components of the cohesive energy for the first transition series (after Gelatt *et al.* 1977)

The position of the minimum in E_t gives a calculated equilibrium WS radius within 2% of the observed value, although E_t itself is somewhat smaller than the observed cohesive energy, as is seen in Fig. 7.4. Turning to the contributions of E_{ren}, E_{sd} and E_{sp} to the cohesion of copper, we recall that the slope of a curve, positive or negative, indicates the exertion of a force, attractive or repulsive

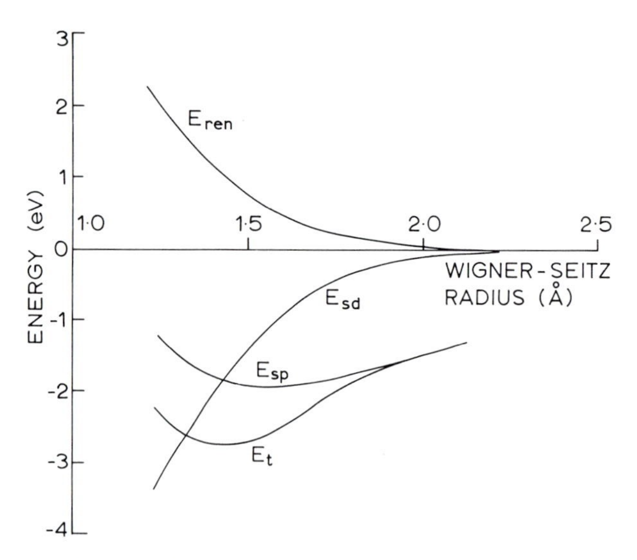

Fig. 7.5 The calculated cohesive energy of copper as a function of atomic radius (after Gelatt *et al.* 1977)

respectively; and that the curvature, positive or negative, indicates a positive or negative contribution to the bulk modulus (equation (5.12)). We see that near the equilibrium radius E_{sd} is attractive, whereas E_{ren} and E_{sp} are both repulsive. When contrasted with the traditional view of copper as a fairly simple 'free-electron' metal, consisting of repulsive hard-sphere ions (due to the filled d shells) held together by an attractive gas of s electrons, this is a remarkable result. At the equilibrium radius the d electrons are pulling the atoms together, through their hybridization energy E_{sd}, while that part of the system which corresponds to the classical free-electron gas is pushing the atoms apart, through the E_{sp} term.

This destroys the traditional explanation – a cornerstone of physical metallurgy – of the f.c.c. structure of copper, silver and gold. In this explanation, which starts from the fact that these are *full* metals (i.e. they have ionic radii almost as large as their atomic radii) it is supposed that the filled shells provide large inert ionic cores which resist being pressed together, much as if they were hard spheres. The free-electron gas, which tries to pack the whole metal into the least possible volume, achieves this most easily when the hard spheres are arranged in a close-packed structure such as f.c.c. In reality, however, the d electrons are strongly attractive at the equilibrium atomic radius, and because this radius is relatively small (e.g. 1·28 Å for copper, compared with 2·35 Å for potassium) the 'free-electron gas' represented by the E_{sp} term is squeezed into less than its optimum volume, to the point where its kinetic energy, through both the Fermi and ground state (*see* Fig. 5.1, at small radii) terms, strongly resists further contraction.

We also see from Fig. 7.5 that E_{sp} and E_{ren} contribute positively to the bulk modulus of copper, whereas E_{sd} weakens this elastic constant. The main contributor is E_{sp}. Again, the contrast with potassium is striking, for the bulk modulus of copper is 40 times greater. Thus we have another indication of the effect of the atoms being pulled close together in copper by the E_{sd} term, squeezing the conduction electron gas so tightly that it strongly resists further compression. The contribution of the Fermi energy to this process can be roughly estimated from free-electron theory, according to which it varies as r^{-2} (equation (5.1)), and so provides an r^{-5} term in the bulk modulus ($\propto r^{-1}\, d^2E/dr^2$). On this basis the modulus of copper should be $(2{\cdot}35/1{\cdot}28)^5 \approx 20$ times greater than that of potassium. We infer that an important additional contribution in copper must come from the rise in ground-state energy (*see* Fig. 5.1) at small radii.

7.4 THE PROBLEMS OF COPPER, SILVER AND GOLD

The classical explanation from electron theory of the structures (and metallurgical properties) of copper, silver and gold is left in disarray by the above arguments. This is ironic, considering that these have long been regarded as ideal subjects for the electron theory of metals and alloys. Why, then, are these metals f.c.c.; and so firmly so (since large amounts of solid-solution alloying elements have to be added to produce other crystal structures)? Because the d band is full and the s band only half-full, we do not expect any strong Brillouin zone effects. We therefore look, as in the old theory, to a 'hard-sphere' repulsion as the basis of the strong preference for the f.c.c. structure. This cannot of course be a filled d-shell repulsion, as considered in the old theory. However, perhaps there is a repulsion from the sp states due to the need to bend the ground-state wavefunction more sharply, so as to fit it into the WS cell, when the nearest-neighbour interatomic distance is small?

There are other problems. From NFE theory we would expect the E_{sp} term to depend on atomic volume, but not otherwise on atomic spacing, so that other structures could be equally competitive at the same volume. There is also a related problem: what gives these metals their elastic resistance to shear deformation? We know that this resistance comes from repulsive forces (Section 5.9) so we might look to E_{sp} as its main contributor. But if E_{sp} depends only on volume it cannot contribute to the shear constants.

There is of course a purely electrostatic contribution to these constants, which is derivable from Table 5.5. The numerical values given in Table 7.1 show that, whereas in the alkali metals the electrostatic term provides practically all the shear resistance, in the copper group (and even more in the b.c.c. transition metals) it is only a minor contributor. Evidently, in the copper group there is some other large factor at work. We thus return to E_{sp} as the possible source of this additional factor, but we are now looking for a dependence on structure. This may lie in the fact that in the WS polyhedron of the f.c.c. structure the smallest distance between polyhedral faces is 1·81 times the WS radius, whereas in b.c.c. it is only 1·76. Since a large contributor to the rise in E_{sp} at small spacings is the ground state energy (*see* Fig. 5.1), which stems from the condition $\partial\psi/\partial r = 0$ across the faces of the polyhedron, it is possible that the ground state energy may provide a central-force type of repulsion acting along the line between nearest-neighbour atoms.

This would introduce a kind of 'hard-sphere' repulsion in the cohesive properties of the copper group, which might explain the close packing and also the large additional factor in the shear constants of these metals. Of course, since such a central-force component could be only a minor part of the total repulsion (the Fermi energy and E_{ren} providing large structure-independent terms) we expect its contribution to the shear modulus in these metals to be small in relation to the bulk modulus. This is so, as is shown in Table 7.2.

7.5 THE SIMPLE METALS OF THE LONG PERIODS

The long periods begin with the simple alkali and alkaline earth metals, in which the d band is empty, and run on through the copper group, in which the band is completely full. As the nuclear charge increases across a long period, the energy

Table 7.1 Electrostatic contribution to the c_{44} shear constant (10^{11} J m^{-3})

	Na	K	Cu	Ag	Au	Fe (b.c.c.)	W
c_{44}	0·049	0·0263	0·753	0·436	0·420	1·16	1·51
Electrostatic contribution	0·05	0·0213	0·239	0·147	0·147	0·25	0·168

Table 7.2 **Ratio of the 'non-electrostatic' part of the shear modulus to the bulk modulus**

Na	K	Cu	Ag	Au	Fe (b.c.c.)	W
0	0·125	0·37	0·29	0·16	0·53	0·45

levels of all the atomic states move downwards. Figure 7.6 shows the 4s and 3d atomic levels across the first transition series (for the $d^{n-1}s$ configuration). We see that the d shell has a higher energy than the s shell in the early members of the series, but that it falls more steeply across the series, especially after copper. This latter stage occurs when the d shell is full so that the further electrons, added in the elements beyond copper, have to go in the outer sp level where they are rather ineffective in screening the d electrons from the increasingly attractive nuclear charge. Thus, in zinc and beyond, the 3d electrons are pulled increasingly tightly into the atomic core, where they take little part in the cohesion of the solid. The same effect occurs in the later periods, and in fact the 4d electrons are pulled in even more strongly, relative to the 5s electrons, on the right-hand side of the second transition series.

We have already discussed the alkali and alkaline earth metals, as well as the gallium group, in Chapter 5, and noted there an explanation of the large axial ratios of zinc and cadmium, as well as of the complex crystal structure of

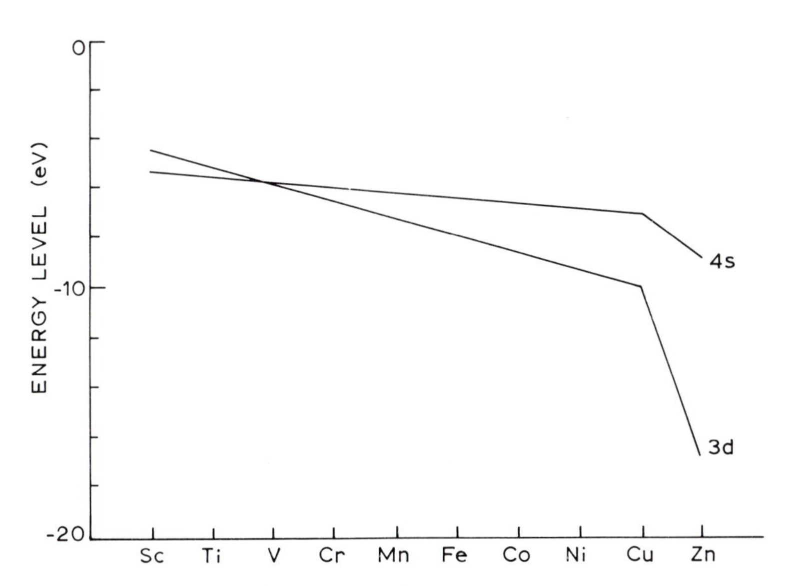

Fig. 7.6 **The energy levels of the 3d and 4s atomic states across the first long period**

mercury. The divalent metals of the zinc group have cohesive energies much lower than their monovalent predecessors, the copper group; quite the opposite of the alkali/alkaline earth relationship. The reason is that the steep drop in energy of the d bands and the corresponding shrinking of the d orbitals into the atomic cores, in going from the copper to the zinc group, removes the spd hybridization which is a major contributor to the cohesion of the copper group. As a result, the zinc group have only the relatively weak effect of the overlap of the p band on the s band to provide their cohesion. As noted by Griffith (1956), they thus rather narrowly miss being inert gases.

7.6 THE FINAL PERIODS

In the atoms of the third long period the filling of the 5d states after the two 6s states have been filled, is interrupted after lanthanum by the 14 *rare earth* or *lanthanide* metals in which the 4f states fill. Despite their high energies, the 4f electrons lie well inside the atom where they are largely shielded from the outside by the 6s and 5p electrons. The 4f band is thus extremely narrow. Its electrons remain localized in their parent atoms and do not contribute to the metallic character of these elements. The rare earths are thus early transition metals, with (usually) three Bloch electrons in a hybridized spd conduction band. Their crystal structures are generally close-packed, although in some cases with stacking sequences more complex than the simple ones found in the f.c.c. and c.p.h. structures. The 4f orbitals of neighbouring atoms are too localized to interact directly, but they interact indirectly via the conduction electrons, which causes their electrons to become magnetically ordered, usually antiferromagnetically.

However, some compounds of rare earth metals such as $CeAl_3$ or $CeCu_6$, and also some which contain actinides, such as UBe_{13} or UPt_3, show a different behaviour: the 4f, 5d and 6s states hybridize together so that all their electrons behave as Bloch electrons. However, because these electrons move so sluggishly through the crystal they appear to have gigantic *effective masses*. These *heavy-fermion compounds* have many unusual electronic properties – for example, a Fermi temperature of only about 100 K. We recall from Appendix 4 (equation (A 4.23)) that the effective mass of an electron is an indicator of the flattening of the *E, k* curve around the origin of *k*-space, as a result of the compression of the electron states into a narrow energy band rather than the free-electron parabola. If the electrons were completely localized in the atoms, their bandwidth would be zero and their effective mass infinite. Correspondingly, if they hop only occasionally from atom to atom, their band width is narrow and their effective mass is finite but very large. The Bloch electrons in these compounds are so nearly localized that their effective masses can reach about 1000 free-electron masses, whereas even in transition metals the effective masses are only about 10 times the free-electron value.

In their valency electronic structures the rare earths all appear to have the same kind of atom (with minor variations when the 4f shell is half- or completely filled). However, the increasing nuclear charge along the series pulls in the atomic orbitals so that there is a progressive reduction in the (trivalent) atomic diameter, from 3·7 Å in lanthanum to 3·4 Å in lutetium. This is the *lanthanide contraction*. Cerium shows some anomalous behaviour: in different temperature ranges it exists in three crystalline forms, two of which are f.c.c., with different atomic diameters, 3·42 Å in the low-temperature f.c.c. structure and 3·64 Å in the high-temperature one. This is a consequence of instability in a singly occupied f shell. At low temperatures the 4f electron in f.c.c. cerium metal is promoted into the spd band where it takes part in the cohesion and reduces the atomic spacing. This effect is delicately balanced, in energy, and moderate heat restores the atomic 4f electron; the high-temperature f.c.c. structure thus has a larger spacing.

The *actinide metals* which follow thorium in the final, incomplete row of the Periodic Table resemble rare earths in that the filling of the 7s, 6d states is interrupted by the filling of the 5f ones. The 5f states in these metals are less completely localized than are the 4f states in the rare earths. As a result, in the earlier actinide metals they hybridize with the 7s and 6d states, which generally contain three electrons per atom, to form a combined conduction band. In this respect the early actinide metals are more like transition metals than rare earths. In the later actinide metals, by contrast, the 5f states become more completely localized within their atomic cores and so these metals (starting with americium) are more like rare earths in their properties and structure.

Because the 5f states are so finely balanced between the alternative forms of Bloch states like the d states in transition metals, and localized states like the 4f states in rare earths, and also because spin–orbit coupling is particularly strong in these heavy metals, the properties and structures of the actinides are very complex (Freeman and Koelling 1974). *Plutonium* is a striking example. Its six different crystal structures begin at low temperature with a complicated monoclinic α phase, followed by a different monoclinic β phase and then a face-centred orthorhombic γ phase, followed at successively higher temperatures by the f.c.c. δ phase, the body-centred tetragonal η phase and finally the b.c.c. ε phase. The 'normality' of these higher-temperature structures is, however, illusory, for the volume *decreases* on going from δ to ε, and the f.c.c. δ phase has a negative coefficient of thermal expansion. Although the electron energy bands have now been deduced for several structures of the actinides (Freeman and Koelling 1974) the explanations of how the 5f electrons bring about the specific complexities of structure and properties in these metals remain tentative and speculative.

REFERENCES

Andersen, O. K., *Solid State Comm.*, **13**, 501, 511, 514 (1973).

Freeman, A. J., and Koelling, D. D, in *The Actinides*, Vol. 1 (ed. A. J. Freeman and J. B. Darby), Academic Press, New York (1974), Chap. 2.

Friedel, J., in *The Physics of Metals, I – Electrons* (ed. J. M. Ziman), Cambridge University Press (1969), Chap. 8.

Gelatt, C. D., Ehrenreich, H., and Watson, R. E., *Phys. Rev. B*, **15**, 1613 (1977).

Griffith, J. S., *J. Inorg. Nucl. Chem.*, **3**, 15 (1956).

Mattheiss, L. F., *Phys. Rev.*, **134**, A970 (1964).

8

The electron theory of alloys

8.1 INTRODUCTION

Since atoms in metals are bonded together only indirectly, by their electrical attractions to the common cloud of free electrons moving among them, a metal remains largely unaffected if some of its atoms are removed and replaced by atoms of a similar metal. This is the basic reason why metals are easily able to form alloys over wide ranges of composition, often in the form of disordered solid solutions. By the same token, because they are so relatively indifferent to one another's presence, metal atoms usually alloy together with rather small heats of formation (Varley 1954).

The heat of formation ΔE_{AB} of an alloy composed of metals A and B, in the (atomic) proportions c and $1 - c$, is the difference between the cohesive energies of the alloy, E_{AB}, and those of its constituent metals:

$$\Delta E_{AB} = E_{AB} - [cE_A + (1 - c)E_B] \tag{8.1}$$

and is typically less than $0 \cdot 1\, E_{AB}$. If the two metals show complete indifference to one another's presence, i.e. if $\Delta E_{AB} = 0$, there is nothing to prevent the entropy of mixing from producing random solid solutions over the entire range of composition, provided A and B have the same crystal structure.

In practice the atoms in an alloy are never totally indifferent; and the small but non-zero heat of formation manifests itself in various ways – for example as partial solid solubility, long-range or short-range ordering in solid solutions, and in more extreme cases by the formation of intermediate phases such as secondary solid solutions and intermetallic compounds. The observed relations between these features of alloys and the atomic properties of the constituent metals have been systematized in the famous *Hume-Rothery rules*. The most important of these rules concerns the *atomic size factor*, which always gives a positive ΔE_{AB} and thus opposes the formation of primary solid solutions. When the atomic diameters differ by more than 15% the primary solubility is small. Next in importance is the *electrochemical factor*, the difference in electronegativity between the constituents. This always gives a negative ΔE_{AB} and so favours the combining of the constituents into an alloy. It indicates a tendency, generally slight, within the metallic state of the alloy, towards the formation of an ionic compound. Third is the *electron–atom ratio* of the alloy. When other factors are small, this often correlates well with the occurrence of particular crystal structures in certain ranges of composition.

8.2 THE ATOMIC SIZE FACTOR

Darken and Gurry (1953) noted that at a temperature T the primary solid solubility is restricted to below about 1 at.% when the energy of solution exceeds $4k_BT$ per atom. As is well known (Friedel 1954, Eshelby 1956, Mott 1962, Cottrell 1975), if we regard a misfitting solute atom as an *elastic sphere* compressed or expanded to fit into a hole of the wrong size in the matrix metal, which is represented as an isotropic elastic continuum, and we estimate the ensuing total strain energy E in both the matrix and solute, then the approximate value

$$E = 8\pi\mu r_0^3\varepsilon^2 \quad (8.2)$$

is obtained for a solute and matrix having the same elastic properties. Here μ is the shear modulus, and r_0 and $(1 + \varepsilon)r_0$ are the unstrained radii of the solvent and solute atoms, respectively. Taking $\varepsilon = 0{\cdot}15$ and $\mu r_0^3 = 0{\cdot}7$ eV, then $E \simeq 4k_BT$ at 1000 K, which gives a simple explanation of Hume-Rothery's size-factor rule.

Mott (1962) established the quantum-mechanical basis of this elastic theory. Starting from a Wigner–Seitz wavefunction $\psi_0(r)$, for an electron in the lowest state in a WS cell (*see* e.g. Fig. 2.3), he expressed the wavefunction for a metallic electron in the alloy approximately as in equation (A 4.11), with $u(r)$ having the forms of an A atom or a B atom WS $\psi_0(r)$ wavefunction, respectively, inside the WS cells of A or B atoms (of the same valency) in the alloy. The WS condition (equation (2.2)) has of course to be satisfied at the boundaries of these cells. Consider a single solute atom B in a dilute solution. If the WS radius r_B of this atom is different from that of the solvent atom r_A, the wavefunction ψ_0 for this atom has to be found for some intermediate radius r, e.g. $r_A < r < r_B$. But this is the same problem as that of finding the bulk modulus of metal B from the Wigner–Seitz theory. Similarly, the expansion of the atomic hole in the solvent A, from r_A to r, is equivalent to a problem of the elasticity of metal A.

Mott extended the quantum-mechanical theory further by taking account of the fact that, when the WS cell of the solute atom is compressed or expanded, there is a renormalization of the wave amplitude in it, similar to that discussed in Section 7.2, which has to be compensated by a slight transfer of electronic charge so as to obtain continuity of ψ, as well as the condition $d\psi/dr = 0$, at the cell boundary.

The importance of the size factor of course extends far beyond primary solubility. Many intermetallic compounds owe their existence to size-factor effects, in particular to the opportunity to fit alloy atoms together exceptionally compactly in complex crystal structures which offer two or more different sizes of sites to accommodate them. The complex crystal structures formed in some pure metals such as manganese and some of the actinides may in fact be 'size factor compounds' between atoms of different sizes, the difference in this case being produced within the same species of atom when the partly filled d and f shells make it possible for electrons to have choices between itinerant and atomically localized orbitals.

Size-factor compounds in alloys are classified into various groups (Hume-

Rothery *et al.* 1969): for example *Zintl compounds* such as LiCd in which the atoms are fairly equal in size, σ *phases* such as FeCr and Mn_3Cr, *Laves phases* such as $MgCu_2$ in which the constituent atoms differ in size by about 22·5% and *interstitial compounds* such as WC and Ta_2C in which the interstitial atom has a radius less than about 59% of that of the metal atom. Frank and Kasper (1958; see also Pearson 1980) have given an elegant explanation of many of these complex crystal structures – which have coordination numbers with often 14, 15 or 16 atoms round large ones and thereby achieve a closeness of packing beyond that achievable in simple structures such as f.c.c. and c.p.h. – in terms of a general trend to pack as many atoms as possible into a given volume.

As Massalski and King (1961) emphasized, for the atomic size factor it is usually the volume per atom that matters rather than the distance between nearest neighbours. This is of course to be expected from Chapter 5, where we saw that the volume-dependent terms are the main contributors to the cohesive energy of a metal. Thus, in an alloy, if the volume available to an atom is the same as in the pure metal of that element, the heat of formation is small.

8.3 ELECTRONEGATIVITY

It is a familiar idea in theoretical chemistry that the approximate wavefunction, for a molecule containing two atoms A and B with atomic wavefunctions ψ_A and ψ_B, can be written as $a\psi_A + b\psi_B$, where $a = b$ for a pure covalent bond, $a = 0$ and $b = 1$ (or vice versa) for a pure ionic bond and $a < b$ for intermediate cases where B is more electronegative than A. The tendency for the electrons in such bonds to spend more time with the more electronegative partner was summarized by Pauling's celebrated table of electronegativity values of the elements (Pauling 1960), from which the degree of ionic character in chemical bonds can be approximately estimated. Hume-Rothery, in his rule of the electrochemical factor, showed that an apparently similar effect operates in alloys (Hume-Rothery *et al.* 1969). The more widely separated in electronegativity are the constituent metals of an alloy, the greater is the tendency for the electrons to concentrate on the more electronegative metal, a tendency which expresses itself through the formation of *ordered solid solutions* when the effect is mild, and of *electrochemical intermetallic compounds*, e.g. Mg_2Si, in the more extreme cases. Such compounds provide an alternative to random solution, as a means for the metal atoms to associate, and so primary solubility is often small in alloys of this type.

The relationship between alloy electronegativity and the electron theory of metals, in particular the pseudopotential NFE theory, has been developed by Inglesfield (1969). He concentrated on the Cd–Hg, Cd–Mg and Hg–Mg systems because the three atoms have the same valency and nearly the same size, so that all alloying factors other than electronegativity are insignificant. An electronic factor, the *alloying potential*, was identified as corresponding to the *electronegativity difference*, for both indicate the greater attraction of one ion than another for valency electrons. The alloying potential is defined by the *virtual crystal method*, which expresses the pseudopotential experienced by the metallic electrons in the alloy crystal as a weighted *mean potential*, the same at every site:

$$\bar{v} = cv_A + (1 - c)v_B \tag{8.3}$$

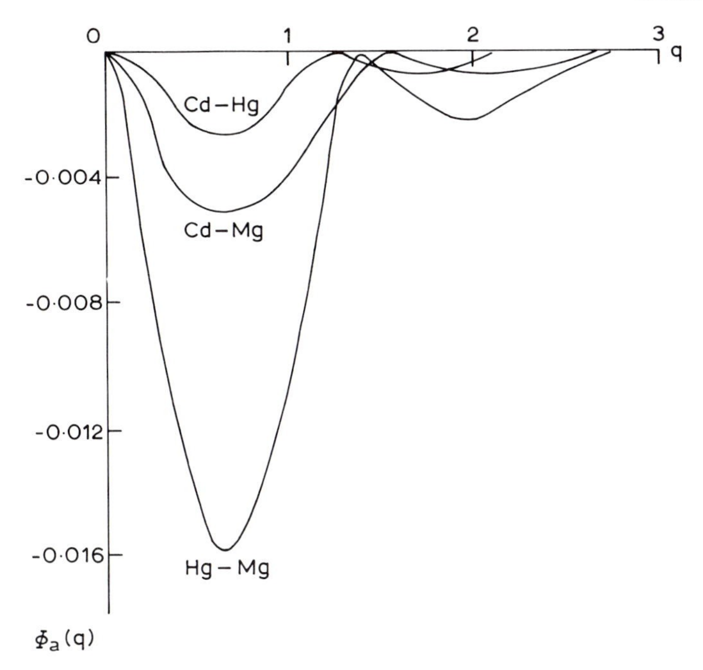

Fig. 8.1 The alloying energy-wavenumber characteristic for three alloys in atomic units, calculated by Inglesfield (1969)

where v_A and v_B are the pseudopotentials of atoms A and B (*see* Chapter 4), c being the atomic concentration of A. The *alloying potentials* are then the differences of the two pseudopotentials from this mean, i.e. $v_A - \bar{v}$ and $v_B - \bar{v}$. Inglesfield expressed the energy of the alloys as the sum of two terms, one which contains the mean potential and depends on the structure of the lattice of *sites*, irrespective of which atoms occupy them, and the other which contains the alloying potential and depends on the distribution of the two kinds of atoms among the sites.

To calculate the pseudopotentials from the atomic properties of the constituent metals, Inglesfield used the model potential proposed by Abarenkov and Heine (*see* Section 4.2) from which he deduced the *alloying energy-wavenumber characteristic*, $\Phi_a(q)$. This is the equivalent for the alloy of the parameter $\Phi(q)$ (*see* Appendix 7, equation (A 7.5)). When multiplied by the square of the structure factor and summed over the wavenumber q, as in equation (A 7.4), it gives the band-structure energy E_{BS}. Figure 8.1 shows the form of $\Phi_a(q)$ which he obtained for the three alloy systems.

To understand the general shape of such a function we first consider (*see* Fig. 8.2) two curves of the *screened pseudopotential form factor*, $w(q)$, similar to those shown in Fig. 4.1, which we take to be representative of the two constituent metals A and B. When their atoms have the same valency and atomic volume the two $w(q)$ curves start off at the same point, where $q = 0$ (*see* equation (4.16)). From the general argument presented in Section 4.2, the more electronegative

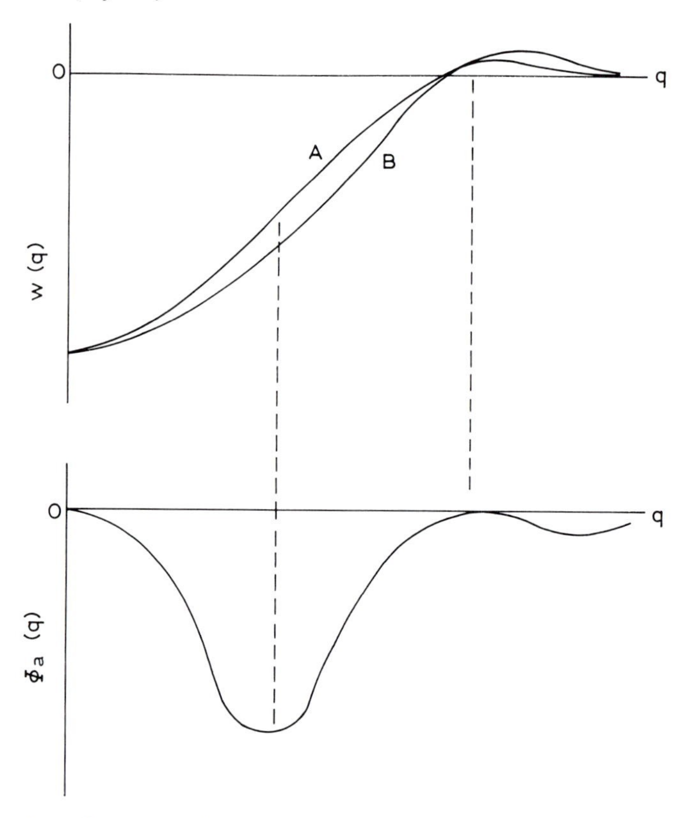

Fig. 8.2 Pseudopotential form factors for atoms A and B and the corresponding energy–wavenumber characteristic

metal, with the stronger Coulomb attraction for the metallic electrons, gives the lower $w(q)$ when q is small, i.e. curve B in Fig. 8.2. From the general relation between $\mathbf{\Phi}$ and w (equation (A 7.5)), we have

$$\mathbf{\Phi}_a(q) \propto -[w_A(q) - w_B(q)]^2 \tag{8.4}$$

where the minus sign comes from the negative multiplying factor, $\chi(q)$, as in equation (A 7.5). This expression leads to the shape of $\mathbf{\Phi}_a(q)$ shown in Fig. 8.1, derived from the *difference* between the $w(q)$ curves of the component atoms. The deep minimum in $\mathbf{\Phi}_a(q)$ at small q is thus a consequence of the electronegativity difference between these atoms.

As we saw in Section 4.4, the important values of the wave-vector $\boldsymbol{q}$ in a crystal are those that coincide with the reciprocal lattice vectors $\boldsymbol{g}$ because only then is the *structure factor* (equation (4.3)) non-zero. The smallest reciprocal lattice vectors of the *site* lattice (e.g. b.c.c. for the ordered MgHg phase, which has the

Table 8.1 Correspondence between Pauling's electronegativity difference and well depth

	Cd–Hg	Cd–Mg	Hg–Mg
Pauling's electronegativity difference	0·2	0·5	0·7
Well depth A (atomic units)	0·1	0·14	0·25

CsCl structure) occur, as in simple metals (*see* Section 5.7), near the q of the first zero in the $w(q)$ curves, i.e. where $\Phi_a(q)$ is small. However, if the alloy is ordered then the extra Bragg reflections from the superlattice (*see* Appendix 10) produce additional reflection wave-vectors at, typically, about half the smallest lattice $\mathbf{g}$, i.e. somewhere in the region of the deep first minimum of $\Phi_a(q)$. The band structure energy (*see* equation (A 7.4)) associated with $\Phi_a(q)$ may then be large and thus, by the argument put forward in Section 5.7, may favour the formation of the ordered structure. This indicates an incipient tendency, within the metallic state of the alloy, towards the formation of an *electrochemical* or *ionic intermetallic compound*.

The first minimum of $\Phi_a(q)$ occurs at about the same value of q for all three alloy systems represented in Fig. 8.1. Because of this Inglesfield was able to express all three alloying potentials in a simple common form, as a spherically symmetric 'square well' potential of radius R_m, the same for all these alloys, with a well depth of energy A, a *different* constant for each alloy system. The value of R_m was chosen so as to place the $\Phi_a(q)$ minimum at the same value of q as in Fig. 8.1, and those of A, which measure the electronegativity difference between the component metals, were adjusted for each alloy system so as to give the same depths of the minima as in Fig. 8.1. These fitted values of A showed a rough correspondence to Pauling's electronegativity differences as is seen in Table 8.1. From equations (8.4) and (A 7.4) we see that the energy gain from the formation of the superlattice increases as A^2, and so is some six times larger in Hg–Mg than in Cd–Hg. In agreement with this, Cd–Hg is observed to be disordered at room temperature whereas Hg–Mg is ordered up to the melting point (900 K). Inglesfield also calculated the ordering energies, and his results were in fair agreement with observation.

As in the case of the parameter Φ_{BS} discussed in Appendix 7 and Section 5.8, $\Phi_a(q)$ can be Fourier-transformed into a *spatial function* $\Phi_a(r)$, in which it represents a *direct interaction* between two ions a distance r apart. Inglesfield showed that this $\Phi_a(r)$ is positive between like ions, indicating a repulsion between them, and negative between unlike ones, so favouring structures in which a maximum of nearest neighbours are unlike ions, as in superlattices and ionic compounds. Although there is thus a strong resemblance between these two classes of structure, it must be remembered that the ordering forces in the alloy stem from the band-structure energy and that it is through this that the electronegativity difference is expressed. There is also an electrostatic energy, as the two types of ion carry slightly different charges in the alloy, but this is small in comparison.

8.4 GENERAL PROBLEMS OF THE THEORY OF ALLOYS

The next question concerns alloys of metals with *different* valencies. We prepare for it by first considering some general problems which face the electron theory of alloys.

The Bloch theorem (Section 1.4), which is the foundation of the electron theory of solids, is an analysis of the quantum-mechanical consequences of *perfect lattice periodicity*; the very existence of that most basic feature of the theory of metals, the wave-vector $\boldsymbol{k}$ as a *good quantum number*, springs from this. In a disordered alloy, however, there is no translational symmetry since different kinds of atoms are randomly distributed among the crystal sites. Strictly, then, there is no $\boldsymbol{k}$ in such an alloy and the vast structure of the electron theory, which has been built upon this concept, appears to be inapplicable to alloys. In one sense the position is even worse than this, and in another sense much better. It is worse in that the crisis we appear to have identified in alloys already appears in pure metals, when their periodicity becomes imperfect as a result of, for example, thermal vibrations, lattice vacancies or dislocations. It is much better in that we know from experience that all these substances remain 'good' metals, even when all semblance of long-range periodicity is lost, in the liquid state; and we know from the theories outlined in Sections 1.7–1.10, as well as the pseudopotential theory of imperfect structures (Section 4.5), that perfect crystallinity and an exact wave-vector $\boldsymbol{k}$ are not essential requirements of the metallic state.

The problem is essentially one of *concentrated, disordered* solutions. There is no difficulty here with superlattices or stoichiometric intermetallic compounds because their crystallinity, although more complex, is in principle as perfect as that of simple metal crystals (*see* Appendix 10). Neither is there any difficulty with *dilute* solutions (or with the effects of thermal vibrations). The occasional local irregularity can be treated, to a good approximation in the limit of extreme dilution, as an isolated scattering centre in an otherwise perfect crystal, as noted in Section 4.5. In the limit, the mean free path l of an electron, between collisions, is so large that the ambiguity $\Delta k \approx l^{-1}$ in the wave-number becomes negligible, i.e. $\Delta k \ll k$ (*see* Section 1.10). Even in concentrated disordered alloys the effect is usually small, $\Delta k < k_F$ (e.g. $l \simeq 100$ Å in primary solutions in copper). In the extreme case of special high-resistivity alloys such as Nichrome (80Ni–20Cr), where the empty d states give the atoms a large scattering cross-section, $l \approx 10$ Å which is approaching the range where major effects of the breakdown of the $\boldsymbol{k}$ concept, such as Anderson localization (*see* Section 1.10), can be expected.

The physical effects of the difference in the atomic characteristics of the components in an alloy are of course reduced by screening (*see* Section 2.7 and Chapter 3). Each metallic electron, in its movements through the entire system, tends to prefer those places where there is a positive Coulomb charge; and so by its transient visits it makes a small contribution to the screening haloes of neutralizing negative charge at such places (*see* Fig. 2.7). This is the general picture of screening in 'good' NFE metals, which we expect to be able to extend to alloys of such metals in which the component atoms have different valencies. It differs, however, from the picture of the pentavalent phosphorus atom in quadrivalent silicon (*see* Section 1.6) where the extra phosphorus electron remained localized in the neighbourhood of its parent atom, being bound to it in a hydrogen-like state (equation (1.22)).

Friedel (1954, 1958) analysed the circumstances under which the screening around a single atom of different charge, in an NFE metal, is either of the metallic type or consists of a localized bound state. We start by assuming that a bound state is formed and calculate it and its energy level, much as in Section 1.6. If this level lies sufficiently far below the bottom of the conduction band, as in Fig. 1.8, then the state remains a bound one at low temperatures, for the reason given in Section 1.6. However, if this 'impurity level' lies close to the bottom of the band, as is to be expected in broad-band metals, then the quantum state of the level hybridizes with those at the bottom of the band (*see* Appendix 5). Eventually the binding is lost, being replaced by an additional itinerant state at the bottom of the conduction band.

Even then, a 'ghost' of the bound state may remain, in the form of a *virtual bound state*. We noted in Section 1.10 that for wavefunctions to combine well they must not only spatially overlap and have similar energy levels; for they must also have similar *symmetries* (*see* Fig. A 2.3). In fact the outer d states of transition metal solutes dissolved in simple metal solvents do not combine very well with the sp states of the conduction bands, and so they preserve to some extent their localized character, even when their energy level lies just above the bottom of the conduction band. A metallic electron in this band, when passing through such a transition metal atom, may thus linger there for a little while as if it were temporarily bound in a localized state. These virtual bound states can be detected from the magnetic moments of transition metal solutes in simple metal solvents.

In applications of electron theory to the structures and properties of alloys the formidable problem of recalculating the energy band structure for each composition is often bypassed by means of the *rigid band approximation*. This assumes that when, for example, a solute of higher valency is dissolved in a metallic solvent, the shape of the energy bands (i.e. the E, k relation) remains unchanged, apart from a scaling brought about by a change in lattice constants. All that changes is the position of the Fermi level, which rises as more electrons are added to the common NFE pool by the higher valency atoms. The justification for this assumption is as follows. Suppose that a localized perturbation $V(\boldsymbol{r})$ is introduced into the electron gas of the host metal; for example, $V(\boldsymbol{r})$ may be the pseudopotential of an alloy atom of different valency. Then, according to perturbation theory (*see* Appendix 4) the resulting change in the energy levels of the (normalized) Bloch states, $\psi_k(\boldsymbol{r}) = u_k(\boldsymbol{r}) \exp(\mathrm{i}\boldsymbol{k}\cdot\boldsymbol{r})$, is given by

$$\langle k|V|k\rangle = \int \psi_k^* V \psi_k \, \mathrm{d}v = \int u_k^* V u_k \, \mathrm{d}v \tag{8.5}$$

(*see* equations (4.5) and (A 4.48)). For free electrons, for which u_k = constant, this change is independent of k – it is the same for every level. The whole energy band is thus shifted rigidly along the energy axis by an amount $\propto$V. For a 'good'

NFE metal, in which u_k is nearly independent of k, the corresponding similar shift is the basis of the rigid band approximation. The shift in the position of the band is brought about by a change in the allowed k values, through the requirement that they satisfy the boundary conditions at the surface of the metal (*see* Appendix 3).

This is of course only an approximation; and improved treatments have been developed, such as the *soft band model* of Cohen and Heine (1958). In general the approximation is reasonable for dilute alloys of 'good' NFE metals, but it is not valid for very narrow bands or for states near the limits of the band structure. For more accurate analyses there are available methods such as the *coherent potential approximation* (Ehrenreich and Schwartz 1976, Ducastelle 1980) in which the material outside a WS sphere is represented by an *effective medium* which is determined self-consistently by a method similar to the KKR treatment of the muffin-tin potential (*see* Section 2.3). The *density functional method* (*see* Section 3.6) has proved particularly useful in the more exact theories (Girifalco and Alonso 1980, Pettifor 1983).

8.5 THE JONES THEORY OF ALLOYS

We come now to perhaps the most famous topic in the electron theory of alloys, Hume-Rothery's discovery of the critical values of electron concentration at which particular phases occur in certain alloys, together with Jones's explanation of these critical values in terms of band-structure energies at a critical point in the filling of Brillouin zones. Hume-Rothery, Raynor and their colleagues proved that when the solvent metals copper, silver and gold are alloyed with metals of higher valency and favourable size factor, such as zinc, aluminium and tin, the boundary of the primary f.c.c. α phase solid solution is reached at an electron per atom (epa) ratio of about 1·4, the composition of the b.c.c. β phase is based on an epa of about 1·5, and the more complex γ and ε phases occur at correspondingly higher epa values. Jones explained this, using the rigid band approximation, by showing that when the Fermi surface, expanding with increasing epa, approaches the Brillouin zone boundary of a given crystalline phase, the increased density of states there due to the decrease in dE/dk near the boundary (*see* e.g. Fig. 1.5) leads to a large band-structure energy, and thus to a greater thermodynamical stability of the phase in this range of alloy composition. By representing the Fermi surface in the free-electron approximation as a sphere, it is then easily calculated that this sphere touches the first zone boundaries (*see* Fig. A 4.6) of the f.c.c. and b.c.c. structures at epa values of 1·36 and 1·48, respectively, thus explaining the observed values.

For a multivalent solute atom contributing only one of its valency electrons to the Fermi distribution, and holding the others back in bound states localized at the atom, the Jones theory remains applicable because the bound quantum states are formed by subtracting itinerant states from the bottom of the distribution so that the effective epa is the same as if all these electrons had been contributed to the distribution (Friedel 1954, 1958).

This delightful theory, given pride of place in all physical metallurgy textbooks, was suddenly brought to the point of disaster when Pippard (1957) showed experimentally that in pure copper, with an epa of 1·0, the Fermi surface already touches the zone boundary. A curious aspect of the history of the theory is that its popular textbook version, in which a free-electron *sphere* expands with increasing epa and eventually touches the zone boundary, is not that originally given by Jones (1937). Jones used a realistically large (4 eV) value for the band gap of copper at the zone boundary, and deduced that the Fermi surface in the pure metal deviated considerably from that of a sphere. In fact he estimated that it would touch the zone boundary at an epa of 1·04, but nevertheless calculated that the f.c.c. structure was more stable than the b.c.c. in the range $1{\cdot}0 \leqslant \text{epa} \leqslant 1{\cdot}43$.

Another problem is that, as we saw in Section 7.4, copper, silver and gold are very different from the simple metals envisaged in the elementary theory, since their cohesion is much greater than the NFE theory predicts for monovalent metals, largely as a result of spd hybridization. It seems, from the consistency of the epa effects observed by Hume–Rothery, that in alloys of this class these metals nevertheless act as if they were simple sp metals.

Despite these difficulties, the empirical Hume-Rothery epa rules have remained as a standing challenge to electron theory, which was not surprisingly taken up again from the new vantage point of pseudopotential theory, particularly by Heine (1969; see also Heine and Weaire 1970). As a start, Heine showed that what matters is not when the actual Fermi *surface* touches the zone boundaries, but when the surface of an equivalent free-electron *sphere* touches them. This is because the energy change associated with the deviation of the Fermi surface from a free-electron sphere is a higher-order function of the deviation of the NFE structure from the free-electron ideal than is the general band-structure energy. Few electrons participate until the main spherical region reaches the boundaries.

This is an encouraging result for the simple form of the Jones theory. However, Heine also showed that there is no special stability to be expected from the band-structure energy precisely when the Fermi sphere touches the boundary. The kinks in the energy curve are very small at this epa. Heine (1969) and Blandin (1966) have suggested that an alternative approach to an explanation might be to start, not from the above Brillouin zone argument, but from the many-electron *screening* effects. We saw in Section 3.4 and Appendix 7 that the dielectric constant $\kappa(q)$ which represents the screening effect of the electron gas varies strongly with the factor $\alpha = q/2k_F$. In the pure metal, with epa = 1·0, the Fermi wavenumber k_F is rather small; α is then large and there is fairly weak screening. When the Fermi sphere touches the zone boundary, i.e. $q = g = 2k_F$, the screening is still rather weak but it is changing rapidly with $q/2k_F$ (as in Fig. 3.1) and becomes large as k_F increases further, i.e. the electron energy then moves towards the free-electron value. The change of screening with $q/2k_F$ thus enhances the effect of the Brillouin zone boundary on the electron energy. Heine showed, for example, that as the epa rises above 1·36, where the sphere first touches the f.c.c. zone boundary, the screening rapidly increases and raises the band-structure energy from its depressed NFE value (*see* Fig. 1.5) to the higher free-electron value. The corresponding effect in the b.c.c. lattice occurs in the neighbourhood of epa = 1·48. A common tangent drawn to these two energy–epa curves touches them at about the epa points 1·36 and 1·48, respectively.

This appears to restore the electron theory of these Hume-Rothery alloy phases. However, questions remain. As Heine and Weaire pointed out, this explanation could still leave the homogeneous f.c.c. solid solution, at compositions in the range of epa ≈ 1·2, unstable against decomposition into two f.c.c. phases, one higher and one lower in epa. There would be an entropy penalty to pay for this of course, in which case the *spinodal temperature* below which the heterogeneity could develop might lie below the temperatures at which diffusion could bring it into being. The theory of these Hume-Rothery alloys is clearly still not fully settled.

Nevertheless, there can be little doubt about the validity of the Jones theory when applied to alloys with the γ-brass structure, which occurs at an epa ≈ 1·6, an example of which is Cu_5Zn_8. This complex cubic structure is made by stacking together a cubic array of 27 b.c.c. cells, then removing the 8 atoms at the large cube corners and also the one at the body centre of the array, and slightly readjusting the positions of the remaining atoms. It has a *Jones zone* (*see* Appendix 4), consisting of 36 small facets. This nearly spherical zone produces correspondingly nearly spherical energy contours in the zone, and the consequently nearly spherical Fermi surface can fit it closely at the appropriate epa. Conditions are thus favourable here for a strong Jones effect. This is confirmed by the density-of-states curve, $N(E)$, which rises to a high peak as the Fermi sphere approaches the zone boundary and then drops away sharply as the epa rises above 1·6 (*see* Fig. 8.3).

8.6 BINARY TRANSITION METAL ALLOYS

The near equality of the atomic radii of most of the transition metals in the same period makes it reasonable, in alloys between these metals, to assume d band

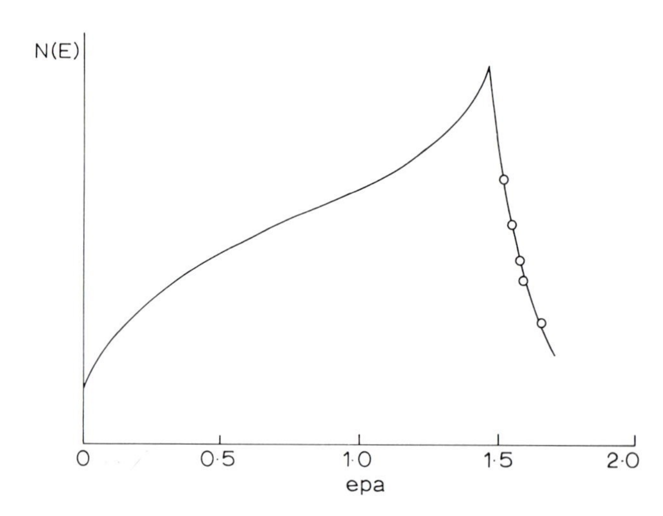

Fig. 8.3 Density of states at the Fermi level as a function of the number of electrons per atom (epa) for the γ brass structure; experimental points are from Veal and Rayne (1963)

effects to be the main basis of their cohesion (Cyrot and Cyrot-Lackmann 1976). There are various treatments of these, of which perhaps the simplest is by Pettifor (1978, 1979), who extended the Friedel theory of rectangular d bands (*see* Section 6.2 and Fig. 6.2). In the development of equation (6.1) for the cohesive energy per atom, E_d, of a transition metal with z_d electrons per atom in the d band, of width w, it was not necessary to consider the value E_0 of the energy level of the atomic d states, about which the band was symmetrically distributed. However, the different metals in the alloy in general have different atomic d levels, and so as an initial step we modify the pure metal equation (6.1) by adding to it a term $z_A E^A$, for metal A, where z_A is the value of z_d in A and

$$E^A = E_0^A + E_{prep}^A + E_{ren}^A \tag{8.6}$$

where E_{prep} and E_{ren} are the terms used in Section 7.2 to convert a free atom into a suitable WS cell for the solid. Hence we rewrite equation (6.1) as

$$E_A = z_A E^A - \frac{w}{20}[z_A(10 - z_A)] \tag{8.7}$$

Consider for simplicity a 50/50 binary alloy of metals A and B, in substitutional solution, interlinked by a network of overlapping d orbitals as in Fig. 6.1. We then have three terms of the above type, i.e. E_A and E_B per atom, for the two pure metals A and B, and E_{AB} per atom of the AB alloy. Thus

$$E_{AB} = z_{AB} E^{AB} - \frac{w_{AB}}{20}[z_{AB}(10 - z_{AB})] \tag{8.8}$$

and the *heat of formation* of the alloy, per atom (equation (8.1)) is

$$\Delta E = E_{AB} - \tfrac{1}{2}(E_A + E_B) \tag{8.9}$$

Then

$$z_{AB} = \tfrac{1}{2}(z_A + z_B) \tag{8.10}$$

and we assume that

$$E^{AB} = \tfrac{1}{2}(E^A + E^B) \tag{8.11}$$

A significant conclusion can be reached by examining the simplest case, where $w_{AB} = w_A = w_B = w$ and $E^A = E^B$. Here equation (8.9) reduces to

$$\Delta E \simeq -\frac{w}{80}(z_B - z_A)^2 \tag{8.12}$$

This brings out, as an alloying effect, the fact which we noted from Fig. 6.3, i.e. that the cohesion is strongest when the d band is half-filled. It thus follows that the heat of alloy formation is particularly favourable, other things being equal, when the constituent metals are chosen from opposite ends of the transition row and give $z_{AB} \simeq 5$.

Usually, of course, $E^A \neq E^B$ and $w_{AB} \neq w_A \neq w_B$. The inequality

$$\Delta E_0 = E_0^B - E_0^A \tag{8.13}$$

for the atomic d levels leads to an electronegativity difference. The molecular orbitals of the tight-binding theory are then no longer purely covalent; there is some ionicity in that the electrons spend more time at the more electronegative end of the orbital, as noted in Appendix 2. By an analysis similar to that given in Appendix 5 Pettifor (1978, 1979) showed that the increased bonding (and antibonding) resulting from this electrochemical effect widens the d band to

$$w_{AB} = w\left[1 + 3\left(\frac{\Delta E_0}{w}\right)^2\right]^{1/2} \simeq w\left[1 + \frac{3}{2}\left(\frac{\Delta E_0}{w}\right)^2\right] \tag{8.14}$$

where

$$w = \tfrac{1}{2}(w_A + w_B) \tag{8.15}$$

Equation (8.9) then becomes

$$\Delta E = -\frac{w}{80}(z_B - z_A)^2 - \frac{1}{4}(z_B - z_A)\Delta E_0 - \frac{3w}{40}\left(\frac{\Delta E_0}{w}\right)^2 [z_{AB}(10 - z_{AB})] \tag{8.16}$$

The first term here is equation (8.12) again, the second is the weighted difference in the atomic d levels, and the third is the increase in cohesion resulting from the difference in electronegativity

To this ΔE Pettifor adds a further term,

$$-\frac{w}{24}(5 - z_{AB})(z_B - z_A)\frac{\Delta\Omega}{\Omega} \tag{8.17}$$

where

$$\Omega = \tfrac{1}{2}(\Omega_A + \Omega_B) \tag{8.18}$$

and

$$\Delta\Omega = \Omega_B - \Omega_A \tag{8.19}$$

Here Ω_A and Ω_B are the respective volumes per atom in the pure metals. This term takes account of the fact that, because the d bandwidths vary with atomic volume (*see* equation (6.17)), the proper values of w_A and w_B for the alloy differ from those of the pure metals when $\Omega - \Omega_A$ and $\Omega - \Omega_B$ are non-zero. This addition, with some simplifying approximations and the assumption that

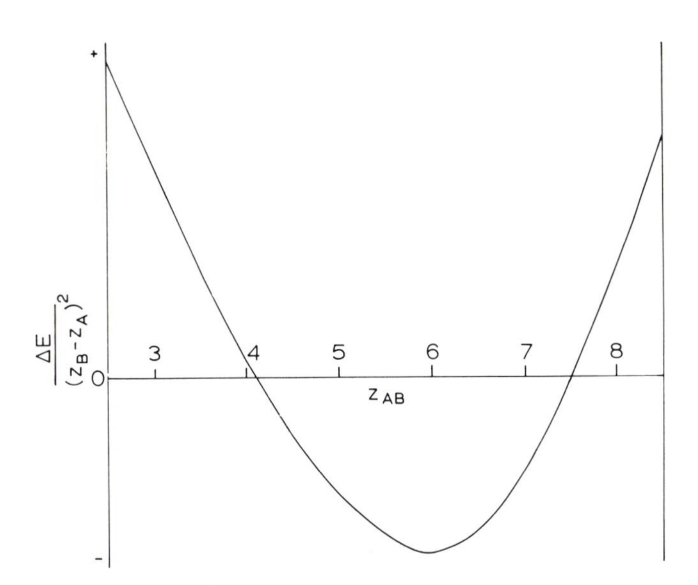

Fig. 8.4 **Change in the heat of formation of alloying, in the second transition series, as a function of the average number of electrons in the alloy d band (after Pettifor 1983)**

$w = 10$ eV (which is appropriate for the second transition row), leads to Pettifor's final expression (Pettifor 1983) for the heat of formation of the alloy:

$$\Delta E = [f_0(z_{AB}) + f_1(z_{AB})]\,(z_B - z_A)^2 \tag{8.20}$$

electron-volts per atom, where

$$f_0(z_{AB}) = \frac{1}{8}\left[1 - \frac{3}{50} z_{AB}(10 - z_{AB})\right] \tag{8.21}$$

$$f_1(z_{AB}) = -\frac{1}{24}(5 - z_{AB})\left(\frac{\mathrm{d}\ln \Omega}{\mathrm{d}\, z_{AB}}\right) \tag{8.22}$$

Figure 8.4 shows the variation of $f_0 + f_1$, i.e. $\Delta E/(z_B - z_A)^2$, with z_{AB}. We see that it is positive at both low and high values of z_{AB}, so that alloys in which both A and B are chosen from the same end of the transition row have unfavourable heats of formation. The strongly negative heat of formation, favouring alloy formation, occurs – as in the simple case of equation (8.12) – when A and B are chosen from opposite ends of the series and $z_{AB} \simeq 6$.

8.7 SEMI-EMPIRICAL ALLOY THEORIES

The difficulties and complexity facing the development of the formal band theory of alloys have encouraged the construction of several *semi-empirical* theories based on assumed prominent features of the electronic structure of alloys. One of the most successful of these is the *Miedema scheme* (Miedema *et al.* 1980) in which a binary alloy of A and B atoms is regarded as an assembly of WS cells. Suppose that initially each cell in the alloy has the same volume as the equilibrium cell in the pure metal of its atom. In general of course, because of their different volumes these cells have to be *deformed* in order to fit together in the alloy, but Miedema *et al.* argue that the energy of this distortion is small and can be ignored, an assumption which is questionable except for open metals such as the alkalis (*see* Fig. 2.3 and Section 7.4). At this stage there is no energy change on forming the alloy. However, there are electronic adjustments which have to be made at the boundaries of the WS cells, and it is from these that Miedema *et al.* deduce the heat of alloy formation.

Two such adjustments are envisaged. First, for the state described above there is a discontinuity $\Delta\rho$ in the density of electrons at the boundaries between A and B cells. This has to be smoothed, which entails changing the volumes of the cells from their equilibrium values and hence leads to an unfavourable, positive term in the heat of formation. Miedema *et al.* argue, on the basis that the boundary surface – where the correction has to be made – goes as the two-thirds power of the cell volume, that this energy term is proportional to $(\Delta\rho)^{2/3}$. Second, there is in general an electrochemical difference between A and B which, as we have seen, leads to a favourable term in the heat of formation. Miedema *et al.* estimate this term by regarding the individual A and B atoms in the alloy as samples of the macroscopic pure metals A and B, with their bulk properties, in which case the electrochemical effect can be dealt with as a problem in *contact potentials* (*see* Chapter 9). The critical quantity introduced for this is the difference $\Delta\phi^*$ in an *electronegativity parameter* ϕ^* which is almost identical to the *work function*, i.e. the minimum energy required to remove an electron from the macroscopic pure metal at zero temperature. In accordance with the general argument presented in Section 8.3, the contribution of this quantity to the heat of formation goes as $-(\Delta\phi^*)^2$.

The total heat of formation can thus be expressed as

$$\Delta E_{AB} = -P(\Delta\phi^*)^2 + Q(\Delta\rho^{1/3})^2 \qquad (8.23)$$

where P and Q are positive constants. As Pettifor (1983) has emphasized, this semi-empirical scheme of Miedema's, which has proved particularly successful in its applications to binary alloys of the transition metals, resembles in some ways the tight-binding theory outlined in Section 8.6. Thus $\Delta\phi^*$ is obviously similar to the term ΔE_0 (equation (8.13)), which leads to the electronegativity difference although, as Pettifor points out, the work function is not the best way to represent this. Again, Miedema's electron density term in equation (8.23) is reminiscent of the Fermi energy of an electron gas (equation (1.3)) although, again as Pettifor points out, here the pseudopotential theory would lead to a Q in equation (8.23) that is not a constant but which changes sign with electron density, being *positive* at low densities where exchange correlation is important, and *negative* at high densities where the kinetic energy is dominant. The theoretical basis of this successful semi-empirical scheme is thus not well established.

Ever since 1932 when Pauling introduced his electronegativity scale, it has been attractive to try to arrange all the chemical elements in a linear series which could serve to systematize the entire family of (binary) compounds. The various factors which control the formation of intermetallic compounds – size factor, electronegativity and electron concentration – have generally proved too much to handle in such a way. However, Pettifor (1984, 1985) has recently succeeded in constructing a *chemical scale*, a linear series of the elements based on the Periodic Table, which acknowledges the chemical similarity of elements from the same group of the Table by placing them in neighbouring positions in the series. For example, the elements of groups IVB, VB and VIB are ordered in the sequence – . . . Pb, Sn, Ge, Si, Bi, Sb, As, P, Po, Te, Se, S – . . . within the series. The serial position of each element is indicated by its *chemical scale number* χ, which ranges from 0 for the most electropositive elements to 4 for the most electronegative.

Binary compounds, e.g. $A_m B_n$, can then be represented as points on a two-dimensional *map*, the axes of which are the chemical scale numbers χ_A and χ_B of the metals in the compounds. Pettifor has constructed such maps for compounds with various structural formulae, AB, AB_2, AB_3, $A_2 B_3$ and so on. The significant feature which this reveals is that compounds of a given crystal structure all tend to occur within one well-defined region of such a map. For example, in the map of all the AB compounds, those with the CsCl (ordered β brass) structure mainly lie in the region $\chi_A, \chi_B \leqslant 1{\cdot}8$, whereas those with the NaCl structure are found only outside this region. The many other types of crystal structure also have preferred localities in such maps. The existence of a general chemical factor, which goes beyond mere electronegativity, is thus indicated by these systematic trends.

REFERENCES

Blandin, A., in *Phase Stability in Metals and Alloys* (ed. P. S. Rudman), McGraw-Hill, New York (1966), p. 115.

Cohen, M. L., and Heine, V., *Adv. Phys.*, **7**, 395 (1958).

Cottrell, A. H., *An Introduction to Metallurgy*, Edward Arnold, London (1975).

Cyrot, M., and Cyrot-Lackmann, F., *J. Phys. F*, **6**, 2257 (1976).

Darken, L. S., and Gurry, R. W., *Physical Chemistry of Metals*, McGraw-Hill, New York (1953).

Ducastelle, F., in *Theory of Alloy Phase Formation* (ed. L. H. Bennett), The Metallurgical Society of AIME, New York (1980), p. 194.

Ehrenreich, H., and Schwartz, L. M., *Solid State Phys.*, **31**, 149 (1976).

Eshelby, J. D., *Solid State Phys.*, **3**, 79 (1956).

Frank, F. C., and Kaspar, J. S., *Acta Crystallogr.*, **11**, 184 (1958).

Friedel, J., *Adv. Phys.*, **3**, 446 (1954); *Suppl. to Nuovo Cim.*, **7**, 287 (1958).

Girifalco, L. A., and Alonso, J. A., in *Theory of Alloy Phase Formation* (ed. L. H. Bennett), The Metallurgical Society of AIME, New York (1980), p. 218.

Heine, V., in *The Physics of Metals, I – Electrons* (ed. J. M. Ziman), Cambridge University Press (1969), Chap. 1.

Heine, V., and Weiare, D., *Solid State Phys.*, **24**, 249 (1970).

Hume-Rothery, W., Smallman, R. E., and Haworth, C. W., *The Structure of Metals and Alloys*, The Institute of Metals and The Institution of Metallurgists, London (1969).

Inglesfield, J. E., *Acta Metall.*, **17**, 1395 (1969).

Jones, H., *Proc. Phys. Soc. A*, **49**, 250 (1937).

Massalaski, T. B., and King, H. W., *Proc. Mater. Sci.*, **10**, 1 (1961).
Miedema, A. R., de Chatel, P. F., and Deboer, F. R., *Physica B*, **100**, 1 (1980).
Mott, N. F., *Rep. Prog. Phys.*, **25**, 218 (1962).
Pauling, L., *The Nature of the Chemical Bond*, Cornell University Press (1960).
Pearson, W. B., in *Theory of Alloy Phase Formation* (ed. L. H. Bennett), The Metallurgical Society of AIME, New York (1980), p. 262.
Pettifor, D. G., *Solid State Commun.*, **28**, 621 (1978); *Phys. Rev. Lett.*, **42**, 846 (1979); in *Physical Metallurgy* (ed R. W. Cahn and P. Haasen), Elsevier, Amsterdam (1983), Chap. 3; *Solid State Commun.*, **51**, 31 (1984); *Solid State Phys*, **19**, 285 (1986).
Pippard, A. B., *Phil. Trans. R. Soc. A*, **250**, 325 (1957).
Varley, J. H. O., *Phil. Mag.*, **45**, 887 (1954).
Veal, B. W., and Rayne, J. A., *Phys. Rev.*, **132**, 1617 (1963).

For further reading, a recent publication is Hafner, J., *From Hamiltonians to Phase Diagrams*, Springer-Verlag, Berlin (1987).

9

The metallic surface

9.1 THE WORK FUNCTION

Although valency electrons are free to move anywhere inside a metal, its surface is a prison wall for them. This feature of the metallic state is measured by the *work function*, which is the minimum energy required to remove an electron from the metal and is thus a latent heat of vaporization of electrons from a metal. In the simple 'potential box' model of a metal, shown in Fig. 9.1, the work function is $\mathbf{\Phi}$, the difference in energy between the Fermi level at 0 K and the potential energy of an electron in the space outside the metal. Typically $\mathbf{\Phi}$ is about one-third of the total depth of the box, and ranges from about 2 eV for the alkali metals to about 6 eV for metals such as platinum.

The work function is related to several measurable properties of metals. It is directly measured by the *photoelectric threshold*, i.e. the lowest frequency ν of an incident photon that can cause an electron to be emitted from the metal as a result of an energy transfer of $h\nu = \mathbf{\Phi}$. At elevated temperatures electrons can also be emitted when they receive an energy of $\mathbf{\Phi}$ or more from thermal fluctuations. The probability of an electron receiving such an energy is of course proportional to $\exp(-\mathbf{\Phi}/k_B T)$. A semi-classical calculation of the kinetics of this process of thermal evaporation leads to the celebrated *Richardson–Dushman* equation for the *thermionic emission current* j leaving unit area of the surface:

$$j = A(1 - \bar{r})T^2 \exp(-\mathbf{\Phi}/k_B T) \tag{9.1}$$

where $A = 4\pi m k_B^2 e/h^3 = 1{\cdot}2 \times 10^6\ \mathrm{A\ m^{-2}\ K^{-2}}$, and $\bar{r}$ is the reflection coefficient which measures the chance (typically $\simeq 0{\cdot}5$) that an electron approaching the

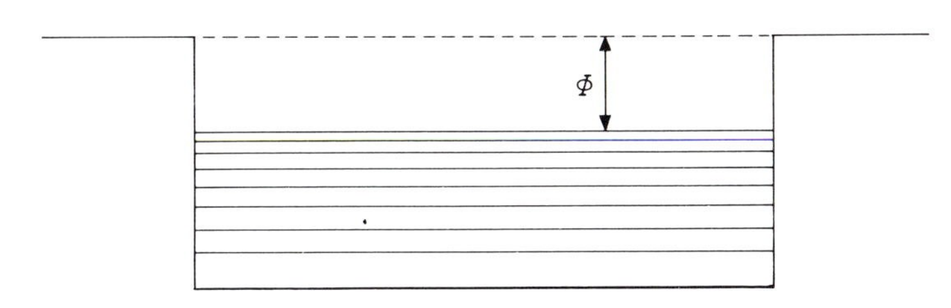

Fig. 9.1 The work function $\mathbf{\Phi}$ in the potential box model of a metal

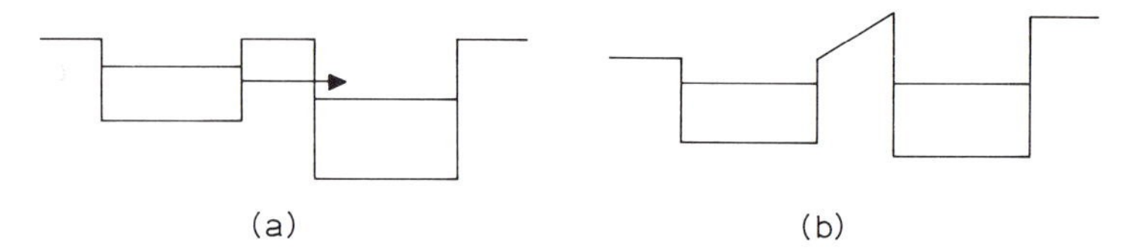

Fig. 9.2 When two metals with different work functions are joined (a), electrons flow from one to the other until a common Fermi level is obtained, producing a contact potential difference (b)

surface from the interior, with the necessary thermal energy $\mathbf{\Phi}$, will be turned back rather than allowed to pass through.

Since the Fermi electron sea, like that of any fluid, seeks the same energy level everywhere, the Fermi levels of two different metals must become equalized when the metals are brought into electrical contact and electrons can flow freely from one to the other. This contact can be made by joining them with a wire, but Fig. 9.2 shows the case where the metals are placed sufficiently close together that the their electrons can *tunnel* through the narrow energy barrier between them. The flow from the metal with the small work function $\mathbf{\Phi}_1$ is not at first compensated by the counterflow through the large barrier $\mathbf{\Phi}_2$. Metal 1 thus becomes positively charged, and metal 2 negatively charged, so that the two potential boxes are displaced relative to each other, as shown in Fig. 9.2(b), until the Fermi level becomes equalized between them. A *contact potential difference*, $\mathbf{\Phi}_1 - \mathbf{\Phi}_2$, is thus established between them, as in *Volta's* classical discovery.

Instead of bringing up a second metal we may bring a single neutral atom up to a clean metal surface, as shown in Fig. 9.3, where the atom is represented by the curve of its potential energy V and by its valency electron level. When the atom comes sufficiently close, electrons can tunnel through the barrier between it and the metal. This electrical contact leads to two further changes. First, the valency electrons in the atom no longer exist in a sharp level: there is some broadening since the atomic valency orbital is no longer a stationary state, because the electrons in it can leak through into the metal. From the uncertainty relation given by equation (A 1.1), a state with a lifetime of Δt has an uncertainty or breadth of energy of the order of $\Delta E \approx h/\Delta t$.

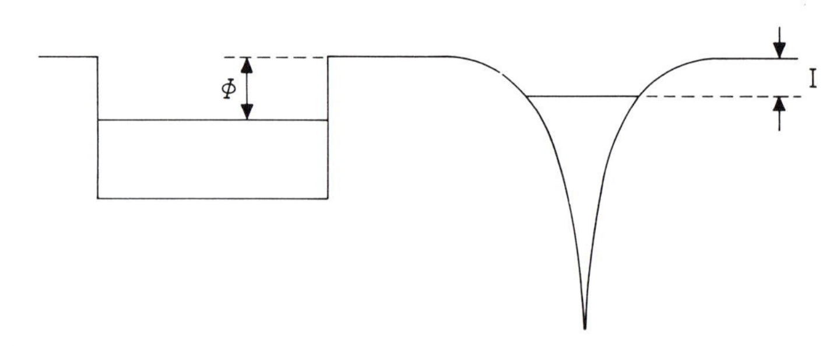

Fig. 9.3 The approach of a neutral atom with ionization energy *I* to a metal with work function $\mathbf{\Phi}$

For simplicity, we shall assume that $\Delta E \ll E$, so that the valency electron can still be regarded as existing in a sharp level E with a sharp ionization energy I. The second effect of the electrical contact is that, when $I < \Phi$, as for example when an alkali atom approaches a transition metal, the valency electron can transfer to the metal. The atom is then converted into a *positive ion*, which adheres to the metal by electrostatic attraction to the *image charge* which its field induces in the metal. The extra negative charge which the metal acquires by the transfer is located preferentially in the surface near the point of attachment of the atom.

There is a corresponding reverse effect. If a positive ion of an atom with a large ionization energy (e.g. an oxygen atom) approaches the metal, so that $I > \Phi$, then electrons can transfer from the metal and neutralize the ion. Metals with clean surfaces can thus act readily as acceptors or donors of electrons, from or to nearby atoms. This property underlies their ability to act as *catalysts* in chemical reactions among such atoms.

Suppose now that a monolayer of easily ionizable atoms, for which $I < \Phi$, becomes attached to the surface of a metal. These atoms become positively ionized and a corresponding sheet of negative charge is accumulated in the metal just below the surface, as shown in Fig. 9.4. These two surface layers carry equal and opposite charges, so that there is no external field from them; but because the positive layer lies slightly further out than the negative one, they jointly form an *electrical double layer*, like a charged parallel-plate condenser with its positive side outwards.

This double layer reduces the work function of the metal. Thus, if we take an electron in the Fermi level of the metal out through the surface, then the electrostatic potential energy of this electron falls as the electron leaves the repulsive, negative inner layer and approaches the attractive, positive outer layer. The potential energy at the surface of the metal is thus changed from the simple box shown in Fig. 9.1 to the form shown in Fig. 9.5(a). Since electrons can easily penetrate such a narrow barrier, the effective work function Φ is reduced, as in the case of *dull emitter thermionic valves* which have a layer of alkali metal atoms on a transition metal filament. For example, a fraction of a monolayer of caesium can lower the work function of a transition metal by up to 4 eV.

The opposite is also possible, as shown in Fig. 9.5(b). This might for example be brought about by the attachment of a layer of oxygen atoms to a metal surface. The electronegative oxygen atoms attract electrons, and so the double layer in this case is formed with its negative side outwards, which increases the work function.

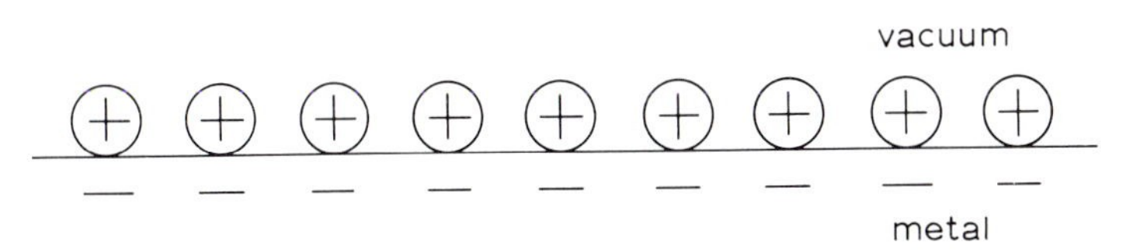

Fig. 9.4 An electrical double layer formed by the attachment of easily ionized atoms to a metal surface

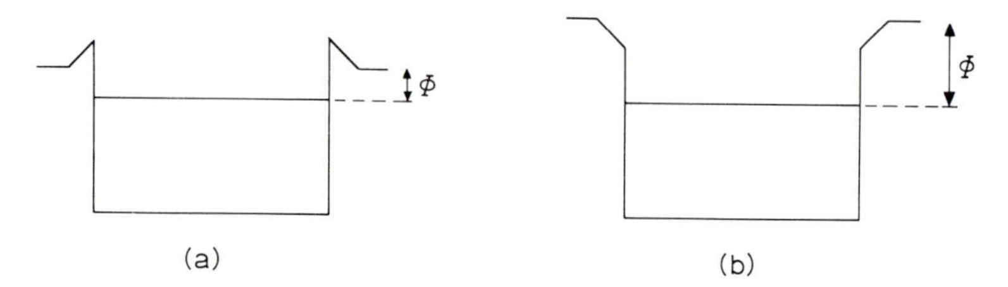

Fig. 9.5 **Change in the potential energy box on the formation of an electrical double layer: (a) positive outwards and (b) negative outwards**

9.2 THE THEORY OF THE WORK FUNCTION

Two distinct effects contribute to the value of the work function of a clean metal. The first is a consequence of the electrical conductivity of the metal, the second a result of the formation of an *intrinsic* double layer, with negative side outwards, at the surface of the metal.

To understand the first effect, we begin with the simplest jellium picture of a metal (*see* Section 3.2), i.e. a solid continuum of uniform positive charge with a

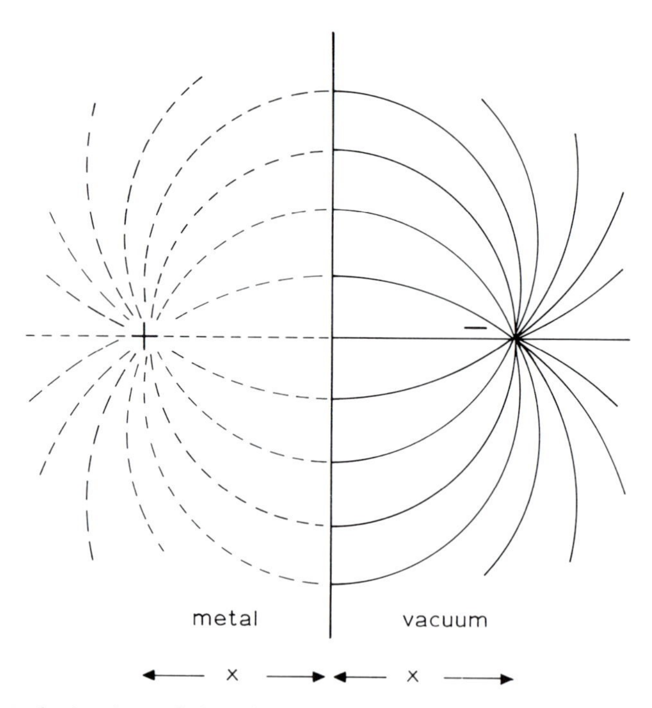

Fig. 9.6 **Polarization of the electron distribution in a metal by an external point charge, leading to an image force**

sharp boundary, in which is dissolved a fluid of mobile, negative neutralizing charge, which also ends sharply at the same boundary. Because of the mobility of this fluid, such a system is attractive to external, electrically charged particles. Suppose, in particular, that we bring a 'firmly held' electron towards such a 'metal'. Figure 9.6 shows the pattern of the electrical field lines which we expect from this electron, according to classical electrostatics. Since the metal is a conductor these lines must approach its surface normally; the mobile electronic fluid in the metal becomes *polarized* by the field into a distribution which gives an *equipotential* along the surface.

This pattern of field lines is of course exactly the same as would be obtained if the metal were removed and a point charge, equal and opposite to that of the electron, were placed in the mirror-image position, $-x$, when the electron is at x. The electron is thus attracted to the polarized metal with the same force, $e^2/4x^2$, as it would be attracted to this image charge. This electron has a corresponding potential energy $-e^2/4x$, which rises from a negative value to zero as it is pulled increasingly far from the metal, and which is part of the work function of the metal.

Obviously, from Fig. 9.6, if the electron were taken sufficiently far from a *finite* piece of metal, the pattern shown there would no longer be valid since it would extend laterally beyond the rim of the metal surface. In fact, the work function is defined so as to avoid this difficulty, i.e. for electrons which are removed only a short way from the surface (short compared with the surface's lateral dimensions).

This classical treatment of the image force also fails when the electron gets sufficiently close to the surface of the metal, both because the assumption of a rigidly held electron cannot then be maintained and also because the particulate structure of the metal then makes itself felt. As $x \to 0$ the potential energy does not sweep down to $-\infty$, but instead approximately levels off to a value characteristic of the interior of the metal, as the electron passes right through the surface and into the metal. What is this interior value? It is in fact the potential due to the electron–electron correlation and exchange effects, discussed in Chapters 3, 4 and 5. The polarization of the electronic fluid, which gives the classical image force when the external electron is outside the metal, is simply the long-range manifestation of these same correlation and exchange effects, i.e. the general effects by which electrons avoid one another as far as possible (*see* Fig. 9.7). There is thus a more complete and rigorous theory of this contribution to the work function, based on correlation and exchange instead of classical electrostatics. We shall discuss it below.

The other source which contributes to the work function, the intrinsic *double layer*, does so because the electronic fluid does not end sharply at the boundary of the metal: it spills out a little beyond the edge of the positive charge, as shown in Fig. 9.8, and so forms an outer layer of negative charge and an inner one which, because the electron density is correspondingly depleted there, has a net positive charge. The existence and thickness of this double layer are determined through a balance between kinetic and potential energies. The kinetic energy favours an electron distribution which fades very gradually, away from the metal, since this avoids the sharp curvatures in ψ which are associated with high kinetic energy; the potential energy of course favours an abrupt end to the

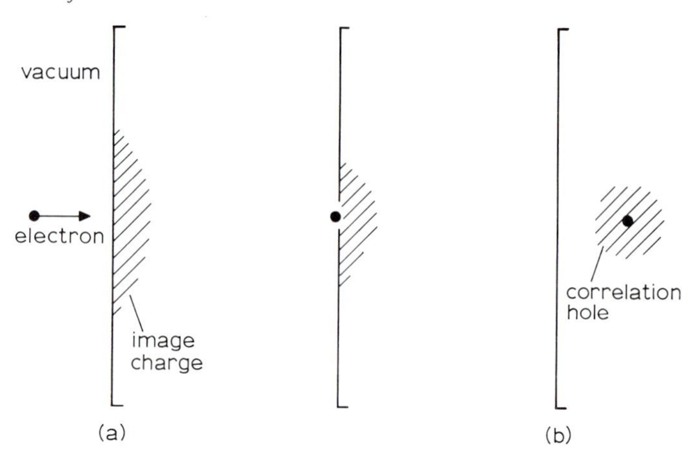

Fig. 9.7 **The image charge (a) induced by an external electron becomes the correlation hole (b) when the electron enters the metal**

distribution at the edge of the jellium, since this keeps the electrons as close as possible to the attractive electropositive material.

As with the monolayer of electronegative atoms shown in Fig. 9.5(b), this double layer increases the work function. We expect the combined effects of this and the image-correlation-exchange contribution to give a total potential energy of an electron as a function of its position x, along a line through the metal and into the space beyond, which is somewhat as shown in Fig. 9.9. As the surface is approached from inside the metal, the potential energy first begins to rise when the electron crosses the double layer shown in Fig. 9.8. Then, outside the metal, it rises further as the image effect becomes dominant. Such a form of potential energy curve represents, more realistically, the simple square-well potential box model shown in Fig. 9.1.

There are two problems facing the calculation of the depth W of this potential energy well and thus of the magnitude of the work function,

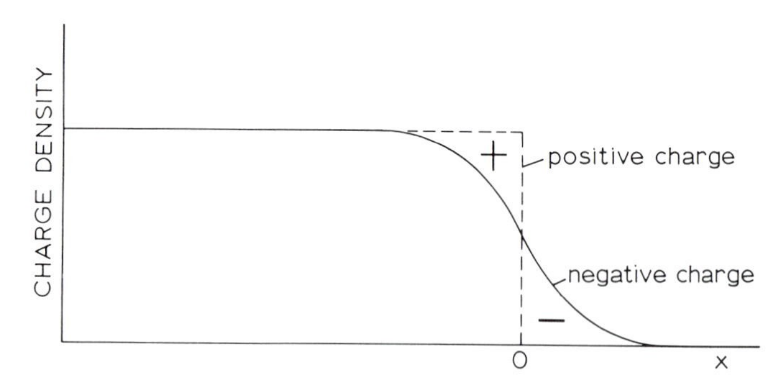

Fig. 9.8 **Charge distributions which represent the positive ions and electronic fluid at the surface of a metal**

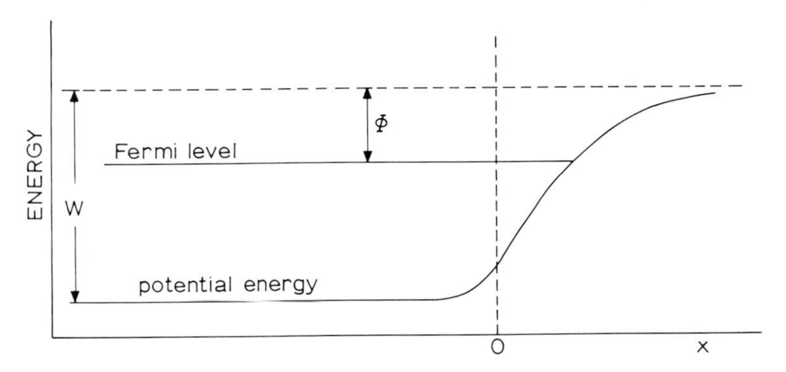

Fig. 9.9 Energy of an electron at the surface of a metal

$$\Phi = W - E_F \tag{9.2}$$

First, W depends strongly on the correlation and exchange energies which, as we have seen, are not easily calculated. Second, the form of this potential energy curve is produced by the distribution of electron density and hence of the amplitude of the wavefunction ψ. But, from Schrödinger's equation, ψ is determined by this potential energy function. Hence we are faced once more with a self-consistent calculation. An initial form of the potential energy curve shown in Fig. 9.9 has to be assumed, derived perhaps from semi-classical arguments, and substituted into Schrödinger's equation to find a $\psi(x)$. From this $\psi(x)$ the electronic density is calculated, as in Fig. 9.8, and used in turn to deduce an improved potential energy curve. This cycle of calculations is then repeated until stable values are produced.

Because the electron density, as in Fig. 9.8, is of central interest and importance in all of this, the density functional method outlined in Section 3.6 has proved particularly useful in such calculations (Lang and Kohn 1970, Lang 1973, Bennett 1974, Smith 1975). The energy $E[r]$ of the interacting electron gas is expressed as a functional (*see* equation (3.21)) of the electron density $n(r)$, and the aim is to find that $n(r)$ which makes $E[r]$ a minimum, through a variational equation such as equation (3.23). We write the first two terms in equation (3.21) as

$$\int V(r)n(r)\,\mathrm{d}r + \frac{1}{2}\int\int \frac{e^2}{r_{12}} n(r_1)n(r_2)\,\mathrm{d}r_1\,\mathrm{d}r_2 = \int \phi(r)n(r)\,\mathrm{d}r \tag{9.3}$$

where $\phi(r)$ is the *electrostatic potential* of the element of the electron distribution, $n(r)\,\mathrm{d}r$, contained in the volume element $\mathrm{d}r$ about the point r, due to (i) its interaction with the positive charge of the lattice (or jellium), represented by $V(r)$, and of density which we shall denote as $n_+(r)$; and (ii) its Coulomb interaction e^2/r_{12} with the other electrons, unmodified by the correlation and exchange effects. This classical electrostatic potential is then related to the net charge distribution $e[n(r) - n_+(r)]$ at this point by *Poisson's equation* (equation (A 2.12)):

$$\nabla^2\phi = -4\pi e[n(r) - n_+(r)] \tag{9.4}$$

In the method used by Kohn and Sham (1965), the exchange and correlation terms are approximated by a potential V_{xc}, as in equations (3.25) and (3.26). We thus rewrite equation (3.25) as

$$V_{\text{eff}}(n) = \phi(r) + V_{xc} \tag{9.5}$$

taking ϕ from equation (9.3) and V_{xc} from equation (3.26). Kohn and Sham used V_{eff} as the potential energy term in a one-electron Schrödinger equation, which is solved to give a set of wavefunctions ψ_i, $i = 1, 2, \ldots$, and hence to give

$$n(r) = \sum_i |\psi_i(r)|^2 \tag{9.6}$$

This Schrödinger equation, together with equations (9.4)–(9.6), enables a cycle of self-consistent calculations to be made. The first step is to assume a suitable $n(r)$, e.g. as in Fig. 9.8. From this $n(r)$ a V_{eff} is then found, as in equation (9.5), obtaining $\phi(r)$ from equation (9.4) and using a suitable approximation such as the local density functional of Section 3.6, to obtain V_{xc}. With this V_{eff} the wavefunctions are then derived from the Schrödinger equation, and finally $n(r)$ is found from equation (9.6). The cycle is then repeated until self-consistent results are obtained.

Figure 9.10 shows the distribution of electron density across the surface which is found by such a method. It confirms the 'intuitive' distribution shown in Fig. 9.8, but also reveals a sequence of slight oscillations in the metal. These are Friedel oscillations which, as we have seen (Section 2.7), appear characteristically in the wave-mechanical analysis of screening electron distributions in the presence of a non-uniform field of positive charge.

The transition region is approximately described by

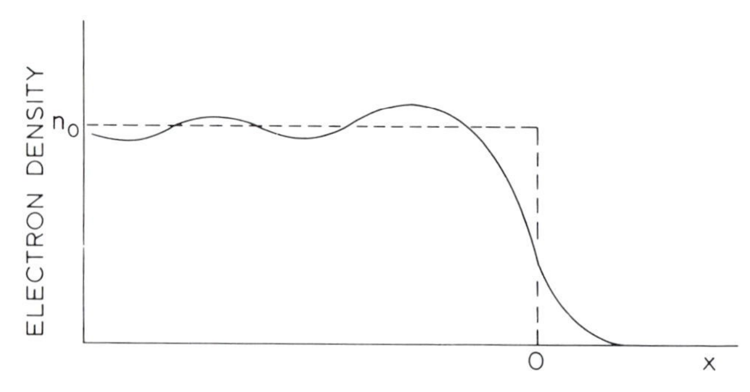

Fig. 9.10 Distribution of electron density across the surface of a metal

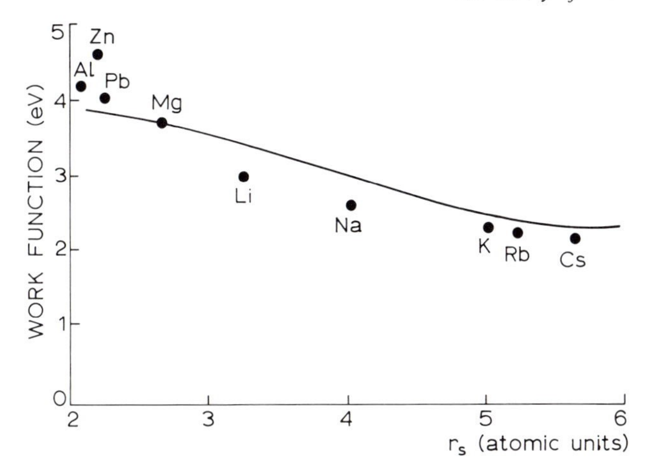

Fig. 9.11 Comparison of observed work functions with theoretical values calculated by Lang and Kohn (1971) and by Smith (1969)

$$n(x) = \begin{cases} n_0[1 - \frac{1}{2}\exp(x/\lambda)] & (x < 0) \\ \frac{1}{2}n_0 \exp(-x/\lambda) & (x > 0) \end{cases} \tag{9.7}$$

as was shown by Smith (1969), who obtained the values $\lambda \simeq 0{\cdot}4$–$0{\cdot}43$ Å for typical metals. By comparison, the Thomas–Fermi screening radius λ_{TF} (equation (3.7)) is of the order of 1 Å. Since the thickness of the double layer is much the same for many metals, the magnitude of the step $\Delta\phi$ in potential across the layer increases with the electron density n_0. Thus Smith showed that in monovalent metals of large atomic spacing, such as caesium, the double layer makes only a minor contribution to the work function, whereas in metals of high electron density such as aluminium it is the major contributor. Figure 9.11 shows theoretical values of the work function, calculated by the method outlined above (Lang and Kohn 1971). Here the parameter r_s, defined by

$$n^{-1} = \tfrac{4}{3}\pi r_s^3 \tag{9.8}$$

is used as a measure of $n(r)$. There is fair agreement with observed values for polycrystalline non-transition metals, as also found by Smith (1969), who used a slightly different version of the method.

Improved results are obtained by taking the atomic structure into account. In place of the sharp edge of the jellium is the dimpled surface of the outer layer of the positively charged ionic spheres. As a result the $n_+(r)$ distribution has, like $n(r)$, a gradual transition through the surface, and the better matching of the positive and negative charge distributions weakens the double layer and so reduces its contribution to the work function. This atomicity effect varies with the structure of the crystal face at the surface. It is least significant on the smooth, densely packed, low index faces that most closely resemble the sharp edge of jellium. High-index faces, in which some ions protrude much further than others, have particularly gradual transitions of positive charge and thus low work functions. For example, the work function on the (310) face of tungsten is only 4·3 eV, whereas it is 4·7 eV on (100) and 5·3 eV on the most close-packed face, (110).

This atomic variation of positive charge on the surface is utilized in the *field-ion* and *electron-tunnelling* microscopy techniques for observing the atomic structure of surfaces. In, for example, the electron-tunnelling method an extremely sharp-tipped probe is moved over the surface, and the intensity of the tunnelling current between it and the surface is measured. A 'contour' map of the intensity distribution shows the high spots of the ionic surface, i.e. the distribution of the atoms in the surface.

A detailed theoretical treatment which takes account of crystallinity and produces a three-dimensional distribution of the electron density has been made by Appelbaum and Hamann (1972). In particular, they obtained a work function of 2·71 eV for the (100) surface of sodium.

We turn now to the *transition metals*, where the work function is associated essentially with the removal of an electron from the Fermi level in a partially filled d band. Because the d states largely retain their atomic character, it is convenient to take a more atomic approach to the problem. The first step is to note the ionization energy I_d needed to remove an electron from a d state in a free, neutral atom (*see* Fig. 7.6). This varies from about 5 to 10 eV across the first transition series, and from about 4 to 12 eV across the second, for atoms in the $d^{n-1}s$ configuration, which best approximates to that in the solid (*see* Section 7.2).

Next, following the renormalization method described in Section 7.2, we squeeze all the electrons of the free atom into the WS cell, and amplify the (otherwise unaltered) atomic wavefunctions inside the cell by the appropriate amount, equivalent to about one electronic charge. The repulsive potential which this added electronic charge brings into the cell raises the energy of the d states by the amount E_{ren} (*see* equation (7.1)), which is typically about 1 eV. The ionization energy from the d state of the renormalized atom is thus $I_d - E_{ren}$.

We now bring a number of such renormalized atoms together, to form the metal. The d level broadens approximately symmetrically into a band, as shown in Fig. 6.2, and so the energy depth of the mid-point of the band, relative to that of a free electron outside the metal, is approximately $I_d - E_{ren}$. The next step is to estimate the difference ΔE_F between the energy level of the Fermi surface in this band and that of the mid-point of the band. Thus

$$\Phi_d = I_d - E_{ren} - \Delta E_F \tag{9.9}$$

is the energy depth of the Fermi surface relative to that of an external electron. According to the elementary approach represented by Fig. 6.2

$$\Delta E_F = \frac{w}{10}(z_d - 5) \tag{9.10}$$

where w is the width of the d band and z_d is the number of electrons per atom in it. Taking $w = 5$ eV, ΔE_F ranges from $-2{\cdot}5$ to $+2{\cdot}5$ eV along a series. Better estimates (Allen 1976) of ΔE_F can be obtained by taking the shape of the d band into account, as in the moment distribution method outlined in Section 6.3. Putting the above numerical values together we obtain $\Phi_d \simeq 6{\cdot}5$ eV for the first series. With a slightly larger bandwidth, a similar value is obtained for the second series.

Finally, there is the contribution of the surface double layer. This is found to be small, and so a much simplified calculation can be used (Allen 1976). At the surface the coordination number, z in equation (6.13), is about three-quarters of its value inside the metal. Hence, from equation (6.13), the width w of the d band is reduced, locally at the surface, by about one-sixth. The Fermi level remains constant through the surface layer, of course. Hence the energy level of the bottom of the band must rise there by about

$$\frac{w}{6}\left(\frac{E_F}{w}\right) \simeq \frac{E_F}{6} \tag{9.11}$$

This rise in the energy level of the bottom of the band represents the increase in the potential energy of the electronic ground state, going outwards through the surface, i.e. the step $\Delta\phi$ in potential across the double layer.

Since typically $E_F/6 \simeq 1$ eV, we see that Φ_d (equation (9.9)) contributes almost all of the work function of a transition metal.

9.3 SURFACE ENERGY

Although it gives sensible numerical values, the method of estimating surface energy by counting the number of 'broken interatomic bonds' across the surface is clearly unsound in principle for metals. This is also true of the more elementary treatments based on electron theory, which makes the theory of the surface energy of metals particularly difficult. The key point is that it is essential to take account of the particulate character of all the participants; this rules out both jellium and also uncorrelated, uniform, electronic charge distributions. The general reason was given by Laplace in the eighteenth century. Surface energy, being an energy per unit area, can only be related to bulk energy by the introduction of a characteristic (atomic) *length*, and so cannot be explained by any continuum approximation.

The theory, due to Lang and Kohn (1971), is in four stages. First is the completely continuum model consisting of jellium and a uniform electronic charge distribution. Suppose that we make two new surfaces by cutting through a block of this material, as in Fig. 9.12. We imagine the process to occur in two stages. First the electron density is held constant right up to the new edge of the jellium where, like the jellium itself, it falls abruptly to zero. There is *no* change at this stage of forming the new surfaces since the local state of every element of the material remains unchanged (apart from the strictly infinitesimal and therefore negligible amount at the new surface). In the second stage we allow the electronic distribution to relax outwards, as in Fig. 9.8. This can only lower the energy, since it releases a constraint on the electrons. In fact, the fall is caused by

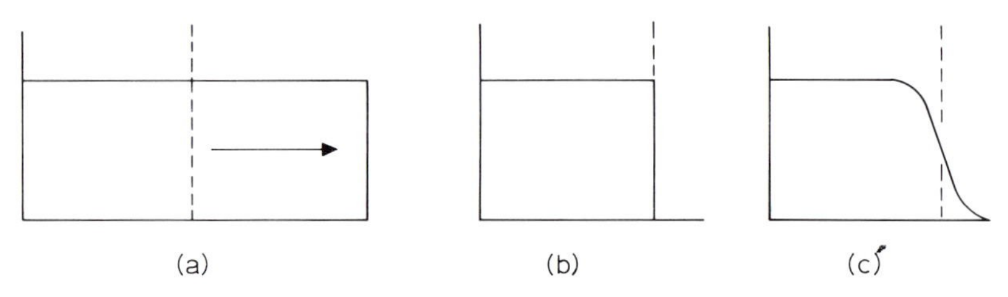

Fig. 9.12 **A new surface is made by cutting a block of jellium (a), removing the right-hand side and initially holding the electron distribution fixed (b), then allowing it to relax (c)**

a drop in kinetic energy as the electron distribution becomes more spread out. Hence the theory at this stage predicts a *negative* surface energy!

The next step is to repeat the process, still with jellium, but using an electron gas in place of the continuous electronic distribution. This brings in the exchange and correlation contributions, so that the surface energy σ_u consists of

$$\sigma_u = \sigma_s + \sigma_{es} + \sigma_x + \sigma_c \tag{9.12}$$

where σ_s and σ_{es} are, respectively, the changes in kinetic and electrostatic energy due to the spilling out of the electron distribution to form the double layer, in Fig. 9.12(c); σ_x and σ_c are the respective changes, per unit surface area, in the exchange and correlation energies resulting from the formation of the new surface. As we have seen, σ_s is negative. The partial separation of the charges in the double layer (as in Fig. 9.8) makes σ_{es} positive, but smaller than σ_s (since otherwise the double layer would not form). The exchange and correlation terms are both positive since they represent the transition from the complete correlation hole, associated with an internal electron, to the image charge, associated with a surface or external one, as in Fig. 9.7. Lang and Kohn, using the density functional method, showed that $\sigma_x \gg \sigma_c$.

Table 9.1 gives some of the results obtained by Lang and Kohn with this model, using two values of the parameter r_s, which relates to the electron density in the metal (equation (9.8)). For alkali metals (e.g. $r_s = 5$), in which the negative kinetic energy term is small, this theory gives a reasonable value for the surface energy, $\sigma \simeq 100$ mJ m^{-2}. But for metals such as aluminium in which the electron density is high (e.g. $r_s = 2$), the huge kinetic energy term overwhelms

Table 9.1 **Contributions to the surface energy, mJ m^{-2} (after Lang and Kohn 1971) (r_s in atomic units)**

r_s	σ_s	σ_{es}	σ_x	σ_c	σ_u
2·0	−5600	1330	3080	180	−1010
5·0	−30	15	95	20	100

the large exchange term, so that the predicted surface energy is still negative.

It thus becomes necessary to improve the theory further by replacing the jellium with a lattice of discrete charged ions, represented by suitable pseudopotentials. Lang and Kohn showed that this introduces large changes in electrostatic energy in both the stages shown in Fig. 9.12. In Section 5.5 we examined the effect of replacing jellium by a lattice of point ions, and showed that this change introduced an *electrostatic energy* E_{el}, representing the interaction of the ions and the electron gas considered as a classical Coulomb system. This E_{el}, which is negative (*see* equation (5.5)) and hence favourable to bonding, was obtained by summing all the Coulomb energies e^2/r between all pairs of particles, over the entire system, using the Ewald–Fuchs method (which is based upon the original Madelung method for making such lattice sums in ionic crystals). The value of E_{el} obtained, given by equation (5.7), was of course for an infinite crystal. Lang and Kohn made a similar calculation for a semi-infinite crystal, bounded by a free surface on one side as in Fig. 9.12(b). Because fewer e^2/r terms enter into the sum in this case, the value E'_{el} obtained is smaller than that for the infinite crystal, and the positive energy given by the difference, when divided by the area of the free surface, represents the contribution δc_{el} of this lattice effect to the surface energy in stage 1 of Fig. 9.12. Lang and Kohn obtained the value

$$\delta c_{el} = \alpha Z n_0 \tag{9.13}$$

where n_0 is the electron density inside the metal, Z is the ionic charge and α is a Madelung number (*see* Section 5.5) equal to 0·00325 and 0·00563 for, respectively, the close-packed faces of f.c.c. (111) and b.c.c. (110) structures. For the f.c.c. (111) face δc_{el} ranges from 7 mJ m^{-2} for caesium to 408 mJ m^{-2} for aluminium.

The second electrostatic contribution of the lattice, δc_{ps}, also positive, comes in stage 2 of Fig. 9.12. In place of the rather small electrostatic term σ_{es} in equation (9.12), we have an electrostatic energy of the pseudopotentials of the ions, in the surface region, with the electrons which become partly separated from them as a result of the spilling out, as in Fig. 9.12(c). Lang and Kohn's δc_{ps} is the difference between the total energy due to this effect and the corresponding energy σ_{es} for the jellium model. Their values ranged from 20 mJ m^{-2} for caesium to 1050 mJ m^{-2} for aluminium.

The total surface energy σ is then

$$\sigma = \sigma_u + \delta c_{el} + \delta c_{ps} \tag{9.14}$$

The values obtained for σ, e.g. 100 mJ m^{-2} for caesium (r_s = 5·63) on the b.c.c. (110) face, and 1030 mJ m^{-2} for aluminium (r_s = 2·07) on the f.c.c. (111) face, are within about 25% of the observed values for the metals considered (Al, Zn, Mg, Li, Na, K, Rb, Cs).

For transition metals, the method outlined at the end of Section 9.2, which is essentially an application of equation (6.15), enables a rough estimate of surface energy to be made, as shown by Allen (1976).

9.4 SURFACE STATES

We turn now to a more fundamental effect of the surface. The Bloch theory, which is the foundation of the electron theory of metals, is based on the assumption of an *infinite* perfect lattice. It is the perpetual repetition of the ionic potential that constrains the solutions of Schrödinger's equation to the infinitely extended Bloch states, e.g. equations (1.14) and (A 4.11), which include the sinusoidal waves of free-electron theory as a special case. However, in a finite crystal the lifting of this constraint allows the equation to provide other kinds of solutions which lack the infinite continuation of the Bloch ones; in particular, solutions that represent states confined to the surface, also known as *Tamm states*.

In the derivation of the Bloch wavefunction, as in Appendix 4, the requirement that the wave-vector $\boldsymbol{k}$ should be composed of *real* numbers, as in equation (A 3.6), stems from the application in Appendix 3 of the *Born–von Karman* cyclic boundary conditions which are a simple way of representing an infinite continuation of the metal. In the finite crystal there is no longer this constraint on real $\boldsymbol{k}$ values. If we introduce a complex wave-vector $\boldsymbol{k} = \boldsymbol{a} + \mathrm{i}\boldsymbol{b}$, where $\boldsymbol{a}$ and $\boldsymbol{b}$ are real, into equation (A 4.11), we obtain

$$\psi(\boldsymbol{r}) = u'(\boldsymbol{r}) \exp(-\boldsymbol{b} \cdot \boldsymbol{r}) \tag{9.15}$$

where

$$u'(\boldsymbol{r}) = u(\boldsymbol{r}) \exp(\mathrm{i}\boldsymbol{a} \cdot \boldsymbol{r}) \tag{9.16}$$

The interesting feature is the term $\exp(-\boldsymbol{b} \cdot \boldsymbol{r})$ in the wavefunction. This decays towards zero in one direction of $\boldsymbol{r}$, and increases towards infinity in the opposite direction. Infinitely large wavefunctions are of course not admissible inside the metal (which of course is why band gaps occur in the energy spectrum, since the solutions of Schrödinger's equation take on this same form at energies which fall in the forbidden ranges of these gaps). At the free surface, however there are additional possibilities because a solution of the equation (9.15) type, with its decaying side pointing into the metal, can be joined at a surface to a solution which also decays outwards, into the vacuum outside the metal. This joining is possible for certain critical values of $\boldsymbol{b}$ and the ensuing wavefunctions, which have significant amplitudes only in the surface layers of the material, represent the surface Tamm states in which electrons are localized in the region of the surface.

These states are important in semiconductors because they can occur at energies in the gap between the conduction and valence bands, so that at the surface electrons get trapped in them. They are often known as *dangling bonds*, from the chemical picture which portrays them as unbonded orbitals that are the remnants of broken interatomic bonds at the surface. Although they also exist in metals, they have less practical significance than in semiconductors because they can generally combine with conduction band states in the same energy range, whereupon their electrons can escape into the metal. Excited surface states can also occur, analogous to the plasma oscillations inside metals (*see* Section 3.3); the quanta of these surface electronic oscillations are known as *surface plasmons*.

9.5 SURFACE RELAXATION AND RECONSTRUCTION

Since surface atoms have no neighbours on one side, they may take up equilibrium positions quite different from those inside the metal. The most general effect of this kind is *surface relaxation*, in which the interatomic spacing perpendicular to the surface is different from that in the interior. Sometimes a more drastic change occurs in which the pattern of the atoms on the surface is quite altered from that inside. This is *surface reconstruction*. The development in recent years of ultra-high vacuum techniques for preparing clean surfaces, and low-energy electron diffraction (LEED) and field-ion microscopy (FIM) for determining surface structures, has enabled these effects to be studied experimentally.

The surface relaxation is usually inwards. We might have expected the opposite since, when the conduction electrons spill out of the surface (Fig. 9.12(c)), the surface ions should follow them by electrostatic attraction. There is, however, a second effect (Finnis and Heine 1974), caused by the dimpled ionic structure of the surface. Electrons outside the crests of the dimples lack the electrostatic attraction of ions on one side and can reduce their energy by spilling off these crests, down into the valley channels between the dimples, where they are nearer to more ions. For the outermost surface ions, particularly on the less closely packed faces, the 'centre of gravity' of electrostatic attraction from the conduction electrons is moved slightly inwards by this effect, and the ions move correspondingly slightly inwards.

The reconstruction which occurs on some clean metallic surfaces involves subtle effects of interatomic interaction which are beginning to come within the scope of the theory. The LEED technique has proved very useful for studying the two-dimensional crystallography of these surface rearrangements, in particular their symmetry. Consider first an unreconstructed layer of a surface crystallographic facet, e.g. a (100) face with a *square net* structure. This provides the substrate on which the top monolayer of atoms sits, in its own reconstructed network pattern. This pattern, although different from that of the (100) net beneath it, is nevertheless crystallographically related to it, since the atoms in the top layer sit in regular positions on those beneath them.

A standard notation is now used to describe this relationship. Suppose that the substrate net is described in a two-dimensional version of the usual crystallographic notation (*see* equation (A 4.27)), by two lattice vectors $\boldsymbol{a}$ and $\boldsymbol{b}$ which are the smallest translations in the plane of the face which carry one atom to the site of another. Then, relative to this two-dimensional lattice, the structure of the reconstructed layer is like a superlattice with lattice vectors $l\boldsymbol{a}$ and $m\boldsymbol{b}$; the reconstruction is said to be an $l \times m$ type. An additional letter p or c, is also used to indicate whether the two-dimensional cell of this superlattice is primitive or whether it has an atom at its centre. Thus the notation is p$(l \times m)$ or c$(l \times m)$. Figure 9.13 shows some simple examples. We note that (1×1) implies that there is no reconstruction, i.e. the pattern of the top layer is identical to that of the corresponding plane inside the crystal. Results obtained by LEED and FIM show that this is so for (110), (100) and (211) facets of clean b.c.c. transition metals, and also for the (111) planes of f.c.c. metals, although there is recent evidence for some surface reconstruction on several metal surfaces at low

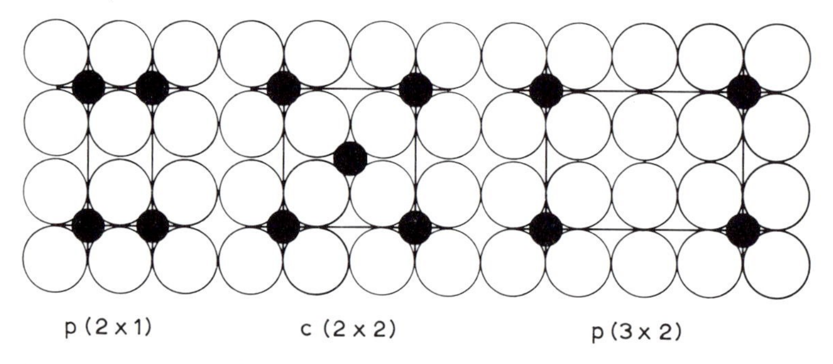

Fig. 9.13 Examples of surface reconstruction

temperatures. The clean (100) planes of Ni, Pd and Cu also have (1 × 1) symmetry, but on the (100) planes of Pt, Au and Ir there is a (5 × 1) symmetry, as a result of an almost hexagonal top layer. In some f.c.c. metals the (110) plane has a (1 × 1) symmetry, but in Pt it has a (2 × 1) reconstruction. Some of these superstructures transform reversibly at high temperatures into the unreconstructed pattern (for reviews *see* Schmidt 1975, Bauer 1975, Ward Plummer *et al.* 1987) There is recent evidence that at high temperatures surface 'melting' takes place, below the melting point of the bulk material.

9.6 ADSORPTION AND CATALYSIS

The chemical interaction of a metallic surface with the environment is, of course, an enormous subject, and one in which the band theory of electronic structure plays only a secondary role. Such interaction begins with the adsorption, usually of a monolayer or partial layer, of foreign atoms or molecules on the clean metal surface. Ever since Langmuir (1918) it has been recognized that there are two kinds of adsorption: *physical*, in which the bond between the adsorbate molecule and the metal is weak (e.g. with an energy below 0·1 eV) and is commonly due to van der Waals forces; and *chemisorption*, in which there is a much stronger bond (e.g. 0·5 eV upwards, sometimes as high as 4 eV) of an obviously chemical nature. Figure 9.14 shows how this is usually represented in an energy diagram. To go from the shallow minimum of physical adsorption to the deeper one of chemisorption, the adsorbed particle has to pass through an energy barrier, which requires thermal activation.

Much information has been provided in recent years on the structures of adsorbed layers by LEED, FIM and other techniques. The subject is a large and complex one because the structure depends on the nature of the adsorbate, of the metal, and of the crystal face exposed to the adsorption. Generally, adsorption occurs more readily on stepped, high-index faces than on close-packed, low-index ones, which is understandable from the most elementary principles of molecular cohesion.

Oxygen on a (100) face of nickel provides a typical example. LEED shows the oxygen layer to have a c(2 × 2) structure. The fourfold nature of the symmetry leads to the conclusion that the adsorption layer has one of the structures shown

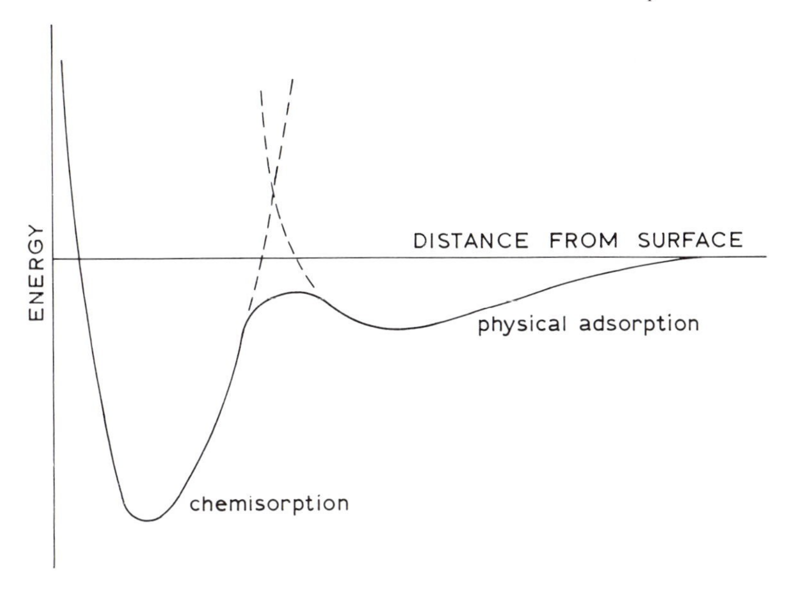

Fig. 9.14 The energy-distance relation for a particle of adsorbate approaching a metal surface

in Fig. 9.15. In Fig. 9.15(a) the adsorbed atoms sit on top of an unreconstructed (100) face of metal atoms; in Fig. 9.15(b) they form a two-dimensional superlattice with a reconstructed layer of metal atoms. This second type might be regarded as an early stage in the transition, from a truly adsorbed layer in which the metal atoms beneath remain part of the bulk metal, to a chemical compound of the adsorbate and surface metal atoms. In general, the less chemically active adsorbate species (e.g. hydrogen, nitrogen, carbon monoxide, metals) leave the metal surface unreconstructed, and the active ones such as oxygen produce reconstruction, leading in higher amounts to surface oxides.

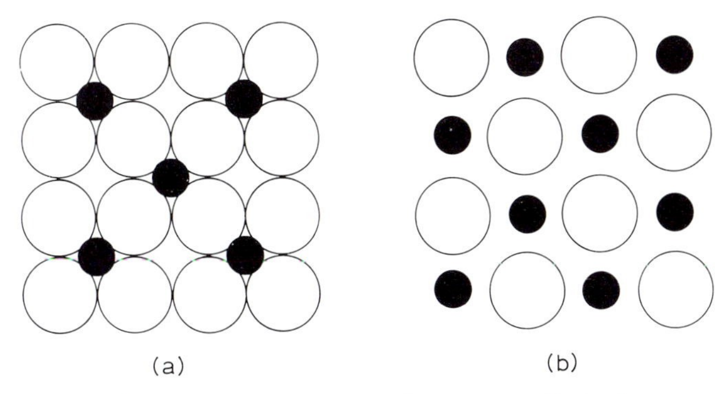

Fig. 9.15 Possible c(2 × 2) adsorption layers (small circles) on a (100) face (large circles)

The $c(2 \times 2)$ structure is the most common form of symmetry for adsorption on (100) planes, e.g. CO on Ni, W, Cu; H_2 on W; Na on Al, Ni; although there are many variations, e.g. CO on Pd and H_2 on Mo give $c(4 \times 2)$, and H_2 on W gives (1×1) at saturation. At very low concentrations the central atom may be missing, e.g. oxygen on nickel gives $p(2 \times 2)$ before going over to $c(2 \times 2)$ at higher concentrations. Other crystal faces of course give different results, e.g. H_2 on W (110) gives (2×1) at half a monolayer and (1×1) at full monolayer; and on W (211) it gives (1×1).

The simplest examples of physical adsorption are provided by inert gas atoms, which retain their neutrality and atomic structures when adhering to the metal surface. The adsorption in this case can be understood in terms of the *van der Waals forces* which exist between all neutral atoms, as a result of correlated motions of the electrons in their atomic orbitals, and also of the *electrostatic polarization* of these neutral atoms in the double-layer electric field of the surface.

Something quite different happens when a metallic atom, e.g. an alkali, is adsorbed on the surface of a metal such as tungsten, as explained by Gurney (1935). Consider a free neutral alkali metal atom with its valency electron in a level whose ionization energy is comparable to the work function of the metal. As the atom approaches the metal surface there is a possibility of its electron tunnelling to and fro, between the atom and the metal, with a characteristic tunnelling time Δt. By the uncertainty principle (equation (A 1.1); *see* also equation (A 1.37)) this gives a spread ΔE in energy of

$$\Delta E \, \Delta t \approx h \tag{9.17}$$

i.e. the sharp valency level of the free atom is broadened into an energy band of half-width ΔE when the atom is close to the metal. For alkali metals on low-index faces of transition metals, $\Delta E \simeq 0{\cdot}1\text{–}1$ eV.

A second effect is that the electron in the atom experiences the image field of the metal surface (*see* Fig. 9.6). There are two images: that of the electron itself which, at a distance x from the surface, is classically given by $-e^2/4x$, and the *negatively* charged image of the positive ion core of the atom, which repels the electron by an electrostatic term of similar form. In fact the repulsive image predominates, with the result that the energy level of the valency electron in the atom is not only broadened by the tunnelling effect; it is also raised, through the image effect, by about $0{\cdot}5\text{–}1$ eV. The general effect of this is to enhance the tendency of the valency electron to transfer to the metal, even when the unmodified valency level lies slightly below the Fermi level of the metal. In fact, an adsorbed alkali atom generally loses 65–90% of its valency electronic charge to the metal, and adheres to it essentially as a positive ion (for a review *see* Roberts and McKee 1978). This implies that the raised band of the atomic valency state lies mainly above, but slightly overlaps, the top of the Fermi distribution in the metal. More generally, when this band is more nearly centred about the Fermi level, the adsorbed atom is bonded to the metal partly through a metallic type of cohesion and partly through an ionic attraction. Finally, if the band lies wholly below the Fermi level there is little ionization, but the filled valency state is incorporated into the general band structure of the metal and the bond is a metallic one.

The same principles apply in the more complex case of the chemisorption of various atoms and molecules on metallic surfaces. The many theoretical treatments of chemisorption range from purely chemical ones, in terms of localized molecular orbitals between the adsorbate particles and the immediately adjacent atoms of the metal surface, to band theory and density functional calculations (for a review *see* Grimley 1975). One of the most complete treatments was provided by Lang and Williams (1976) for a single atom chemisorbed on a jellium. They showed that a situation somewhat similar to the *virtual bound states* or *resonances* described in Section 8.4 can exist. The atomic orbital of the adsorbed atom becomes part of the band structure of the metal, but the itinerant electrons of this band structure spend more time in the neighbourhood of the atom, so that there is a local *peak* in the density of states of the metal, localized in space to the vicinity of the atom and localized in energy to the vicinity of the atomic valency level. Lang and Williams applied the theory to lithium, silicon and chlorine atoms adsorbed on aluminium. They showed that the lithium atom is almost completely ionized and so clings to the metal as an Li^+ ion; that the chlorine 'resonance' is almost completely occupied by two electrons, coming from an atomic 3p level which lies well below the Fermi surface, so that the atom exists as a Cl^- ion; and that for silicon the centre of the broadened band of the 3p level lies close to the Fermi level, so that this band provides a bond through only its lower half being occupied.

The enormous subject of *heterogeneous catalysis* on metal surfaces is mainly a matter of molecular chemistry, but there are some aspects that bring in the band structure. In fact, because its processes operate on an individual atom-to-atom scale, catalysis reveals aspects of metallic structure that tend to be underemphasized in the band theory.

It is a familiar idea that metals are good catalysts because their free electrons serve as an easily accessible reservoir which can readily give or take electrons to or from adsorbed atoms and molecules, according to whether the valency atomic levels of the adsorbate lie below or above the Fermi level. This facility for ionizing chemical reactants and also for partially screening the ionic charges by the image effect can break down chemical barriers to reactivity. It is also well known that transition metals are generally outstandingly good catalysts, which suggests that a partly filled d band, with a high density of states at the Fermi surface – and which can thus supply or receive large numbers of electrons at constant Fermi energy – is particularly effective in catalysis.

However, there are complications. When gold is added to palladium the ability of the material to catalyse, e.g. to participate in reactions in which hydrogen is transferred between molecules, drops sharply to almost zero at 50 at.% Au. This is in agreement with the simple *rigid band* picture (*see* Section 8.4), according to which the extra electron in the gold atom raises the Fermi level to the top of the band at about this composition. But in the apparently equivalent example of the addition of copper to nickel, the corresponding effect does not appear until some 75% Cu is added. It would appear that in these alloys the nickel atoms retain their d state holes (about 0·5 electrons) even in the presence of large numbers of copper atoms. The rigid band assumption is not valid in this case and a better picture may be gained by using, instead, the *coherent potential approximation* (*see* Section 8.4), according to which the atoms retain more of their pure-metal characteristics in the alloy.

There are other possible complicating effects. In particular, the surface composition of an alloy, in equilibrium, is generally expected to differ from that in the interior; for example, copper concentrates at the surface in Ni–Cu alloys. There is also the question of the detailed structural chemistry at the active site on the surface where the reaction takes place. If one type of metal atom (e.g. Pd) is more catalytically active than another (e.g. Au) then its effectiveness in an alloy may depend upon the extent to which such atoms occur together in small clusters at active sites, which in turn depends on the composition and thermodynamic characteristics of the alloy. (For a review of electronic and alloy effects in catalysis, *see* Ponec (1975).)

REFERENCES

Allan, G., in *Electronic Structure and Reactivity of Metal Surfaces* (ed. E. G. Derouane and A. A. Lucas), Plenum Press, New York (1976), p. 45.

Applebaum, J. A., and Hamann, D. R., *Phys. Rev. B*, **6**, 2166 (1972).

Bauer, E., in *Interactions on Metal Surfaces* (ed. R. Gomer), Topics in Applied Physics Vol. 4, Springer-Verlag, Berlin (1975), p. 225.

Bennett, A. J., in *Critical Reviews in Solid State Sciences* (ed. D. E. Schuele and R. W. Hoffman), CRC Press, (1974).

Finnis, M. V., and Heine, V., *J. Phys. F*, **4**, L37 (1974).

Grimley, T. B., in *Electronic Structure and Reactivity of Metal Surfaces* (ed. E. G. Derouane and A. A. Lucas), Plenum Press, New York (1976), p. 35.

Gurney, R. W., *Phys. Rev.*, **47**, 479 (1935).

Kohn, W., and Sham, L. J., *Phys. Rev.*, **140**, A1133 (1965).

Lang, N. D., *Solid State Phys.*, **28**, 225 (1973).

Lang, N. D., and Kohn, W., *Phys. Rev. B*, **1**, 4555 (1970); *Ibid.*, **3**, 1215 (1971).

Lang, N. D., and Williams, A. R., *Phys. Rev. Lett.*, **37**, 402 (1976).

Ponec, V., in *Electronic Structure and Reactivity of Metal Surfaces* (ed. E. G. Derouane and A. A. Lucas), Plenum Press, New York (1976), p. 573.

Roberts, M. W., and McKee, C. S., *Chemistry of the Metal–Gas Interface*, Oxford University Press (1978).

Schmidt, L. D., in *Interactions on Metal Surfaces* (ed. R. Gomer), Topics in Applied Physics, Vol. 4, Springer-Verlag, Berlin (1975), p. 63.

Smith, J. R., *Phys. Rev.*, **181**, 522 (1969); in *Interactions on Metal Surfaces* (ed. R. Gomer), Topics in Applied Physics Vol. 4, Springer-Verlag, Berlin (1975), p. 1.

Ward Plummer, E., *et al.*, in *Advancing Materials Research* (ed. P. A. Psaras and H. Dale Langford), National Academy Press, Washington D.C., (1987), p. 283.

10

Superconductivity

10.1 INTRODUCTION

Superconductivity is a dramatic subject. Its discovery in 1911 by Kamerlingh-Onnes came as a surprise to a generation still imbued with classical ideas of the impossibility of perpetual motion. The effect is spectacular: a sudden fall, on cooling a suitable material to its *critical temperature* T_c, of the d.c. electrical resistivity to zero (*see* Fig. 10.1), or so nearly zero that an electrical current induced in a ring of the material will continue to run round it, unaided and with undiminished intensity, for years. Superconductivity is a manifestation in the macroscopic world of those peculiarly quantum-mechanical features which are normally seen only in the atomic world. It has great practical possibilities for both heavy-current electrical engineering and computer electronics. There is now intense scientific excitement in the whole subject, following the recent discovery of ceramic superconductors with exceptional properties.

Most metals are superconducting at low temperatures, the only possible exceptions being some of the monovalent metals (Li, Na, K, Cu, Ag, Au) and the

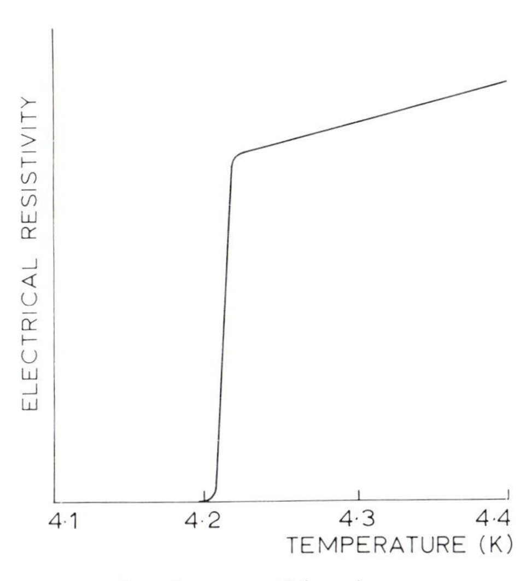

Fig. 10.1 **The superconducting transition in mercury**

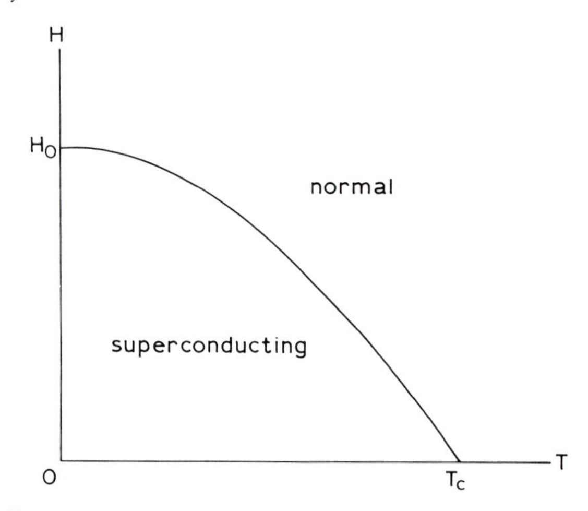

Fig. 10.2 Effect of a magnetic field on the transition temperature

ferromagnetics. Critical temperatures T_c range from 23·2 K for the alloy Nb_3Ge down to about 0·01 K for tungsten. Non-magnetic impurities in a superconducting metal have little effect on T_c, but magnetic atoms drastically reduce the critical temperature.

At any temperature below T_c the application of an external magnetic field of *critical strength* H_c causes the superconducting material to become 'normal' again. This critical field H_c varies with temperature, as shown in Fig. 10.2, reaching a maximum value H_0 at zero temperature. The transition between the normal and superconducting states, brought about by changing T or H or both, is completely reversible in both directions for the 'soft' (Type I) superconductors (*see below*). This magnetic limit imposes a corresponding limit on the amount of current that a superconductor can carry. A steady current passing along a wire produces its own magnetic field, in the form of circular field lines round the wire; for a thick wire the critical current is that which produces the field H_c at the surface of the wire.

A Type I superconductor does not permit any magnetic field to exist inside it, except within a narrow surface layer whose thickness is indicated by the *penetration depth*, typically about 500 Å. We might think that this exclusion of magnetic fields is simply a consequence of the perfect, resistanceless conductivity of the material in the superconducting state, since it follows from electromagnetism that, when an external magnetic field is brought up to a perfect conductor, eddy currents are induced in the surface layer which oppose the penetration of this field into the interior. However, the discovery of the *Meissner effect* in 1933 showed that a superconductor is not merely a perfect conductor. Suppose that the material, at a temperature where it is in the normal state, is exposed to a magnetic field. Because of the finite electrical resistance in the normal state the eddy current decays and the field then penetrates into the interior. If the material, containing this penetrated field, is cooled below T_c it *expels* the field. This is the Meissner effect and it is a unique and most characteristic feature of the superconducting state, which is a state of *perfect diamagnetism*. It proves that, whatever its history, the internal state of the superconductor below T_c is always the same, i.e. always without an internal magnetic field. This, together with the evidence of reversibility, means that the superconducting state is a *thermodynamically distinct state* of the electronic gas in such materials. It has its own characteristic free energy, which is less than that of the normal state at temperatures below T_c. The existence of the critical magnetic field H_c can be understood on this basis. When an external field H is excluded, the diversion of the field lines round the material leads to an increase in magnetic energy, proportional to H^2. The critical field H_c is that at which this increase fully offsets the difference in the free energy of the material between its normal and superconducting states. The energy of the critical excluded field, $H^2/8\pi$, indicates that the energy difference between the superconducting state and the normal state at absolute zero is of the order of 10^{-8} eV per atom, which is minuscule compared with, for example, the Fermi energy.

The above description strictly applies only to 'soft' Type I superconductors, and only those that are large compared with the penetration depth λ. The size effect occurs because, even in the superconducting state, a magnetic field can penetrate substantially into a slender filament or film of thickness $d \leqslant \lambda$. Superconductivity can then continue up to a higher critical field

$$H_d \approx \frac{\lambda}{d} H_c \tag{10.1}$$

As well as Type I superconductors with characteristics as described above, there are 'hard' Type II superconductors (e.g. niobium, vanadium, intermetallic compounds, semiconductors) which behave differently. When the applied magnetic field reaches the critical value at which perfect diamagnetism begins to fail, some lines of magnetic flux enter the material. But this transition to the normal state develops only gradually, progressing only with further increase in the applied field, and is not completed until a much higher limiting field is reached. The filamentary structure of the normal and superconducting regions developed in this 'mixed' state enables Type II superconductors to carry d.c. (but not a.c.) currents of up to about 10^{10} A m^{-2} in magnetic fields of the order of 10 T, which makes them useful for building powerful electromagnets. An important role in these 'hard' materials is provided by foreign particles and lattice defects (e.g. from work hardening) which 'pin' the magnetic filaments and so help to retain the superconducting state in the face of intense applied fields.

The distinction between Types I and II depends on the ratio of the penetration depth λ to the *Pippard coherence length* ζ of the material. The interface between normal and superconducting regions has a finite thickness, measured by the coherence length, which originates from the spatial spread Δx of the superconducting current carriers in accordance with the uncertainty principle (equation (A 1.1)). Whereas $\zeta > \lambda$ for Type I superconductors, Type II superconductors are characterized by $\zeta < \lambda$. We can understand this by noting that when the interfaces are very thin it is possible to have many alternating strands of normal and superconducting material within a single region of allowed magnetic field variation. It has been established that both λ and ζ depend on the mean free path l of the conduction electrons in the normal state of the material. In fact $\lambda \propto l^{-1/2}$ and $\zeta \propto l^{1/2}$, so that a high resistivity in the normal state, as the result of a small l (e.g. as in a concentrated alloy or very impure metal), favours Type II superconductivity.

Further information about the nature of the superconducting state has been provided by specific heat capacity measurements. If the accurately known specific heat capacity due to lattice vibrations is subtracted from the total, the remainder is the specific heat capacity of the electron gas. As noted in Appendix 3, in a free-electron gas the electrons within an energy of the order of $k_B T$ from the Fermi surface can absorb thermal energy, because there are empty quantum states available for them at energies only slightly above the Fermi level. As a result, a free-electron gas should have a specific heat capacity proportional to temperature, and this is confirmed by measurements on metals in their normal state. The behaviour of metallic superconductors is quite different. The electronic specific heat capacity is extremely low at low temperatures, but rises exponentially on heating in a manner which clearly suggests that thermal activation is required before an electron can absorb thermal energy. This in turn implies the existence of an *energy gap*, a thin band of forbidden energies which envelops the Fermi surface and separates it from the band of free-electron or Bloch states above it. This energy gap is of a very different nature from that which was considered in the zone theory of insulators (*see* Section 1.5). There the gap was produced by Bragg reflections from the crystal lattice, and so was fixed in k-space by the Brillouin zone boundaries which represent these reflections. In superconductors the gap is unrelated to lattice reflections, and is *not* fixed in k-space, but is a gap that envelops the Fermi surface *wherever that surface may be*. It follows that the entire system, i.e. the Fermi distribution together with its surface gap, can be displaced *as a whole* in k space by an applied electric field. Being then asymmetrically positioned, relative to the centre of k-space, such a displaced system represents a current flowing through the material. Although the whole structure can be moved rigidly in this way, it is impossible for electrons to change their states independently of this movement unless they are supplied with an activation energy, because they cannot otherwise jump across the energy gap. The state of the *Fermi distribution plus gap* is thus self-preserving. If it is displaced to an asymmetrical position it will, at temperatures below T_c, hold this position even when there is no longer any external electric field applied. The gap holds the system in position and prevents individual electrons from breaking free, which would cause the asymmetrical state to decay. It is thus possible to understand how the persistent currents of the superconducting state can be maintained in the absence of applied voltages. The energy gap has been confirmed and measured by tunnelling experiments. It ranges from about $3k_B T_c$ when $T = 0$ to zero when $T = T_c$.

10.2 THE BOSON STATE

The frictionless behaviour of the electron gas at low temperatures identifies it as a *quantum superfluid*. Another quantum superfluid, similar in many respects is liquid helium at low temperatures. Helium remains a liquid down to 0 K, and on cooling below 2·18 K it changes its state, to helium II, in which it flows without viscous resistance. This low-temperature transition to a superfluid state occurs because the helium atom (or, strictly, the atom of the abundant ^{4}He isotope) is a *boson* particle.

To explain this we return to fundamental quantum mechanics. We recall from Appendix 2 (equations (A 2.2)–(A 2.8)) the problem of constructing a wavefunction to represent two electrons. We now do this more generally, for any pair of identical particles. Let $\psi(1, 2)$ be the wavefunction for the state in which one particle has the coordinate 1 and the other coordinate 2. Because these two identical particles are physically indistinguishable, $\psi(2, 1)$ – in which the particles have exchanged their coordinates – must represent the *same* state, so that $\psi(1, 2) = c\ \psi(2, 1)$, where c is some number to be determined. Let us exchange the coordinates a second time:

$$\psi(1, 2) = c\psi(2, 1) = c^2\psi(1, 2) \tag{10.2}$$

so that $c = \pm 1$. This consideration divides the universe's particles into two great classes, the antisymmetric *fermions* for which $c = -1$, and the symmetric *bosons* for which $c = +1$. Electrons, protons and neutrons are all fermions, whereas photons, phonons and some other elementary particles are bosons. Composite structures such as helium atoms, considered as single particles, can also be classified as bosons or fermions. The rule is that if they contain an odd number of fermions they are themselves fermions, whereas if they contain an even number they are bosons. Thus the ^{4}He atom, with six fermions (two protons, two neutrons and two electrons), is a boson, whereas the rare isotope ^{3}He (only one neutron) is a fermion.

This differentiation into two classes of particles is related to *spin*. It is a general law of nature, which has been explained by relativistic quantum mechanics, that all bosons have integer values of their intrinsic spin quantum number, whereas all fermions have half-odd-integer spins. For example, it is noted in Appendix 1 that the spin of an electron is $m_s = \pm\frac{1}{2}$, depending on whether the direction of the spin axis is 'up' or 'down'. This $\frac{1}{2}$ means that the angular momentum of the spin, with respect to this axis, is $\pm\frac{1}{2}\hbar$. Because of this role of the spin we must interpret the wavefunctions ψ in equation (10.2) as *total* wavefunctions which include a spin wavefunction, as in equation (A 2.8).

Fermions and bosons differ in their ability to occupy quantum states. We have already seen, from the Pauli exclusion principle, that when states are defined by total wavefunctions then only *one* electron is allowed to occupy such a state (more conventionally, in terms of states defined only by their orbital components of the wavefunction, two electrons of opposite spins are allowed to occupy one orbital state). The same is true of all fermions. For bosons there is no such restriction. In fact, the more bosons there are in a given quantum state, the more attractive is this state for yet more bosons.

In order to see this consider, as in equation (A 2.2) onwards, two identical particles 1 and 2, and two (total) quantum states A and B for them. For example, $\psi_A(1)\psi_B(2)$ represents the joint state in which 1 is in A and 2 is in B. If the particles were distinguishable there would be four recognizably different possible arrangements: $\psi_A(1)\psi_A(2)$, $\psi_B(1)\psi_B(2)$, $\psi_A(1)\psi_B(2)$ and $\psi_A(2)\psi_B(1)$. Giving equal weights to each of these, as we would if 1 and 2 were classical particles such as billiard balls and A and B were pockets for them, we would then find a probability of $\frac{1}{4}$ that they were both in pocket A, a probability of $\frac{1}{4}$ that both were in B, and a probability of $\frac{1}{2}$ that there was one in each pocket. But if they were fermions, having two in one pocket (i.e. in the same total state ψ_t) would be impossible and, as in equation (A 2.7), the only allowed state for them would be $\psi_A(1)\psi_B(2) - \psi_A(2)\psi_B(1)$. Hence there would be unit probability that there was one in each pocket. This of course simply re-expresses the exclusion principle. For bosons, however, there are *three* allowed possible arrangements: $\psi_A(1)\psi_A(2)$, $\psi_B(1)\psi_B(2)$, and $\psi_A(1)\psi_B(2) + \psi_A(2)\psi_B(1)$ as in equation (A 2.6). Giving equal weight to each of these, we find that the probability of having both in pocket A, both in B, or one in each pocket is $\frac{1}{3}$ in each case.

Identical bosons are thus more likely than classical particles to be in the same state. They appear to attract one another, as a consequence of their physical indistinguishability and freedom to go into the same quantum state. Further analysis shows that, the more bosons there already are in this quantum state, the greater is its attraction. In fact, when there are N identical bosons already in a given quantum state, the probability of a further identical boson entering this state is increased by a factor $N + 1$. In a macroscopic sample of cold liquid ^{4}He, the hugeness of N causes all the atoms to 'condense' into a single quantum state, i.e. into the ground state of lowest energy. This is known as a Bose-Einstein condensation. Because the bosons then all belong to one and the same wavefunction and are linked together by (weak) interactions, there is an extraordinary coordination in their behaviour. They all tend to move in unison. This is a 'rigid' motion in the sense that the particles are collectively commanded to move all in the same way. The separate turbulent motions of individual particles going their own independent ways are suppressed. This is the superfluidity of liquid helium, which has its origin in the quantum condensation that is a feature of a many-particle boson system at low temperatures.

It is significant that ^{3}He, a fermion system, does not behave in this way. It is still more significant that a Bose-Einstein condensation occurs even in this form of helium, at 0·002 K, because at this extremely low temperature weak attractive forces between the ^{3}He atoms are sufficient to bind *pairs* of atoms together to form boson 'molecules'. This behaviour of ^{3}He would have provided the strongest possible clue to the nature of superconductivity, were it not for the historical fact that this nature had already been elucidated before the ^{3}He discoveries were made.

10.3 COOPER PAIRS

The similarity between superfluid helium and superconductivity was clearly recognized by London (1950), who used electrodynamical arguments to infer, mainly from the characteristics of the Meissner effect, that the electrons might be coordinated over great distances through their all belonging to one single 'macroscopic' quantum state. This of course would be impossible if the electrons remained as fermions. The crucial advance – the recognition that the electrons might be reorganized into boson-like entities – came when Cooper (1956) introduced the notion of *bound electron pairs*. He showed that if there is a net attraction, however weak, between two electrons in states just above the Fermi surface, these electrons can jointly act as a *pair* which, as a single entity, behaves as a boson and so is able, like the atoms in liquid ^{4}He, to condense with other such pairs into a single ground state wavefunction.

An outline analysis of the formation of this state is given in Appendix 11 where it is shown that, when there is an attractive interaction, a pair of electrons can become *bound* in the sense that, even though each has a wave-vector which corresponds to a free-electron energy slightly *above* the Fermi level, the energy of the pair is *less* than twice the Fermi energy. It may seem unlikely that a very weak attractive interaction could lead to this result, but we are dealing with electrons selected from the top of the Fermi distribution. The filling of the Fermi sphere by the remaining electrons, together with the exclusion principle, prevents the two electrons being chosen from those free-electron states with wavenumbers less than k_F. The energy gain from the weak interaction can bring the total energy of the pair below (twice) the Fermi level, in which state the electrons are then 'bound' in so far as the pair cannot disintegrate into two independent electrons without increasing their energy.

If such pairing takes place between electrons at the Fermi surface, the pairs then condense into a single quantum state

$$\psi = \psi_0 \exp(i\theta) \qquad (10.3)$$

with phase θ, where $|\psi_0|^2$ is the number of pairs per unit volume, each pair characterized by its charge $2e$ and mass $2m$. These conclusions have been confirmed experimentally in studies of superconducting rings. When a supercurrent flows round such a ring the wavefunction has to join up smoothly, round the ring, which means that its phase must change by $2\pi n$ in the course of one circuit, where $n = 0, \pm 1, \pm 2 \ldots$. As a result, the magnetic flux threading the ring and produced by this current is *quantized* in units which have been shown to be characteristic of current carriers of charge $2e$.

10.4 THE BCS THEORY

To complete the theory two further steps were necessary. First a physical mechanism had to be provided for the inter-electronic attraction. Second, Cooper's analysis had to be generalized from the case of a single bound pair, in an otherwise normal Fermi distribution of free electrons, to that in which a large number of such pairs is formed. Both steps were taken in the BCS theory (Bardeen, Cooper and Schrieffer 1957).

A strong clue to the first step was provided by the discovery of the *isotope effect* (Maxwell 1950, Reynolds *et al.* 1950) which showed that, when a different isotope of a superconducting substance is used, the critical temperature is proportional to the inverse square root of the isotopic mass. This inverse square root is precisely the relationship for the vibrational frequency of a mass oscillating on an elastic spring. This effect in superconducting metals confirmed a conjecture by Fröhlich (1950) that the mechanism responsible for superconductivity in these materials is an interaction between the electrons and the *phonons*, which are the quanta of lattice vibrations, since the energy of a phonon $\hbar\omega$ is proportional to the vibrational frequency and hence to the inverse square root of the vibrating mass. The combination of this idea with that of Cooper pairs was the central achievement of the BCS theory.

Two electrons, of course, strongly repel each other by their electrostatic interaction. However, we are not concerned with isolated electrons but with two electrons in an environment of many other electrical charges, from ions and other electrons, in a metal. The positive ions provide overall electrical neutrality and, as we have seen in Chapter 3, the other electrons provide very effective long-range screening of electrical interaction. This leaves the way open for more subtle and minor effects to make themselves felt, including the electron–phonon–electron interaction. The general picture is that as one electron moves through the lattice, its Coulomb attraction to the positive ions between which it passes draws these ions slightly together as they try to close in on the electron. There is thus, in the wake of the electron, a slight increase in positive charge density, which can be represented as a packet of phonons. This positive charge in turn attracts another nearby electron; and so, in this indirect way, the second electron is, as it were, attracted towards the first one. This is the basis of the effective electron–electron attraction in the BCS theory, which under favourable circumstances leads to these electrons forming Cooper pairs.

In the general quantum theory of fields the problem of such interactions, i.e. the problem of *action at a distance*, is solved by the idea of *virtual particles*. In the present case these are virtual phonons. The uncertainty principle (equation (A 1.1)) allows a system a brief violation of the principle of conservation of energy by acquiring an extra energy ΔE, provided this energy disappears within a time Δt, where $\Delta E\,\Delta t \leqslant \hbar$. It follows that an electron can emit a phonon of energy $\hbar\omega$, where ω is the frequency of the lattice vibrations which constitute the phonon, provided this phonon is absorbed within a time $\Delta t \leqslant \omega^{-1}$. In the BCS theory one electron of initial wave-vector $\boldsymbol{k}_1$ emits a virtual phonon of energy $\hbar\omega$ and wave-vector $\boldsymbol{q}$ (where $|\boldsymbol{q}| = \omega/c$ in the simplest case, c being the speed of sound waves in the material) and so, by conservation of 'momentum', changes its own wave-vector to $\boldsymbol{k}_1' = \boldsymbol{k}_1 - \boldsymbol{q}$. A second electron, with initial wave-vector $\boldsymbol{k}_2$, absorbs this phonon and so changes its wave-vector to $\boldsymbol{k}_2' = \boldsymbol{k}_2 + \boldsymbol{q}$.

The theory of specific heat capacity due to lattice vibrations shows that the characteristic energy of a phonon is related to the Debye temperature Θ_D, through $k_B\,\Theta_D = \hbar\omega$, where Θ_D is typically of the order of 300 K (*see* e.g. Cottrell 1975). Although $\Theta_D \gg T_c$ in metals, nevertheless $k_B\,\Theta_D$ is small compared with the Fermi energy. Electron–electron interactions in which phonons of this energy are created and annihilated are thus similar, in thermal magnitude, to the room-temperature electron–electron collisions discussed in Section 3.5 and, as with those collisions, the same limitations apply, imposed by the Pauli principle. As a result, the states $\boldsymbol{k}_1, \boldsymbol{k}_1', \boldsymbol{k}_2$ and $\boldsymbol{k}_2'$ range in energy about $E_F \pm \hbar\omega$. For simplicity it is assumed in the BCS theory that, as a result of the above exchange of virtual phonons, the electron–electron interaction energy is a (negative) *constant*, $-V$, for values of $\boldsymbol{k}$ lying within this range, and zero otherwise (*see* equation (A 11.12)).

It is also assumed in the theory that the electrons in the normal state can be represented by pairs of free-electron wavefunctions ψ_k, and that the ground state electron-pair wavefunction ϕ can be constructed as a sum of the free-electron wavefunctions:

$$\phi = \sum c_k \psi_k \tag{10.4}$$

as in equation (A 11.6). The Hamiltonian H consists of the normal free-electron Hamiltonian plus the above interaction term, $-V$ (*see* equations (A 11.3), (A 11.4) and (A 11.12)). The energy of the system is given by the standard matrix element formula (equation (A 1.31)) as

$$E = \langle \phi | \mathrm{H} | \phi \rangle = \sum_{k_1, k_2} c_{k_2}^* c_{k_1} \langle \psi_{k_2} | \mathrm{H} | \psi_{k_1} \rangle \tag{10.5}$$

The aim in the BCS theory is to make E as low as possible by choosing the coefficients c_k so that only negative values of $\langle \psi_{k_2} | \mathrm{H} | \psi_{k_1} \rangle$ are admitted into the sum. Arbitrary values of k_1 and k_2 are generally equally likely to give positive or negative values of this, but when k_1 and k_2 are taken in pairs, both occupied or both empty, only negative values of the matrix element are obtained. We need as many such terms as possible in the sum and, since momentum must be conserved when the electrons go from $\boldsymbol{k}_1, \boldsymbol{k}_2$ to $\boldsymbol{k}_1', \boldsymbol{k}_2'$, the energy is lowest when every such pair has the same momentum:

$$\boldsymbol{k}_1 + \boldsymbol{k}_2 = \boldsymbol{k}_1' + \boldsymbol{k}_2' = \ldots = \boldsymbol{k}_i + \boldsymbol{k}_j = \ldots = \text{constant} \tag{10.6}$$

A supercurrent flows when the constant is non-zero. When it is zero the system is in its lowest energy state (as assumed in Appendix 11) and in this case the two electrons in any such pair have *opposite* momenta, e.g. $\boldsymbol{k}_2 = -\boldsymbol{k}_1$. Because of the exchange effect, it also pays for them to have opposite spins.

When these conditions are satisfied the paired electrons lie in an energy level which, because of the low value of E, is below the Fermi level even though the original free-electron states ψ_k in equation (10.4), from which their wavefunction is constructed, belong to energy levels just above the Fermi surface. It then follows that, at zero temperature, as many electrons as possible which satisfy the various conditions summarized above form pairs of opposite momentum and spin. In this condensed state the pairs can move in a unified way, like superfluid helium, in which the whole system cooperates to prevent any of its members from being scattered by the lattice irregularities. Persistent currents are then possible.

In Appendix 12 the energy W of this superconducting state at zero temperature is estimated as

$$W \simeq -2N(0)(\hbar\omega)^2 \exp[-2/N(0)V] \tag{10.7}$$

relative to the energy of the normal state at zero temperature. Here $-V$ is the constant which represents the electron–phonon–electron interaction, and $N(0)$ is the density of one-electron states of one spin at the Fermi surface in the normal metal. Also estimated in Appendix 12 is the magnitude of the superconducting energy gap, i.e. the energy of the lowest excited state of the superconductor, relative to the ground state, at zero temperature. This minimum excitation consists of the removal of one electron of a pair to an unpaired state, so producing two unpaired electrons. Equations (A 12.9) and (A 12.10) give this gap, in terms of the binding energy Δ (equation A 11.18) of a single Cooper pair, as

$$2\Delta \simeq 4\hbar\omega \exp[-1/N(0)V] \tag{10.8}$$

We expect the critical temperature T_c to be the temperature at which the thermal energy $k_B T_c$ can overcome this gap, and in fact the BCS theory gives the result

$$2\Delta = 3{\cdot}52 k_B T_c \tag{10.9}$$

In terms of the Debye temperature Θ_D we thus have

$$T_c \simeq \Theta_D \exp[-1/N(0)V] \tag{10.10}$$

This expression brings out one of the important successes of the BCS theory. Although the theory involves phonons of energy characteristic of the Debye temperature, T_c is nevertheless much lower than Θ_D, because of the exponential term. From a knowledge of the critical magnetic field it can be deduced that, typically, $N(0)V \simeq 0{\cdot}3$, in which case $T_c \simeq \Theta_D/30 \simeq 10$ K.

Corresponding to this value of T_c is a gap of the order of 10^{-3} eV. From equations (10.7) and (10.8) we have $-W \simeq 0{\cdot}1\ N(0)(2\Delta)^2$, and hence the number of superconducting electrons is of the order of $-W/2\Delta \simeq 0{\cdot}4\ N(0)k_B T_c$. Since, from equations (A 3.8) and (A 3.9), $N(0) \simeq N/E_F$ where N is the total number of conduction electrons per unit volume, and $E_F \simeq 10$ eV typically, we deduce that in the fully superconducting state about 10^{-4} of the electrons are paired. It then follows that the spread of k values at the Fermi surface, within which the pair states are formed, is of the order of $10^{-4}\ k_F$. The corresponding spread of momentum, $\Delta p \simeq 10^{-4}\hbar k_F$, leads via the uncertainty principle to a spatial spread

$$\Delta x \simeq \hbar/\Delta p \simeq 10^4/k_F \simeq 1000\ \text{Å} \tag{10.11}$$

This is a rough measure of the spacing of a pair and correlates with Pippard's coherence length (*see* Section 10.1). Thus the pairs should not be thought of as close 'molecules'; they are widespread entities and the region occupied by any one pair overlaps those of millions of other pairs. This Bose-Einstein condensation is not like the condensation of gaseous molecules into a liquid – it is a 'condensation of momentum', and the spatial dimensions of the condensing pairs remain uncondensed. The intermingling of large numbers of such pairs, in one another's shared space, is a characteristic feature of the superconducting state in BCS metals.

10.5 THE SEARCH FOR HIGHER CRITICAL TEMPERATURES

The vision of new electrical technologies based on superconductivity has inspired great efforts to develop materials with high and more easily accessible critical temperatures. The BCS theory has not been encouraging, however. As we have seen, T_c is brought low by the exponential factor in equation (10.10). A widely held view is that the BCS mechanism is unlikely ever to give a T_c above about 40 K, for if the electron–phonon coupling is made strong, as is required for a high T_c, it is likely to produce instead a transition to a different crystal structure.

Moreover, when an electron is strongly coupled to the lattice motion its effective mass is increased by the inertia of its interlinked ions, and it becomes less mobile. This effect shows itself as a tendency towards the formation of *polarons*. These are quasiparticles, usually considered in the theory of ionic and semiconducting crystals, in which electrons in the conduction band become strongly associated by electrostatic interaction with nearby ions. Positive ions are pulled towards such an electron; negative ones are pushed away. The electron then does not move as a free particle, but as a combination of electron plus lattice distortion, and thereby moves sluggishly. It is this combination of electron and ionic *configuration* that is the polaron.

We see from equation (10.10) that, with the BCS mechanism, a high T_c requires large values of all three contributors, Θ_D, $N(0)$ and V. The effect of Θ_D is less obvious than appears at first sight, because the large phonon energy $\hbar\omega$ required for a large Θ_D also makes the creation of virtual phonons a more difficult process, and so tends to reduce V. The assumption of a simple V term in the BCS theory is of course an oversimplification, and later theories (Morel and Anderson 1962, Carbotte 1969) have calculated instead the electron–phonon interaction, which depends both on the phonon energy and on the pseudopotential of the lattice ions through which the conduction electrons interact with the lattice and so generate and absorb the phonons. Elastically 'soft' lattice vibrations with small phonon energies, and a large pseudopotential, are expected from the theory to favour a high T_c. Thus we do not look for such materials among the 'ideal' metals with very weak pseudopotentials (e.g. $N(0)V = 0{\cdot}175$ and $T_c = 1{\cdot}2$ K for aluminium). By contrast, $N(0)V$ is fairly large in lead (0·39) and mercury (0·35), but this is offset by low Debye temperatures (96 K for lead and 70 K for mercury) so that the critical temperatures are only modest (7·2 K for lead and 4·2 K for mercury).

The requirement for a high density of states $N(0)$ may explain why some transition metals have moderate T_c values, e.g. 8·8 K for niobium and 4·9 K for vanadium. The structure of the density-of-states distribution (*see* Fig. 6.12), with the pronounced dip in the region of the Cr, Mo, W group, also correlates qualitatively with the low T_c of tungsten (0·01 K). However, the electronic structure of the transition metals, with their partly filled, tight-binding d orbitals, is so different from that envisaged in the BCS theory that the applicability of the theory to these metals is doubtful. It is known, for example,

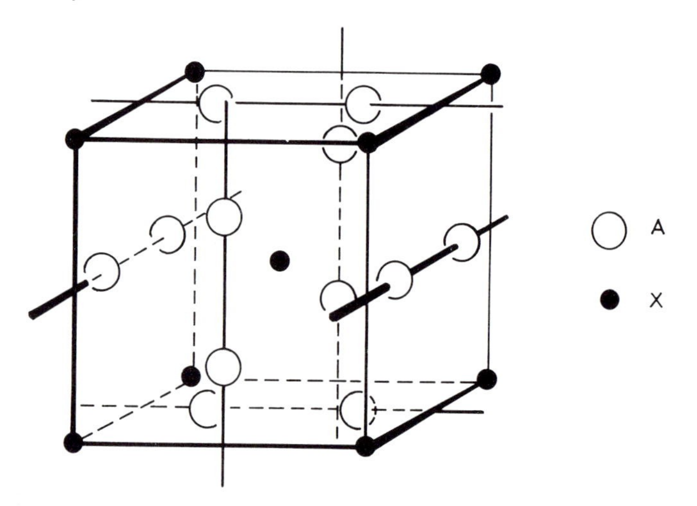

Fig. 10.3 The A15 structure of an A_3X compound

that the isotope effect does not occur in some of these metals, e.g. ruthenium and osmium.

A hopeful sign appeared in 1950 with the discovery that the compound NbN has a T_c of 15 K. This was followed in 1953 by an even more encouraging discovery – that some compounds of composition V_3X or Nb_3X (where X is a non-transition element) and with the A15 type of crystal structure, have high critical temperatures. The first discovery was V_3Si, with T_c = 17 K, then Nb_3Sn with T_c = 18 K; in the early 1970s Nb_3Ge was found to have T_c = 23·2 K. Since then, however, no examples with higher critical temperatures than this have been found, and it is now considered that the A15 alloys have an upper limiting T_c of about 25 K.

The A15 structure (also known as the β tungsten structure) is an extraordinary one (*see* Fig. 10.3). In the Nb_3X and V_3X alloys the transition metal atoms are arranged in straight, non-intersecting strings threading across the faces of a b.c.c. matrix of X atoms. The A15 structures with the highest T_c are metastable, and in some cases undergo a martensitic transformation to a slightly tetragonal form at a temperature just above T_c. This structural instability is a sign of elastic 'softness' which, together with a high density of states, may be the reason for the high T_c.

The growing conviction in the 1960s that really high critical temperatures could never be reached in conventional metals and alloys led several people to examine the possibility of superconductivity in other materials. Theoreticians speculated on alternative mechanisms which might circumvent the limitations of the BCS phonon process. Little (1964) proposed that one-dimensional organic metallic conductors (*see* Section 1.1) might become superconducting at low temperatures. This was subsequently confirmed for some organic salts, although always with $T_c \simeq 1$ K. Inorganic superconductors were also discovered, e.g. $TaSe_3$ with $T_c \simeq 2·3$ K, and some heavy-fermion superconducting compounds such as UBe_{13} and UPt_3 (*see* Section 7.6), but they all had a low T_c. It all proved disappointing, until very recently.

10.6 CERAMIC SUPERCONDUCTORS

In 1986 Bednorz and Müller achieved $T_c > 30$ K in a mixed oxide of copper, lanthanum and barium (Bednorz and Müller 1986). Guided by the discovery in the 1970s of fairly high superconducting temperatures in $BaPb_{1-x}Bi_xO_3$ (13 K) and $LiTi_2O_4$ (13·7 K), they were led to study La–Ba–Cu–O compounds of the type La_2CuO_4 with a little of the lanthanum replaced by barium and also with a small deficiency δ in the oxygen content, e.g. $La_{2-x}Ba_xCuO_{4-\delta}$, $x \simeq 0{\cdot}15$. Quite soon a T_c of 40 K was reached by replacing the barium with strontium, and a feverish worldwide search for still higher critical temperatures began. By early 1987 Chu had achieved $T_c > 90$ K in a compound in which the lanthanum was replaced by yttrium (Chu *et al.* 1987). This was the second crucial advance because it made superconductivity accessible to liquid nitrogen cooling. Chu's material was originally multiphase, but the phase with the high T_c, $YBa_2Cu_3O_7$, has since been identified and prepared as a pure compound. There has now been an extensive exploration of similar compounds in which other chemically similar elements, e.g. various rare earths and alkaline earths, as well as bismuth and thallium, have been introduced in place of those above and a T_c of 125 K has been achieved. Hints of superconductivity at 240 K or even at room temperature have frequently been announced. It is possible that thin, delicate filaments of a really high-T_c phase may be formed in some of these materials, which are easily destroyed by the strains of thermal cycling.

These ceramic superconductors are of Type II and can withstand high magnetic fields (e.g. 180 T at 0 K) before their mixed state disappears, which could make them useful for d.c. magnets. However, their transition from the pure diamagnetic to the mixed state occurs at rather low fields and so they are unsuitable for a.c. currents. Some natural feature in them is effective in pinning the magnetic filaments, so that the special measures used in Type II alloy magnets to introduce pinning centres by, for example, cold working may be unnecessary. Their current-carrying capacities are generally small, typically of the order of 10^6 A m^{-2}, far below what is required for many practical applications. It appears that this is because there are non-superconducting grain boundaries in the polycrystalline forms of the materials. Single crystals have been shown to carry currents of up to about 10^{10} A m^{-2}, and similar values have been obtained in oriented thin films (epitaxial layers). This last observation appears to be related to the *anisotropy* of the materials. As described below, these mixed oxides are *layer structures*. There is evidence that the superconducting current flows mainly within the layers.

10.7 THEORY OF THE HIGH-T_c SUPERCONDUCTORS

The parent material for these mixed-oxide superconductors is the compound La_2CuO_4, in which, on an ionic interpretation, the oxygen atoms collect eight electrons to form four O^{2-} ions, taking three from each of the two lanthanum atoms, which become La^{3+} ions, and two from the copper, which becomes a Cu^{2+} ion. While the La^{3+} ions retain no electrons outside their filled 5s, 5p shell, the Cu^{2+} ion is left with one hole in its 3d shell which, from simple band theory, might be expected to make the compound a metal. In fact, the stoichiometric compound turns out to be an insulator. Two explanations of this have been offered.

The first (Anderson 1987, Anderson and Abrahams 1987) goes back to the Mott–Hubbard theory of the metal–insulator transition, outlined in Chapter 1. The spacing between the copper ions is too large to allow the electrons to make uncompensated transfer between one another, as would be required for metallic conductivity from their partly filled d shells. The repulsive energy given by equation (1.32) for such transfers is greater than the gain from the ensuing formation and broadening of a partly filled d band. However, the Mott–Hubbard energy gap which separates this insulating state from the 'ionized' states (in which, for example, $Cu^{2+} + Cu^{2+} \rightarrow Cu^{3+} + Cu^{+}$) is small, so that La_2CuO_4 behaves as a semiconductor. Moreover, the replacement of a small proportion of the trivalent lanthanum atoms by divalent alkaline earth ones, as in the *Bednorz–Müller* compound, converts the material into a metal through the creation of some Cu^{3+} or O^{-} ions (or, in the language of band theory, through the lowering of the Fermi level and the creation of empty states at the top of the band). With this development of metallicity the material becomes a high-T_c superconductor.

The other explanation (Mattheiss 1987) is based on Brillouin zones and crystal structure. The structure of all these ceramic superconducting compounds is that of a *layered perovskite*, described in Appendix 13. The unit crystal cell is orthorhombic, being a slight distortion of a tetragonal cell. Band structure calculations suggest that for the tetragonal structure one of the bands straddles the Fermi level; on band principles alone (ignoring the electron correlations responsible for Mott–Hubbard insulators), the material should be a metal. However, the tetragonal structure distorts into the orthorhombic, with a small rotation in opposite directions of successive octahedral groupings of oxygen ions round copper ions, and this produces a small crystallographic energy gap at the Fermi surface, which makes the material a semiconductor.

This distortion and consequential gap formation is considered to occur by a particular kind of *Jahn–Teller effect* known as a *Peierls instability*. The Jahn–Teller principle, first developed for ionic crystals, is an example of *symmetry breaking*. Suppose, as a simple example, that a Cu^{2+} ion is at the centre of a regular octahedron, at the corners of which are six identical negative ions, which lie along the x, y, z axes of a cubic crystal structure. Suppose also that the empty d state of this ion is in the $d_{x^2-y^2}$ orbital (*see* Fig. A 1.3), the lobes of which lie along the x and y axes. Along these axes the electron concentration in the copper ion is therefore particularly low, so that the negative ions on these axes are attracted inwards more closely than those on the z axis, and the cubic structure thereby deforms into a less symmetrical tetragonal one. The general Jahn–Teller principle applies whenever a symmetrical wave-mechanical configuration does not correspond to stable equilibrium in a molecular structure.

The Peierls instability (Peierls 1955) depends on the band-structure energy discussed in Chapter 5 (*see* Fig. 5.2). Consider, for example, a linear chain of metal atoms with uniform spacing a. The first Brillouin zone boundary occurs at $k = \pm \pi/a$. If the metal is monatomic this zone would be half-filled, up to

$k = \pm \pi/2a$. Suppose that every second atom is slightly displaced, by exactly the same amount. The structure is now a superlattice of spacing $2a$, and the weak pseudopotential field of period $2a$ generates additional zone boundaries (*see* Appendix 10) the first of which occurs at $k = \pm \pi/2a$, i.e. at the position of the Fermi surface. According to Fig. 5.2 the band structure energy favours such a distortion; however, the new zone boundary forms a *gap* at the Fermi surface, so that the material is no longer a metal but an insulator or semiconductor. The quasi-spherical shapes of Fermi surfaces in three-dimensional metals are generally unsuitable for such an effect, but in a structure where large areas of the Fermi surface are flat – and where *nesting* can therefore occur (*see* Section 6.7) – a transformation to a slightly distorted and less symmetrical crystal structure may be brought about by the Peierls instability.

It is believed by some theoreticians that the tetragonal → orthorhombic transformation in La_2CuO_4 is brought about by such an effect. As shown in Appendix 13, in the oxygen-deficient superconducting oxides such as $YBa_2Cu_3O_{7-\delta}$ the transformation can also be related to an ordering process in which the oxygen atoms, in those copper sheets in which the oxygen sites are only half-filled, align themselves into parallel chains of alternate copper and oxygen atoms.

It is well established that the critical temperature in such materials is very sensitive to the oxygen content (Robinson 1987). The high T_c of over 90 K is obtained when $\delta \simeq 0$, whereas at $\delta = 0{\cdot}4$ it is only 55 K. The tetragonal structure appears to give this T_c of 55 K, whereas the orthorhombic structure is required for the higher temperature. In contrast to this critical effect of oxygen content, T_c seems to be wholly insensitive to the choice of the rare earth metal in the structure.

Measurements (Gough *et al.* 1987) of the quantized magnetic flux, threading rings of these superconducting oxides, have confirmed (as with the earlier superconductors – *see* Section 10.3) that the current carriers have a charge of $2e$, which is indicative of Cooper pairs. This is hardly surprising in view of the general theory, outlined in Sections 10.2 and 10.3, that superconductivity is a characteristic property of electrically charged bosons. Nevertheless, there is much evidence that the mechanism in these oxides is *not* that of the BCS theory. The very high values of T_c are one indicator of this, but there are others, notably that replacing the standard ^{16}O atom by its heavier isotope ^{18}O produces little or no change in T_c, even though it leads to the expected change in the oxygen vibrational frequency (Batlogg *et al.* 1987). For this and other reasons it is generally considered that phonons play a small part, if any, in the mechanism of superconductivity in these high-T_c oxides, and that purely electron–electron interactions must be mainly responsible.

Several tentative theories of such electron–electron mechanisms have been proposed. The most highly developed is Anderson's (Anderson 1987, Anderson *et al.* 1987, Pauling 1987), which links the Mott–Hubbard theory with Pauling's theory of the resonating valence bond (RVB) outlined in Section 5.3, and is also closely related to the theory of antiferromagnetism. In Pauling's theory the atoms in, for example, lithium are linked together in pairs by means of covalent bonds, each of which consists of two electrons of opposite spins, much as in the theory of the hydrogen molecule (*see* Fig. A 2.2). Each monovalent atom in its neutral state provides one bonding electron and so can bond, at any instant, with only one neighbour. However, since the number of neighbours is greater than the number of bonds, the bond attached to any one atom can 'pivot' from one neighbour to another, a process known as *resonance*. So long as every atom is strictly limited to one bond – no more, no less – this resonance can occur only in a *synchronized* way amongst all the bonds; as one bond swings from a given atom, so must another one swing on to that atom. Pauling attributed the metallic properties, not to this, but to an additional *unsynchronized* resonance in which the bonds pivot to some extent independently of one another, so that a given atom may occasionally have two bonds (and thus temporarily become Li^-) or none (Li^+). This process requires electron transfers, e.g. $Li + Li \rightarrow Li^+ + Li^-$, of the type considered in the Mott–Hubbard theory; it is opposed by the Hubbard energy (*see* equation (1.32)), which comes mainly from the electrostatic repulsion between the electrons on the same atom (e.g. Li^-); and is favoured by the band energy (*see* Section 1.8), which is a drop in kinetic energy when electrons are released to wander through the material. Pauling's energy of unsynchronized resonance is thus equivalent, in a chemical picture, to this drop of energy on band formation; and on this basis the Mott–Hubbard theory can be reinterpreted as an RVB theory, as suggested by Anderson.

In his discussion of La_2CuO_4 from this point of view, Pauling (1987) points out that copper and oxygen differ on his electronegativity scale by 1·6 units, on which basis the copper–oxygen bond in the structure is expected to be half covalent, half ionic. The insulating or semiconducting character of La_2CuO_4 shows that the repulsive Hubbard energy slightly exceeds the band or resonance energy, so that in the pure stoichiometric compound the conductivity processes outlined above cannot occur. It is thus essential to have non-stoichiometry, e.g. by replacing some lanthanum atoms with divalent ones, so as to produce some 'holes' in the network of electrons in the RVB structure, throughout the crystal. The presence of these holes provides the freedom, e.g. through transfers such as $Cu^{3+} + Cu^{2+} \rightarrow Cu^{2+} + Cu^{3+}$, which enables the bond resonance to become unsynchronized.

Since the two electrons in a covalent bond have opposite spins we are led, by regarding these electrons as 'belonging', one to each of the two atoms joined by such a bond, to view such a molecule as an antiferromagnetic ordered pair of atoms. Generalizing, we see that the Mott–Hubbard or RVB structures are also antiferromagnetic. Because of the resonance in the RVB, however, this is not a rigidly fixed pattern of antiferromagnetic spins. According to Anderson, the spin of an electron on a given atom fluctuates between up and down, in antiparallel synchronization with that of the electron with which it happens to be paired in the particular covalent bond that temporarily joins this atom to one of its neighbours. This fluctuating arrangement appears to be the most stable form of antiferromagnetic state in low-dimensional systems (Anderson 1987, Bethe 1931). In the two-dimensional alternation of copper and oxygen atoms in the layered perovskite structure, the spins of successive Cu^{2+} ions are antiferromagnetically ordered. This arrangement results from an indirect interaction between these ions (known generally as *superexchange*) which in fact occurs through the correlation of the spins of each neighbouring pair of copper ions with those of the electrons (in p orbitals) of the oxygen ion between them.

Metallic conductivity and high-T_c superconductivity appear in La_2CuO_4 when a small proportion of the trivalent La atoms is replaced by divalent alkaline earths, reducing the electron concentration. In the RVB picture, unsynchronized resonance can then take place and a current can flow, even though the coupling into electron pairs remains. At low temperatures the now mobile RVB pairs form a condensed boson fluid which, being charged, provides superconductivity. There is an important distinction to be made between this effect of *doping* in La_2CuO_4 and that of, say, doping silicon to produce electron holes in the conduction band (*see* Section 1.6): here the doping produces, not a drop in the Fermi level, but a reduction in the number of O^{2-} ions and RVB bonds.

A variant of this theory, linked to antiferromagnetism rather than to RVB, has been given by Hirsch (1987), related to an earlier theory (Emery 1987). He argued that the reduction in electron concentration upon doping expresses itself through the formation of electron holes on the oxygen ions (i.e. some ions become O^- instead of O^{2-}). Suppose that such a positive hole passes from oxygen to oxygen (via copper, along a chain of alternating ions in one of the copper–oxygen layers of the structure). Because the hole is a vacancy in an otherwise completed shell of spin-paired electrons in the oxygen ion, it has an associated spin. In passing along the chain it exchanges spin directions with each copper ion through which it passes. It thus leaves in its wake a trail of reversed copper spins, i.e. *an antiphase domain string* in the antiferromagnetic copper superlattice, rather as the passage of a vacant atomic site would produce an antiphase string in, for example, ordered β brass. The extra energy of the antiphase boundary of this string opposes the forward movement of the hole, but it also attracts a second hole of opposite spin since this, following closely behind the first, reverses the copper spins once more and so restores them to their original perfect antiferromagnetic order. There is thus an attractive coupling between such pairs of charged holes of opposite spins, as a result of which they move together as a charged boson at low temperatures. The condensed fluid of these pairs provides the superconducting state.

Although such mechanisms as these are not yet established as actually operating in the high-T_c superconductors, they do offer the prospect of a means of escape from the low-temperature limitations of the BCS mechanism. In principle, there appears to be no reason why substantially higher critical temperatures, up to room temperature and perhaps beyond, should not now be obtainable.

REFERENCES

Anderson, P. W., *Science*, **235**, 1196 (1987).
Anderson, P. W., and Abrahams, E., *Nature*, **327**, 363 (1987).
Anderson, P. W., *et al.*, *Phys. Rev. Lett.*, **58**, 2790 (1987).
Bardeen, J., Cooper, L. N., and Schrieffer, J. R., *Phys. Rev.*, **108**, 1175 (1957).
Batlogg, B., *et al.*, *Phys. Rev. Lett.*, **58**, 2333 (1987).
Bednorz, J. G., and Müller, K. A., *Z. Phys. B*, **64**, 189 (1986).
Bethe, H. A., *Z. Phys.*, **71**, 205 (1931).

Carbotte, J. P., in *Superconductivity*, Vol. I (ed. P. R. Wallace), Gordon and Breach, New York (1969), p. 491.
Chu, C. W., *et al.*, *Phys. Rev. Lett.*, **58,** 405 (1987).
Cooper, L. N., *Phys. Rev.*, **104,** 1189 (1956).
Cottrell, A. H., *An Introduction to Metallurgy*, Edward Arnold, London (1975).
Fröhlich, H., *Phys. Rev.*, **79,** 845 (1950).
Gough, C. E., *et al.*, *Nature*, **326,** 855 (1987).
Hirsch, J. E.., *Phys. Rev. Lett.*, **59,** 228 (1987).
Little, W. A., *Phys. Rev.*, A **134,** 1416 (1964).
London, F., *Superfluids*, Vol. I, Constable, London (1950) (reprinted by Dover, New York. (1961)).
Mattheis, L. F., *Phys. Rev. Lett.*, **58,** 1028 (1987).
Maxwell, E., *Phys. Rev.*, **78,** 477 (1950).
Morel, P., and Anderson, P. W., *Phys. Rev.*, **125,** 1263 (1962).
Pauling, L., *Phys. Rev. Lett.*, **59,** 225 (1987).
Peierls, R. E., *Quantum Theory of Solids*, Oxford University Press (1955).
Reynolds, C. A., *et al.*, *Phys. Rev.*, **78,** 477 (1950).
Robinson, A. L., *Science*, **236,** 1063 (1987).

APPENDIX 1

Summary of quantum mechanics

GENERAL PRINCIPLES

In all the processes of nature, including the processes of measurement, the action involved is quantized in units of *Planck's constant* ($h = 6{\cdot}626 \times 10^{-34}$ J s; $\hbar = h/2\pi = 1{\cdot}054 \times 10^{-34}$ J s). Action has the dimensions $\mathrm{M\,L^2\,T^{-1}}$, and Planck's constant can be written in the form of products of 'conjugate' quantities as

$$h = \Delta p\,\Delta x = \Delta E\,\Delta t = \Delta L\,\Delta\theta \quad \text{etc.} \tag{A 1.1}$$

where p is momentum, x position, E energy, t time, L angular momentum and θ angular orientation, and Δ denotes the amount of each of these quantities necessitated by the quantum of action.

When we make a good measurement, of for example the position of an object, we try to get as sharp a value of x as possible: that is, we set up the method of measurement so as to make Δx very small. But this inevitably makes Δp large, so that a well-positioned object does not have a sharp value of momentum; and vice versa. This is *Heisenberg's indeterminacy* or *uncertainty principle* ($\Delta p\,\Delta x \geqslant \frac{1}{2}\hbar$). In practice, because $\hbar$ is so small only the very lightest of particles, such as electrons, show the effect strongly.

The energy–time uncertainty relationship has a special interpretation in which Δt refers to the *lifetime* of a particular state. In practice we are often interested in the *stable* states of physical systems, i.e. states of motion which continue unchanged for long times. Because Δt is very large, such states have well-defined energies. In the limit $\Delta t \rightarrow \infty$ they are *stationary states*, in which systems exist in sharp *energy levels*, and which can be analysed by a simpler time-independent form of the theory.

Because their sharpness – or even existence – depends on the conditions of observation, variables of motion such as momentum, position, energy and angular momentum lose their classical meaning and are known in quantum mechanics as *observables*. The implication is that an electron does not generally 'possess' any such properties, but may be made to display a sharp value of one of them if it is observed under conditions which give that one a small Δ.

Since observables generally have spreads of values the theory deals with the *probability* of finding particular values when observations are made. For example, the probability of finding a given electron in a certain small region of space, $dx\,dy\,dz$ ($\equiv dv$), centred on the point x, y, z, can be written as $\rho(x, y, z)\,dv$. Since the electron must be somewhere, the scale of ρ must be chosen so that

$$\int \rho \, \mathrm{d}v = 1 \tag{A 1.2}$$

where the integration is taken over the whole volume in which the electron exists. This scaling process is called normalization.

The probability laws of quantum mechanics differ from the classical ones, as can be seen from the effect of *interference* as in *electron diffraction*. When electrons pass through diffraction slits, as in a crystal, the probability densities of overlapping emerging beams do not add together as they would if the electrons were classical bullets. Instead, the beams interfere with one another, sometimes reinforcing to give extra intensity, sometimes cancelling to give a dark band in the diffraction fringes. The basic feature is thus not ρ itself, which of course is always a positive quantity, but something which can be either + or −, so that reinforcement (+ + or − −) or cancellation (+ − or − +) is possible, according to the 'phases' of the overlapping beams. Of course $\sqrt{\rho}$ can be either + or −, and in fact it is this square root, known as the *probability amplitude*, that is the basic feature in quantum mechanics. This is the *wavefunction*, symbolized by ψ (or Ψ for time-dependent phenomena), also known as the *state function* or *state vector*. The wavefunction is the source of all information about the dynamical condition of a quantum-mechanical particle.

The language of quantum mechanics is modelled on that of experiment, observation and measurement. Measured values of observables are obtained by 'operating' on systems in given initial states. It follows that three different kinds of mathematical symbol appear in quantum mechanics. First is the function ψ, which represents the system we are measuring (e.g. an electron) in its initial state of existence. This ψ is always taken to be a mathematically *single-valued* and *continuous* function of the coordinates. Although we often speak of 'a system', such as one particular electron, it is often better to think of a large number, or *ensemble*, of identical copies of the single system, i.e. to think of the same experiment being carried out in an identical way on each of a number of identical systems, and the results averaged. On this basis we can think of $\psi^2(x, y, z)$ as the *probability density* of an electron in the space x, y, z, a quantity that is measured for example in the electron density maps of crystals, observed by X-ray diffraction techniques.

The second quantum symbol represents the *operation* of measuring an observable on a system, i.e. the interaction of the measuring apparatus with the system. It is a mathematical *operator*, (A, B, etc.) which acts on ψ to make it give a value of the observable in question. Thus $\mathrm{A}\psi$ represents mathematically the operation of A on ψ; and physically the measurement of the particular observable which has the operator A. The three operators, along the x axis, that we shall need are

$$\left.\begin{aligned} \mathrm{A}_x &= x \\ \mathrm{A}_p &= \frac{\hbar}{\mathrm{i}}\frac{\partial}{\partial x} \\ \mathrm{A}_E &= \mathrm{H} = \frac{1}{2m}\left(\frac{\hbar}{\mathrm{i}}\frac{\partial}{\partial x}\right)^2 + V(x) = -\frac{\hbar^2}{2m}\left(\frac{\partial^2}{\partial x^2}\right) + V(x) \end{aligned}\right\} \tag{A 1.3}$$

The first, for the measurement of position, simply means 'multiply by x', i.e. $A_x\psi \equiv x\psi$. The second, the *momentum operator*, is a differentiation of ψ with respect to x. Momentum is thus given by the *gradient* of ψ. Despite the presence of i ($=\sqrt{-1}$), quantum mechanics always works out to give real numbers for measured quantities. Operators that do this are said to be *Hermitian*. The third operator is H, the *Hamiltonian* or *energy* operator. The expression for H given above is for a particle of mass m and potential energy $V(x)$ at position x. It is the quantum equivalent of the classical expression

$$E = \frac{p^2}{2m} + V(x) \qquad \text{(A 1.4)}$$

for the energy of a particle. In three dimensions the Hamiltonian becomes

$$\mathrm{H} = -\frac{\hbar^2}{2m}\nabla^2 + V(x, y, z) \qquad \text{(A 1.5)}$$

$$\nabla^2 = \frac{\partial^2}{\partial x^2} + \frac{\partial^2}{\partial y^2} + \frac{\partial^2}{\partial z^2}$$

The third quantum symbol is simply the real number produced from the operation as the measured value, called the *eigenvalue*, of the observable. We denote it by a, b, E, and so on.

There are two different circumstances under which a measurement can be made. In the first, the ensemble to be measured is already in a quantum state that gives a sharp value of the observable in question (an *eigenstate*). We may, for example, wish to measure the energy required to ionize the electron of a hydrogen atom in its stationary ground state (i.e. in a sharp energy level). The ψ of the system in this case is said to be an *eigenfunction* of the operator being used, and the effect of the operation is expressed by

$$\mathrm{A}\psi = a\psi \qquad \text{(A 1.6)}$$

which means that the operation delivers the measured value a of the observable represented by A and (in an ideal measurement) leaves the system continuing in the same state ψ. This a is the *eigenvalue* of the eigenfunction ψ under the operation of A. For an energy measurement on a stationary state, which is an eigenstate of H, we have

$$\mathrm{H}\psi = E\psi \qquad \text{(A 1.7)}$$

and find the energy to be E. Written out in full, this is the time-independent Schrödinger equation

$$\frac{d^2\psi}{dx^2} + (E - V)\psi = 0 \qquad \text{(A 1.8)}$$

in one dimension, or

$$\nabla^2 \psi + (E - V)\psi = 0 \tag{A 1.9}$$

in three. Here we have simplified the expressions by measuring position and energy in *atomic units*, i.e. the *Bohr radius* a_o and the *Rydberg* R_y:

$$a_o = \hbar^2/me^2 \tag{A 1.10}$$

$$1\ R_y = me^4/2\hbar^2 \tag{A 1.11}$$

A measurement can also be made on a system which is *not* in an eigenstate of the operator for the measurement in question. There may for example be an electron travelling through an apparatus with a fairly sharp momentum. If we want to find where it is, we must apply the position operator to a state function which is spread out over a long stretch of the x axis. After this operation we know (momentarily) where the electron is. The operation has done two things: it has delivered a particular value of x for the position of the electron, but it has also *changed* the wavefunction of the electron from one spread out along the x axis to one which is a narrow peak around the measured position. What has happened is

$$A\phi \rightarrow A\psi \rightarrow a\psi \tag{A 1.12}$$

where ϕ is the original state, ψ is an eigenfunction of A and a is the measured value. Thus $A\phi$ is the new state function ψ of the electron multiplied by a.

Successive measurements are represented by successive operations, e.g.

$$BA\phi = B(A\phi) \rightarrow B(a\psi) = a(B\psi) \rightarrow a(B\beta) \rightarrow ab\beta \tag{A 1.13}$$

where β is an eigenfunction of B and a and b are the eigenvalues obtained by operating first with A and then with B. The order of such operations is important. Suppose that $A = x$ and $B = \partial/\partial x$. Then

$$BA\psi = \frac{\partial}{\partial x}(x\psi) = \psi + x\frac{\partial \psi}{\partial x} \tag{A 1.14}$$

whereas

$$AB\psi = x\frac{\partial \psi}{\partial x}$$

so $BA \neq AB$. Such operators are said to be *non-commuting* and correspond to pairs of conjugate observables, as in equation (A 1.1). Some operators do commute, i.e. $BA = AB$; momentum and energy are an example.

To each observable, e.g. A, there belongs not just one eigenfunction but an entire set ψ_i, where $i = 1, 2, \ldots,$ together with a corresponding set of eigenvalues: a_1 which goes with ψ_1, a_2 which goes with ψ_2, and so on. For example, the energy operator H is associated with the complete set of stationary states and all the energy levels E that go with them. In some cases the set, or part of it, is a finite set of *discrete* states and eigenvalues, e.g. the energy levels of bound states in atoms; in others the set is infinite and *continuous*, e.g. the set of allowed states of motion of a free particle in infinite space.

The operation (A 1.12) should thus be written as

$$\mathrm{A}\phi_i \rightarrow \mathrm{A}\psi_j \rightarrow a_j\psi_j \qquad \text{(A 1.15)}$$

to indicate that the system is initially in some particular ϕ state, denoted by ϕ_i, and that under the operation of A it goes into some particular eigenstate ψ_j in which the particular observed value a_j is delivered. An important quantum feature appears at this point. When A operates on a given ϕ_i the particular ψ_j that emerges – and hence the value of a_j that comes with it – is chosen at random from the set of all the ψ_j. If we make an ensemble of such measurements, $\mathrm{A}\phi_i$, sometimes we get ψ_1 with eigenvalue a_1, sometimes ψ_2 with a_2, and so on. Under these circumstances there are two important observable features of the ensemble ϕ_i when operated on by A: (i) the *probability* of getting the particular eigenstate ψ_j and (ii) the *average value* of the observable, $\langle a \rangle$, found in this ensemble of measurements. Quantum mechanics gives expressions for both these quantities; they will follow below after we have noted some necessary properties of the wavefunction.

First, the wavefunction is generally a *complex* quantity:

$$\psi = u + \mathrm{i}v \qquad \text{(A 1.16)}$$

where u and v are real functions; it has a *complex conjugate*

$$\psi^* = u - \mathrm{i}v \qquad \text{(A 1.17)}$$

with the understanding that the 'square of ψ', i.e. the probability density, is given by

$$\rho = \psi^*\psi = |\psi|^2 = u^2 + v^2 \qquad \text{(A 1.18)}$$

As an example of this, consider the solution of the Schrödinger equation (A 1.8) for the simplest case, $V = 0$. This is a sinusoidal equation, but the solution is not $\psi \sim \sin kx$ or $\cos kx$ because these functions would, by themselves, give a *periodic* distribution of electron density along the x axis, which does not make sense for a free electron on an infinite line where there is nothing in V to favour one position rather than another. Instead, the solution is

$$\psi(x) = c \exp(\mathrm{i}kx) = c(\cos kx + \mathrm{i} \sin kx) \qquad \text{(A 1.19)}$$

with the welcome result that

$$\psi^*\psi = c^2(\cos^2 kx + \sin^2 kx) = c^2 = \text{constant} \qquad \text{(A 1.20)}$$

as expected intuitively. Here c is a normalization constant and

$$k = \frac{(2mE)^{1/2}}{\hbar} = \frac{2\pi}{\lambda} \qquad \text{(A 1.21)}$$

is the *wavenumber*, related to the *wavelength* λ.

Second, it is a general result that if ψ_i and ψ_j are two different eigenfunctions of a common Schrödinger equation, with eigenvalues a_i and a_j respectively, then the integral of their product over the whole of the spatial region in which they both exist is zero:

$$\int \psi_i^* \psi_j \, \mathrm{d}v = 0 \qquad \text{(A 1.22)}$$

where $\mathrm{d}v = \mathrm{d}x\,\mathrm{d}y\,\mathrm{d}z$. This is known as *orthogonality*. In the most usual case, orthogonality results from the different *symmetries* of different wavefunctions. These cause them, even when overlapping, to have the same signs in some volume elements $\mathrm{d}v$ and opposite signs in others, so that the positive and negative forms of $\psi_i^* \psi_j$, in different volume elements, cancel out when integrated over the whole volume. If the wavefunctions are also *normalized*, i.e. if

$$\int \psi_n^* \psi_n \, \mathrm{d}v = 1 \qquad \text{(A 1.23)}$$

the two conditions can be combined into a general *orthonormal* condition

$$\int \psi_i^* \psi_j \, \mathrm{d}v = \delta_{ij} \qquad \text{(A 1.24)}$$

where $\delta_{ij} = 0$ when $i \neq j$ and $\delta_{ij} = 1$ when $i = j$.

Finally, returning to equation (A 1.12) (or dropping the subscript i on ϕ_i in equation (A 1.15)), we note that any wavefunction such as ϕ can be regarded as a *sum* of *all* the n eigenfunctions ψ_j of an operator A:

$$\phi = c_1\psi_1 + c_2\psi_2 + \ldots + c_n\psi_n = \sum_{j=1}^{n} c_j\psi_j \equiv \sum_{n} c_n\psi_n \qquad \text{(A 1.25)}$$

where the constants c_n are in general complex numbers which, for normalized functions, satisfy the sum rule

$$\sum_n c_n^* c_n = \sum_n |c_n|^2 = 1 \qquad \text{(A 1.26)}$$

as follows from the application of equation (A 1.24) to equation (A 1.25).

The first observable feature of the ensemble ϕ, when operated on by A, can now be stated. The probability in this case of getting the particular eigenstate ψ_j (or, in practice, of measuring the eigenvalue a_j of the observable belonging to the operator A) is given by

$$\text{prob}\,(\psi_j \text{ in } \phi) = c_j^* c_j = |c_j|^2 \qquad \text{(A 1.27)}$$

The second observable feature can be obtained as follows. Because

$$\mathrm{A}\phi = \mathrm{A}(c_1\psi_1 + c_2\psi_2 + \ldots) = a_1c_1\psi_1 + a_2c_2\psi_2 + \ldots \qquad \text{(A 1.28)}$$

where the a_j are the eigenvalues, and because of the orthonormal properties of the eigenfunctions ψ_j, then

$$\int \phi^* \mathrm{A}\phi \,\mathrm{d}v = \sum_n c_n^* c_n a_n = \langle a \rangle \qquad \text{(A 1.29)}$$

gives the *average value* $\langle a \rangle$ of many measurements of the observable A on an ensemble of systems all in the initial state ϕ. If ϕ is not normalized the corresponding expression is

$$\langle a \rangle = \int \phi^* \mathrm{A}\phi \,\mathrm{dv} \Big/ \int \phi^* \phi \,\mathrm{d}v \qquad \text{(A 1.30)}$$

This important expression can be written in other ways. Often used in these is Dirac's notation, of which the symbol $\langle a \rangle$ is an example. The *bra* $\langle \phi |$ means ϕ^* and the *ket* $| \psi \rangle$ means ψ, and

$$\left.\begin{aligned} \langle \phi | \mathrm{A} | \psi \rangle &= \int \phi^* \mathrm{A}\psi \,\mathrm{d}v \\ \langle \phi | \phi \rangle &= \int \phi^* \phi \,\mathrm{d}v \end{aligned}\right\} \qquad \text{(A 1.31)}$$

The first of these is the number (or *matrix element*) obtained by integrating the product of ϕ^* with the wavefunction $\mathrm{A}\psi$ obtained by operating on ψ with A. The numerator of equation (A 1.30) is thus a particular form of this matrix element, i.e. $\langle \phi | \mathrm{A} | \phi \rangle$.

The practical value of the procedures (A 1.25)–(A 1.30) is that they enable solutions of Schrödinger's equation to be found, in complicated situations where direct solutions are often impossible, as *sums* of simpler, known solutions.

Although we are mainly concerned in this book with stationary states, we can note here the *time-dependent* form of Schrödinger's equation, such as is needed for dealing with, for example, radiation processes. This form is obtained by making the substitution

$$E \rightarrow -i\hbar \frac{\partial}{\partial t} \qquad \text{(A 1.32)}$$

in the time-independent equation. We use Ψ for the time-dependent wavefunction. For stationary states,

$$\frac{1}{\Psi}\frac{\partial \Psi}{\partial t} = \frac{iE}{\hbar} = \text{constant} \qquad \text{(A 1.33)}$$

with the solution

$$\Psi = \psi(x, y, z) \exp(iEt/\hbar) = \psi(x, y, z) \exp(i\omega t) \qquad \text{(A 1.34)}$$

where $\psi(x, y, z)$ is the space-dependent solution of the time-independent Schrödinger equation and ω is the *wave frequency*, with

$$E = \hbar\omega \qquad \text{(A 1.35)}$$

The frequency factor is of little interest for a single stationary state but is important for states which consist of sums of stationary states of different energies. For example

$$\Psi = \psi_1 \exp(iE_1 t/\hbar) + \psi_2 \exp(iE_2 t/\hbar) \qquad \text{(A 1.36)}$$

where $E_1 \neq E_2$. We rewrite this as

$$\Psi = \exp(iE_1 t/\hbar)\{\psi_1 + \psi_2 \exp[i(E_2 - E_1)t/\hbar]\} \qquad \text{(A 1.37)}$$

from which we can see that at $t = 0$ its spatial form is $\psi_1 + \psi_2$, whereas at $t = \pi\hbar/(E_2 - E_1)$ it is $\psi_1 - \psi_2$ (since $\exp(i\pi) = -1$), and at $2t$ it is $\psi_1 + \psi_2$ again. The wavefunction Ψ thus oscillates between the forms $\psi_1 + \psi_2$ and $\psi_1 - \psi_2$ with frequency $(E_2 - E_1)/2\pi\hbar$. This could represent various physical situations, such as an electron tunnelling from one atom to another and back again, in a molecule.

Equation (A 1.36) can be generalized to a larger, even infinite, sum of component terms with different energies. Such a time-dependent Ψ state does not have a sharp energy. This gives a meaning to $\Delta E\, \Delta t \approx h$ in equation (A 1.1). Thus Δt is the time within which a non-stationary state of energy spread ΔE changes significantly. It also explains the *tunnel effect*, i.e. the penetration of a narrow potential energy barrier by a particle whose kinetic energy is, on a classical basis, smaller than the energy height of the barrier. Quantum mechanics allows an extra energy ΔE ($\approx h/\Delta t$) to be 'borrowed' by the particle, while passing through the barrier, provided it is not retained for a time longer than Δt.

ATOMIC WAVEFUNCTIONS

In the hydrogen atom a single electron a distance r from the nucleus is attracted to the nuclear charge $+e$ by a potential $V = -e^2/r$. Schrödinger's equation is then

$$\nabla^2\psi + \left(E + \frac{e^2}{r}\right)\psi = 0 \qquad \text{(A 1.38)}$$

in atomic units. Since V has spherical symmetry about the nucleus, the simplest solutions of this equation, i.e. the s states of the atom, also have spherical symmetry. For these ∇^2 reduces to a purely radial operator

$$\nabla^2 = \frac{1}{r^2}\frac{\partial}{\partial r}\left(r^2\frac{\partial}{\partial r}\right) \qquad \text{(A 1.39)}$$

which gives the ground state solution

$$\psi_{1s} = \frac{1}{\pi^{1/2}}\exp(-r) \qquad \text{(A 1.40)}$$

in atomic units, at a (negative) energy level of 1 R_y. This is the 1s state of hydrogen. Fig. A 1.1 shows some representations of it. The probability density for the position of the electron $\psi^*\psi$, i.e. ψ^2, decreases exponentially as $\exp(-2r)$ with increasing distance. The function $4\pi r^2\psi^2$ gives the *radial density*, i.e. the probability that the electron is somewhere in the radial *shell* of radius r. It

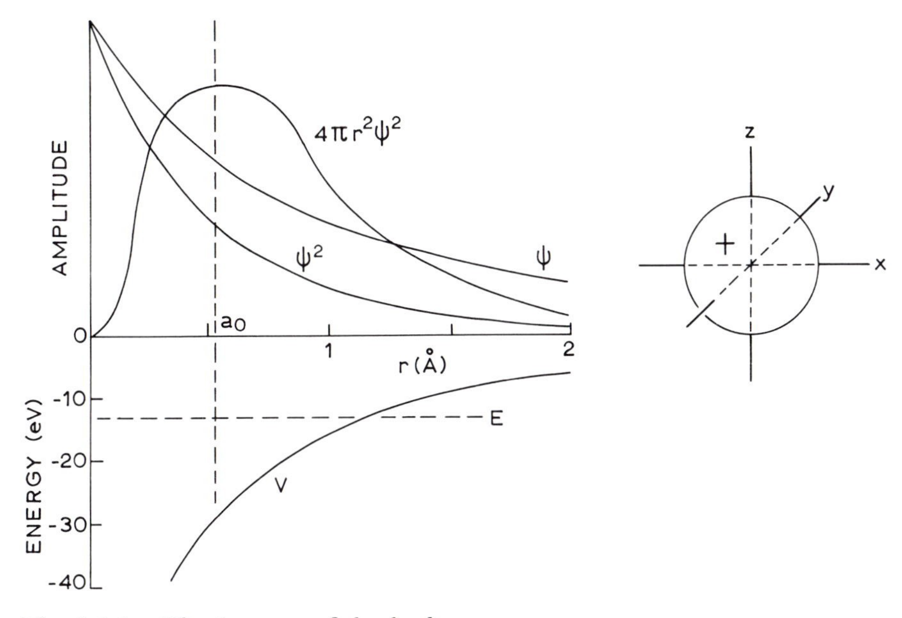

Fig. A 1.1 The 1s state of the hydrogen atom

has a maximum at the Bohr radius (equation (A 1.10)), which can roughly be regarded as the 'radius' of the hydrogen atom in its ground state. The state can be represented schematically by a spherical 'boundary surface', centred on the nucleus, in which ψ has the same sign (positive) everywhere; this is the 1s *orbital*.

The many other solutions of equation (A 1.38), which form the infinite sequences 2s, 3s, . . . , *n*s, . . . , 2p, 3p, . . . , *n*p, . . . , 3d, 4d, . . . , *n*d, . . . , 4f, 5f, . . . , *n*f, . . etc., where n is the *principal quantum number,* have $n - 1$ *nodes* in their atomic orbitals, i.e. surfaces on which $\psi = 0$ and across which ψ has opposite signs. The more nodes there are, the more sharply ψ has to bend to and fro to fit into them; and so the shorter is its wavelength and the higher is the energy. The simplest state higher than 1s is the spherically symmetrical 2s state,

$$\psi_{2s} = \frac{1}{4(2\pi)^{1/2}}(2 - r)\exp(-\tfrac{1}{2}r) \qquad \text{(A 1.41)}$$

This falls off with increasing r at only half the rate of equation (A 1.40). It also has one *spherical node*, since $\psi = 0$ when $r = 2$, situated at twice the Bohr radius. If ψ is positive inside this node, then it is negative outside it. The higher s states are more widely spread out and the *n*s orbital contains $n - 1$ concentric spherical nodes.

The less symmetrical solutions of equation (A 1.38) the p, d and f states, have nodal surfaces of other kinds. The simplest are the p states, with one *nodal plane* through the nucleus; thus *n*p has $n - 2$ spherical nodes in addition to one planar node. The simplest p states are the $n = 2$ group, $2p_x$, $2p_y$, $2p_z$, which have the form

$$\psi_{2p_x} = \frac{1}{4(2\pi)^{1/2}}x\exp(-\tfrac{1}{2}r) \qquad \text{(A 1.42)}$$

with similar expressions for ψ_{2p_y} and ψ_{2p_z}. The exponential decrease with r is the same as for 2s, but in place of the spherical node there is a planar node, since $\psi = 0$ at $x = 0$, through the nucleus and perpendicular to the x axis. Figure A 1.2 shows the boundary surfaces of the three 2p orbitals. By symmetry, these

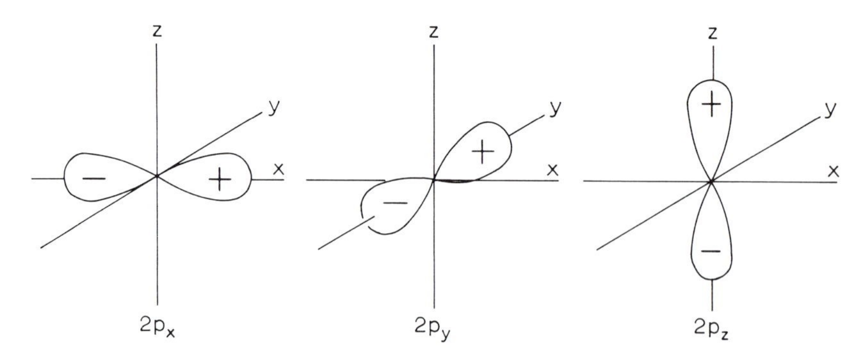

Fig. A 1.2 The boundary surfaces of the 2p orbitals

three states obviously all have the same energy – we say that they are *degenerate*. If superposed, they all add up to make a spherically symmetrical distribution. Thus, *any* three orthogonal x, y, z axes can be chosen in this sphere to represent the three component p states. In practice the choice is usually dictated by external circumstances, e.g. the positions of neighbouring atoms in a molecule or crystal, or an external magnetic field.

The d states have two nodal surfaces passing through the nucleus; thus nd has $n - 3$ spherical nodes. For each n ($\geqslant 3$) there are five independent d states. The 3d wavefunctions are

$$\psi_{3d} = \frac{1}{81}\left(\frac{2}{\pi}\right)^{1/2} r^2 f \exp\left(-\tfrac{1}{3}r\right) \quad \text{(A 1.43)}$$

where $f = f(x, y, z, r)$ has one of the following five forms

$$f = \frac{yz}{r^2}, \quad f = \frac{zx}{r^2}, \quad f = \frac{xy}{r^2}, \quad f = \frac{x^2 - y^2}{2r^2}, \quad f = \frac{3z^2 - r^2}{2\sqrt{3}(r^2)} \quad \text{(A 1.44)}$$

Their boundary surfaces are shown in Fig. A 1.3 (d_{z^2} refers to the orbital with the shape $3z^2 - r^2$). As with the p orbitals, the complete set of the five 3d orbitals adds up to a spherical distribution, and from this sphere orthogonal axes can be chosen in arbitrary orientation (or to suit external symmetries) for the x, y, z coordinates of the orbitals. Despite appearances, the orbitals are all equivalent. For example, d_{z^2} can be regarded as a normalized superposition of two similar to $d_{x^2-y^2}$, since $3z^2 - r^2 = (z^2 - x^2) + (z^2 - y^2)$. In the spherically symmetrical

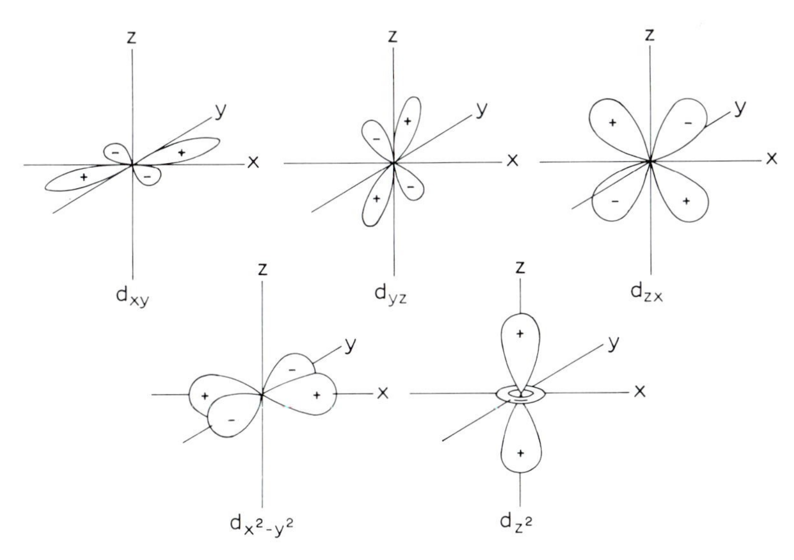

Fig. A 1.3 The boundary surfaces of the 3d orbitals

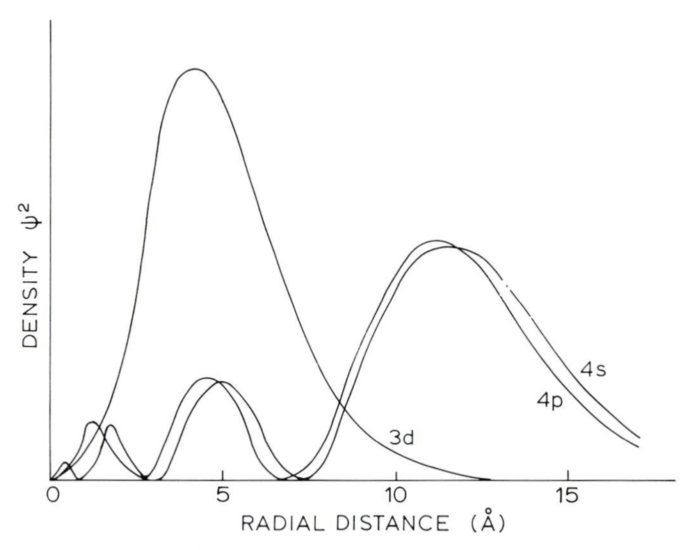

Fig. A 1.4 Densities of hydrogen atomic orbitals

field of the hydrogen nucleus these orbitals all have the same energy, but this degeneracy is removed in a less symmetrical environment such as a crystal lattice field. In such cases d_{xy}, d_{yz} and d_{zx} commonly remain as one degenerate group, often denoted as d_ε or t_{2g}; $d_{x^2-y^2}$ and d_{z^2} form a second degenerate group symbolized by d_γ or ε_g.

Because np has one spherical node fewer than ns, it does not spread quite so far from the nucleus. This effect is more marked for nd, which has two fewer spherical nodes. An example is shown in Fig. A 1.4 where, because of their importance for the first row of transition metals in the Periodic Table, the squared *radial components* of the 3d, 4p and 4s orbitals are compared.

The nodal surfaces are also important for the *angular momentum* of an electron in an orbital. Returning to Fig. A 1.2, let us take a circular path, at distance r from the nucleus, round the z axis in the p_x diagram. In one circuit, of path length $2\pi r$, we pass once through each of the positive and negative lobes; so this path is one wavelength, λ, long. Since $\lambda = h/p$ the momentum along the path is $p = h/2\pi r$, and the angular momentum rp is $h/2\pi$, i.e. $\hbar$. The same argument applies to p_y, so that both states have unit angular momentum about the z axis. By contrast, p_z has circular symmetry about this axis and thus no angular momentum about it.

Because of its spherical symmetry an s state has zero angular momentum. Going on to Fig. A 1.3, we see that d_{xy} and $d_{x^2-y^2}$ have two units of angular momentum about the z axis, d_{yz} and d_{zx} have one each, and d_{z^2} has none.

The electronic states of atoms are classified according to their nodes and angular momentum. Four quantum numbers are used: n, l, m_l and m_s. The *principal quantum number n* determines the *shell* of the state and the number $n - 1$ of nodes. The number of states in a shell is n^2, i.e. there is one state in $n = 1$ (1s), four in $n = 2$ (2s, $2p_x$, $2p_y$, $2p_z$) nine in $n = 3$, and so on. A unique feature of the hydrogen atom, which stems from its simple potential $V = -e^2/r$, is that all the states in the same shell have the same energy, given by

$$E_n = -\frac{1}{n^2}\frac{me^4}{2\hbar^2} \qquad \text{(A 1.45)}$$

This degeneracy does not extend to other atoms, however. The second quantum number l measures angular momentum and can take any integral value from 0 to $n - 1$. In the third shell, for example, there are states with $l = 0$, 1 and 2 (3s, 3p and 3d, respectively). The third quantum number m_l measures the component of angular momentum about some particular axis, e.g. that of an applied magnitude field. It may take any integral value from $+l$ to $-l$, including 0. We have already noted values 0, 1, 2, about one of the axes in Fig. A 1.3. The fourth quantum number m_s is not concerned with the orbital motion of the electron, but with its internal motion – i.e. with *spin*. It can be $+\frac{1}{2}$ or $-\frac{1}{2}$, according to the direction of the spin.

The state of an electron in an atom is thus completely specified by giving the values for each of the four quantum numbers of that state. By the *Pauli exclusion principle*, only one electron can occupy one state specified in this way; or two, with opposite spins, if only the three orbital quantum numbers are specified.

THE HYDROGEN MOLECULAR ION

The simplest molecular problem is that of the hydrogen molecular *ion*, H_2^+, i.e. a single electron moving in the field of two protons. As in Fig. A 1.5, let the two protons A and B be fixed a distance R apart and let the electron e be at distances r_A and r_B from them. The Schrödinger equation for this problem is obtained by substituting the potential

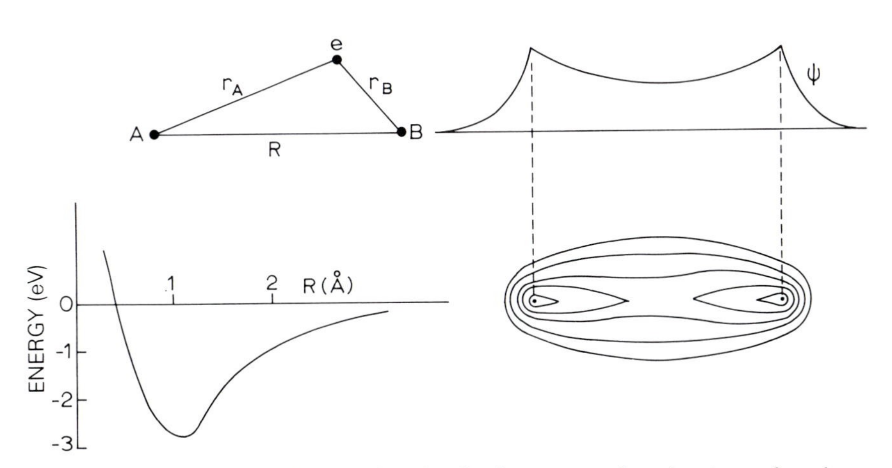

Fig. A 1.5 **Burrau's solution for the hydrogen molecular ion, showing the wavefunction, the total energy of the molecule and contours of probability density for the electron distribution**

$$V = -\left(\frac{e^2}{r_A} + \frac{e^2}{r_B}\right) \tag{A 1.46}$$

It can be solved exactly to give the orbital ψ, shown in Fig. A 1.5, and also the electron energy E for any given value of R. The total energy of the molecule is then obtained by adding to E the proton–proton repulsive energy e^2/R. The minimum which is thus obtained, 2·8 eV below the energy of a separate hydrogen atom and a bare proton, and the molecular spacing $R = 1{\cdot}06$ Å, are in agreement with observed values.

This bond is easily understood. When the electron lies in the space between the nuclei it is electrostatically attracted to both. Since in this region $r_A \simeq r_B \simeq \frac{1}{2}R$, these attractions are stronger than the electrostatic repulsion between the nuclei. However, the molecule cannot collapse to zero volume under these attractions because the momentum, and hence the kinetic energy of the electron, rise when the probability cloud shrinks, as follows from equation (A 1.1).

An assumption made in the above analysis was that the two nuclei remain fixed in position as the electron moves among them. This is the *Born–Oppenheimer approximation*, an essential step in most molecular and solid-state analyses, although there are certain problems for which it cannot be used. Mostly, however, it is justified because nuclei are heavy and slow compared with electrons. Usually an electron reacts so rapidly to the changing field of moving nuclei in molecules and solids that the nuclei can be considered to be instantaneously frozen in their positions when calculating the behaviour of the electron. One obvious case where this approximation cannot be made is in *radiation damage*, in which a displaced atom crashes its way violently through the solid, often moving at speeds greater than those of the valency electrons. It then travels as a positive ion, with some electrons stripped off and unable to keep up with it.

APPENDIX 2

Atoms and molecules

THE PROBLEM OF TWO ELECTRONS

In a system of two or more electrons, the electron–electron interactions make the exact solution of Schrödinger's equation virtually impossible. The problem already appears in the simplest example, the helium atom, in which two electrons move in the field of a nuclear charge $+2e$ and in each other's fields. The equation for this is

$$\nabla_1^2 \psi + \nabla_2^2 \psi + \left(E + \frac{2e^2}{r_1} + \frac{2e^2}{r_2} - \frac{e^2}{r_{12}} \right) \psi = 0 \tag{A 2.1}$$

where the subscripts 1 and 2 refer to the coordinates of electrons 1 and 2, and r_{12} is the distance between the electrons. We are now seeking the *joint* probability of finding one electron at one position and the other at another position; ψ is thus a formidable function of the coordinates of *both* electrons, i.e. $\psi = \psi(1, 2) = \psi(x_1, y_1, z_1, x_2, y_2, z_2)$, in contrast with the one-electron problems examined in Appendix 1. But there is another difficulty. To find ψ we must first know V; but to find the $V_{12} = e^2/r_{12}$ term in equation (A 2.1) we must first know ψ, since this determines r_{12} (probabilistically) for each given position of one electron. The most drastic way out of this vicious circle is the *independent electron* approximation, which simply deletes e^2/r_{12} from the equation. Each electron then moves only in the field of the nucleus and its probability distribution is independent of the presence of the other electron. It also follows that each electron then has its own one-electron wavefunction. The *many-body* ψ of equation (A 2.1) can now be represented as the product of two one-electron functions:

$$\psi(1, 2) = \psi_A(1)\psi_B(2) \tag{A 2.2}$$

where $\psi_A(1)$ represents electron 1 in quantum state A, and similarly for $\psi_B(2)$. With these changes equation (A 2.1) now splits into two separate one-electron Schrödinger equations, like equation (A 1.38) but with $2e^2$ in place of e^2, one equation each for $\psi(1)$ and $\psi(2)$, which can be solved in the same way as the hydrogen equation. Thus the 1s state of helium, on this approximation, is

$$\psi(1, 2) = c \exp(-2r_1) \exp(-2r_2) \tag{A 2.3}$$

with an energy E of -8 Ry. The observed value is $-5{\cdot}8$ Ry.

To improve the result we make a *first-order perturbation*, involving e^2/r_{12}, in which the above wavefunctions $\psi(1)$ and $\psi(2)$ are left unchanged. We can think of them as providing two overlapping clouds of electricity, and the perturbation as limited to calculating the electrostatic interaction energy of each piece of one cloud, e.g. that in volume dv_1 with each piece of the other, in dv_2. This *perturbation energy* is

$$\Delta E = C \int\int \frac{e^2}{r_{12}} [\exp(-2r_1)]^2 [\exp(-2r_2)]^2 \, dv_1 \, dv_2 \qquad \text{(A 2.4)}$$

where C is a normalizing factor. The result, $\Delta E = +2{\cdot}5$ Ry, improves the calculated E to $-5{\cdot}5$ Ry. Further improvements can then be made by estimating the *correlation energy* which results from the electrons tending to keep out of each other's way.

CORRELATION AND EXCHANGE

The obvious cause of correlation is the electrical repulsion e^2/r_{12} between electrons. This *charge correlation* leads to a lower e^2/r_{12} energy compared with the value calculated from the independent electron approximation in which no allowance is made for the electrons keeping out of each other's way. There is also a second and extremely important cause of correlation, of a purely quantum nature. This is *spin correlation*, also known as the *exchange effect*. Consider, as an example, a helium atom with one electron in 1s and the other in 2s, i.e. the atom is in the configuration $(1s)^1\,(2s)^1$. We put A = 1s and B = 2s in equation (A 2.2). We could, though, equally well have in place of equation (A 2.2)

$$\psi(2, 1) = \psi_A(2)\psi_B(1) \qquad \text{(A 2.5)}$$

in which the roles of electrons 1 and 2 are exchanged. Since the electrons are physically indistinguishable, there can be no real change if they exchange places, i.e. $\psi^2(1, 2) = \psi^2(2, 1)$. Hence $\psi(1, 2) = \pm\,\psi(2, 1)$. This requirement is satisfied if we use, in place of the two wavefunctions $\psi(1, 2)$ and $\psi(2, 1)$, the combinations

$$\psi_G = \psi_A(1)\psi_B(2) + \psi_A(2)\psi_B(1) \qquad \text{(A 2.6)}$$

$$\psi_u = \psi_A(1)\psi_B(2) - \psi_A(2)\psi_B(1) \qquad \text{(A 2.7)}$$

This is simply a mathematical way of recognizing that the distinctive labelling of electrons 1 and 2 is physically unwarranted. The *symmetrical* wavefunction ψ_G keeps its sign unchanged when the electrons are exchanged, whereas the *antisymmetrical* wavefunction ψ_u changes sign.

Because of this there is correlation in the positions of the electrons, *even when e^2/r_{12} is ignored*. The correlation is related to the *Pauli exclusion principle*. To include the internal spin motion in the total behaviour of an electron, the wavefunction is generalized to the total one

$$\psi_t = \psi_o \psi_s \tag{A 2.8}$$

where ψ_o is the orbital part that depends on x, y, z, as above, and ψ_s is an additional component that indicates the internal state of the electron, i.e. its spin direction. For an electron, the total wavefunction is always antisymmetric. Obviously, the exchange of two electrons with parallel spins can have no effect on the sign of ψ_s. The antisymmetry in this case must lie in ψ_o; that is, ψ_u in equation (A 2.7) is the orbital wavefunction for electrons with parallel spins. Similarly, when they have opposite spins, ψ_G is their orbital wavefunction. An alternative way to express this is to use only equation (A 2.7), *not* equation (A 2.6), for the electron pair, but to regard it as a statement of ψ_t, not ψ_o. The antisymmetry of ψ_t means that equation (A 2.7) is always valid for electrons on this interpretation. This form of equation (A 2.7) can be generalized to a system of N electrons, in which case it becomes a *Slater determinant*. The multi-electron total wavefunction $\psi(1, 2, \ldots, N)$ is then written as the product of N one-electron functions, $\psi(1)\,\psi(2)\ldots\psi(N)$, and the generalized equation (A 2.7) consists of a normalized sum, each term of which is one of these N one-electron ψ products prefixed by $(-1)^p$, where p is the number of exchanged electron pairs in that term, relative to the prototype term at the head of the sum. Thus equation (A 2.7) is an example of this for $N = 2$.

Returning to our original $\psi = \psi_o$ interpretation of equations (A 2.6) and (A 2.7), we have $\psi_u = 0$ when A = B. Since the spins are parallel in this case, $\psi_u = 0$ can be interpreted to mean that two electrons with parallel spins cannot occupy the same (orbital) state, which is a familiar statement of the Pauli principle. Even when the orbitals A and B are different, $\psi_u = 0$ when the electrons have the *same space coordinates* since in this case $\psi_A(1) = \psi_A(2)$ and $\psi_B(1) = \psi_B(2)$. Thus, two electrons having the same spin cannot occupy the same point in space.

This *spin correlation* or *exchange* effect has many consequences. It is the basis of the Fermi–Dirac distribution in the electron theory of metals, as well as other (e.g. astrophysical) applications. Another consequence can be seen in Fig. A 1.2. Because of the geometric orthogonality of the p orbitals, three electrons of parallel spins, occupying one orbital each by the Pauli principle, are fairly well separated from each other and so have a small electrostatic repulsive energy. A similar argument applies to the d states. Such states of spin correlation are thus energetically favoured, which is the basis of *Hund's rules* of atomic structure and also of *ferromagnetism*.

It also alters energy levels. Taking $\psi = \psi_o$, the perturbation energy ΔE due to the term e^2/r_{12} in equation (A 2.4) is

$$\begin{aligned}\Delta E &= C \iint \frac{e^2}{r_{12}} [\psi_A^2(1)\psi_B^2(2) + \psi_A^2(2)\psi_B^2(1) \\ &\qquad \pm 2\psi_A(1)\psi_B(2)\psi_A(2)\psi_B(1)]\, dv_1\, dv_2 \\ &= I_1 \pm I_2 \end{aligned} \tag{A 2.9}$$

The first two terms, which give I_1, are the *Coulomb energy*, i.e. the classical interaction energy of two charged electrical clouds of density ψ_A^2 and ψ_B^2. But the third term, the *exchange energy*

$$I_2 = C \int\int \frac{e^2}{r_{12}} [2\psi_A(1)\psi_B(2)\psi_A(2)\psi_B(1)] \, dv_1 \, dv_2 \qquad \text{(A 2.10)}$$

makes the total perturbation energy ΔE differ from the classical value. It produces *two* energy levels, $I_1 - I_2$ and $I_1 + I_2$. Usually $I_1 \approx I_2$ so that the effect can be large. In the particular case where $\psi_A = \psi_B$ (as in the ground state of helium) $I_1 - I_2 = 0$; and there is only one energy level, $\Delta E = I_1 + I_2$.

STATIC FIELD APPROXIMATIONS

The next improvement to the one-electron atomic theory is made by assuming that each of the electrons in an atom moves in a *static field* of the nucleus (charge $+ Ze$) and the other $Z - 1$ electrons. This static field $\phi(x, y, z)$ of the *average* electrostatic potential due to these other particles provides the potential

$$V = -e\phi \qquad \text{(A 2.11)}$$

in the one-electron Schrödinger equation. A one-electron solution is then calculated. This is repeated for every electron, and the multi-electron wavefunction is constructed as a Slater determinant of these one-electron functions.

The simplest static field theory, the *Thomas–Fermi* method, does not even go this far, but treats the electrons in the atom semi-classically as a (Fermi–Dirac) gas which behaves as an electrified fluid, the charge density ρ of which is related to the electrostatic field potential ϕ by the classical Poisson equation

$$\nabla^2\phi = -4\pi\rho \qquad \text{(A 2.12)}$$

The analysis of this leads to the Thomas–Fermi equation

$$\nabla^2\phi = A\phi^{3/2} \qquad \text{(A 2.13)}$$

with

$$A = (32\pi^2 e/3\hbar^3)(2me)^{3/2} \qquad \text{(A 2.14)}$$

the Thomas–Fermi coefficient, which can be solved to give the electron distribution. It gives a fair approximation for atoms of large Z and is a useful starting point for more accurate methods.

The most important static field method is that of the *self-consistent field*, due to Hartree, Fock and Slater. It is accurate to about 5% for the lighter atoms. The first step is to *guess* a set of Z one-electron wavefunctions, using the Thomas–Fermi solution together with any other information available. The better the initial guesses, the easier is the later work. Then a static V for the

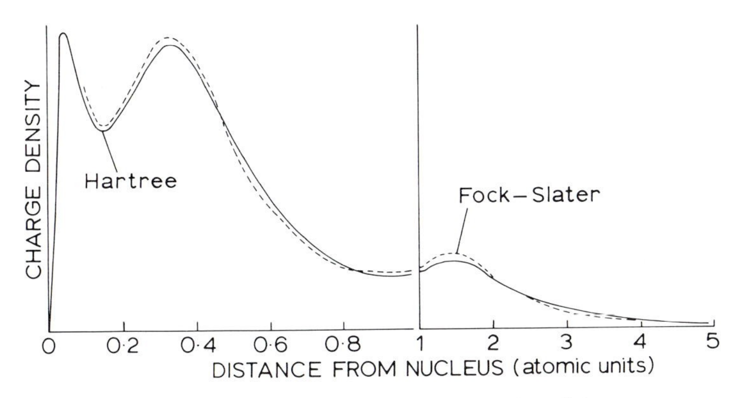

Fig. A 2.1 The charge distribution in Cl^- calculated by two self-consistent field methods

nucleus and $Z - 1$ electrons is calculated from the guessed wavefunctions; and Schrödinger's equation is solved to give the ψ of the remaining electron. This ψ is then used to give a better V, and the process is repeated to give the ψ of another electron. This is repeated for the various electrons until the V and ψ functions show no significant further improvements, at which stage they are taken to be the self-consistent solution to the problem. Heavy calculations are necessary, but many atoms and ions have now been described by this means.

Hartree did not take account of exchange. Fock adapted the method to antisymmetrical total wavefunctions ψ_t, but the analysis then became even more arduous until Slater introduced some labour-saving, approximate methods for taking account of exchange. A comparison is given in Fig. A 2.1 of the Hartree and Fock–Slater charge distributions for the Cl^- ion.

The guessing of wavefunctions can be done most effectively by preparing a *trial function*

$$\psi = c_1\psi_1 + c_2\psi_2 + \ldots \qquad \text{(A 2.15)}$$

where $\psi_1, \psi_2, \ldots$ are known functions, and $c_1, c_2, \ldots$ are *variable parameters*. The *method of variations* is then used to find those values of $c_1, c_2, \ldots$ which make the energy of the trial function a minimum. This gives a good ψ from which to start the self-consistent procedure.

THE PERIODIC TABLE

The multi-electron effects discussed above leave the s, p, d, f pattern of the atomic states intact, as would be expected from the independent electron approximation, but they alter the energy levels by removing some of the degeneracies of the hydrogen atomic levels. In particular, since nodes through the nucleus reduce the electron density near the nucleus, electrons in states of this kind are forced to spend more time further away, where they are to some extent *shielded* or *screened* from the nuclear attraction by other electrons in states nearer the nucleus. Thus, in any given n shell the energy levels of the states increase in the order ns, np, nd, nf.

This effect, together with the principle that electrons occupy states of lowest energy with no more than two (of opposite spins) in each, explains the Periodic Table (Fig. 1.1), in which a new row begins (at an alkali metal) whenever a new shell begins to be occupied by one electron going into the s state of that shell. The order of the energy levels explains the two *short periods*, in which the s and p states of $n = 2$ and 3 fill; the three *long periods* in which the filling of s and p states of the $(n + 1)$th shell is interrupted by the filling of the d states of the nth shell, in the *transition metals*; and the series of *rare earths* and *actinides* in which the filling of f states interrupts other fillings.

The chemical *valency*, on which the periodic structure of the Table is based, reflects the extent to which an electron in the outermost n shell is shielded from the nuclear attraction by the other electrons of the atom. The inert gases have zero valency because, in their outermost shells of completely filled s and p states, each electron is on average about as near to the nucleus as any of its companions in the same shell, and so is not densely screened by any of them. Similarly, the +1 valency of an alkali metal (an *electropositive* or *electron donor* element) stems from its lone $(n + 1)$s valency electron which is densely screened by electrons of the n shell beneath it, and so is weakly bound to the atom and can be easily removed by ionization. Equally, the −1 valency of a halogen (an *electronegative* or *electron acceptor* element) stems from the single hole in the otherwise full outer ns, np shell, which can provide a home for an added electron with a good participation in the attractive field of its nucleus. The formation of *ionic compounds* between electropositive and electronegative elements of course follows from this.

The weak *van der Waals* attraction between inert-gas atoms (and all other atoms) is a more subtle effect of screening. Although completely filled outer sp shells are stable, the movements of the electrons in these orbitals produce *fluctuations* of the electrical charge distribution. Thus, two fairly close neutral atoms become slightly attractive by correlating their fluctuations in such a way that the outer electrons of one surge a little towards the other just when the near side of this other, by its own fluctuations, happens to be slightly less screened than usual.

THE LCAO METHOD FOR MOLECULES

In a multi-electron molecule the electrons in the inner shells of an atom can be regarded to a good approximation as remaining in their free-atom orbitals, undisturbed by the presence of other atoms and uncombined with the orbitals of these atoms. The disturbed electrons, which make the chemical bond, are in the outermost, valence-electron shell. Even here, however, the orbitals partly retain their free-atom form. This has led to a useful method in theoretical chemistry in which it is assumed that the *molecular orbital* ψ_{AB} between two atoms A and B (*see* Fig. A 1.5) can be constructed as a *linear combination of atomic orbitals* (LCAO);

$$\psi_{AB} = c_A \psi_A + c_B \psi_B \tag{A 2.16}$$

These are the *free-atom* orbitals ψ_A and ψ_B, so the LCAO method has the advantage that no further solutions of Schrödinger's equation are needed once the free-atom wavefunctions are known.

For the ground state of H_2^+ we take the LCAO orbital as

$$\psi_{AB} = \psi_A + \psi_B \tag{A 2.17}$$

where ψ_A and ψ_B are the 1s atomic hydrogen functions (equation (A 1.40)) with $r = r_A$ and r_B respectively, in the notation of Fig. A 1.5. The Hamiltonian (equation (A 1.5)) is

$$H = -\nabla^2 - \frac{e^2}{r_A} - \frac{e^2}{r_B} + \frac{e^2}{R} \tag{A 2.18}$$

where the final term, the internuclear repulsion, is included for completeness. The mean energy E is calculated from equation (A 1.30), taking $\psi^* = \psi$ since the atomic orbitals are real functions (*see* equation (A 1.40)). Thus

$$E = \int (\psi_A + \psi_B) H (\psi_A + \psi_B)\, dv \Big/ \int (\psi_A + \psi_B)^2\, dv \tag{A 2.19}$$

We suppose that ψ_A and ψ_B are normalized (but not orthogonal, since they belong to *different* free-atom Hamiltonians), and also make use of symmetry, i.e.

$$\left.\begin{aligned} \int \psi_A^2\, dv &= \int \psi_B^2\, dv = 1 \\ \int \psi_A H \psi_A\, dv &= \int \psi_B H \psi_B\, dv = H_{AA} \\ \int \psi_A H \psi_B\, dv &= \int \psi_B H \psi_A\, dV = H_{AB} \end{aligned}\right\} \tag{A 2.20}$$

to obtain

$$E = \frac{H_{AA} + H_{AB}}{1 + S} \tag{A 2.21}$$

where S is the *overlap integral*

$$S = \int \psi_A \psi_B\, dv \tag{A 2.22}$$

Using equation (A 2.18) we deduce

$$E = E_0 + \frac{e^2}{R} - e^2 \left(\int \frac{\psi_A^2}{r_B}\, dv + \int \frac{\psi_A \psi_B}{r_B}\, dv \right) \Big/ (1 + S) \tag{A 2.23}$$

where $E_0 = -1$ Ry is the energy of the 1s ground state. Considered as a function of R, $E \rightarrow E_0$ as $R \rightarrow \infty$ and shows a minimum at $R = 1{\cdot}32$ Å, which gives a dissociation energy of 1·76 eV, about 60% of the observed value, 2·79 eV. This shows both the strength and the weakness of the LCAO method. The strength lies in the simplicity of the method, which avoids having to calculate new wavefunctions *ab initio* from Schrödinger's equation; the weakness lies in the inaccuracy of the result. More accurate calculations have been made, using modified free-atom wavefunctions which concentrate the electron distribution more strongly in the region between the nuclei.

The LCAO method achieves its semi-quantitative success through mathematical serendipity. The real cause of the molecular bond is the electrostatic distortion of the free-atom probability distribution by the presence of the other nucleus, corresponding to the attraction of the electron into the region between the nuclei, whereas in the LCAO method *undistorted* atomic orbitals are used. Fortunately, however, where the undistorted ψ_A and ψ_B functions overlap with the same sign, in the region between the nuclei, their sum $\psi_A + \psi_B$, when squared, $(\psi_A + \psi_B)^2$, represents a 'heaping up' of electron density in this region, as shown in Fig. A 2.2.

As indicated in Fig. A 2.2, instead of the *bonding* combination ψ_+, given by equation (A 2.17), we could equally well have combined the two atomic orbitals in the *antibonding* arrangement

$$\psi_- = \psi_A - \psi_B \qquad \text{(A 2.24)}$$

which, by a repetition of the above analysis, leads to an increase in energy when the atoms come together. Thus the overall result is that, as R is reduced from infinity, the initially degenerate energy level E_0 of the free atom splits into two levels, one higher and one lower, when the two atomic orbitals begin to overlap. In the antibonding superposition a *nodal plane* N is formed symmetrically between the nuclei, so that the electron avoids this region, whereas in the

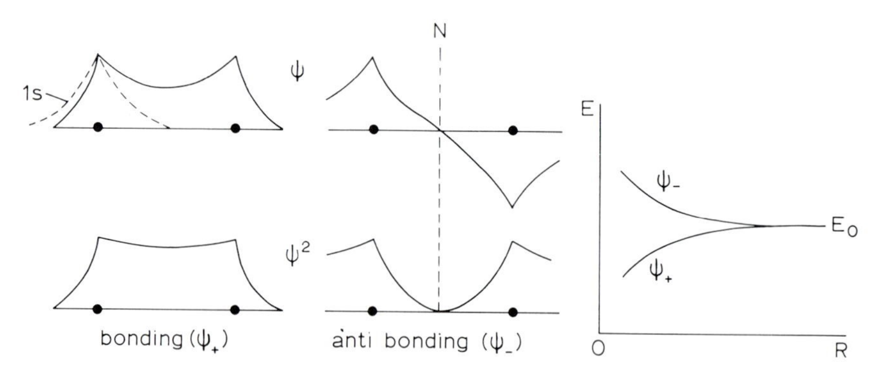

Fig. A 2.2 Bonding and antibonding molecular orbitals formed by the superposition of 1s atomic orbitals, showing the resulting splitting in the molecule of the degenerate free-atom energy level

bonding superposition there is no such node and the wavefunction extends, without change of sign, over the whole space between the nuclei.

The neutral hydrogen molecule H_2 raises the two-electron problem again. The wavefunction is represented as a symmetrized product of two one-electron functions, as in equations (A 2.6) and (A 2.7). Using the H_2^+ functions for these one-electron functions, we obtain for the molecular orbitals of H_2

$$\begin{aligned} \psi_{\pm} &= [\psi_A(1) \pm \psi_B(1)]\,[\psi_A(2) \pm \psi_B(2)] \\ &= \psi_A(1)\psi_A(2) + \psi_B(1)\psi_B(2) \pm [\psi_A(1)\psi_B(2) + \psi_A(2)\psi_B(1)] \end{aligned} \qquad \text{(A 2.25)}$$

in which the two electrons are labelled 1 and 2. The various terms in this expression have individual physical meanings. Thus $\psi_A(1)\psi_A(2)$ represents an *ionic* configuration in which both electrons are centred on nucleus A, whereas $\psi_A(1)\psi_B(2)$ represents a configuration in which electron 1 is on A and 2 is on B.

This indicates a weakness of the molecular orbital approach. It ignores charge correlation and so gives too much opportunity for ionic configurations in the molecule. A drastic but simple way of allowing for charge correlation is provided by the *valence-bond* method, which stems from the original *Heitler–London* (1927) theory of the hydrogen molecule. In this theory the two ionic terms are simply omitted, i.e. the ground state wavefunction is assumed to be

$$\psi_s = \psi_A(1)\psi_B(2) + \psi_A(2)\psi_B(1) \qquad \text{(A 2.26)}$$

This gives a binding energy of about two-thirds of the observed value. Again, the method can be improved to give more accurate values.

In general the molecular orbital method is better for molecules in which the atoms are very close, so that the outer wavefunctions overlap and spread extensively over the whole molecule; the valence-bond method is better for more separated atoms in which the electrons remain fairly *tightly bound* to their parent atoms in clearly recognizable atomic orbitals.

The two electrons in a bonding orbital have opposite spins. If a third H atom approaches the molecule, its electron may attempt to join them. If it exchanges with the electron of opposite spin, this leads to two electrons of parallel spin in the molecular orbital, giving an unstable molecule. If it exchanges with the one with the same spin as itself, the exchange effect gives a repulsion. Thus H_3 cannot form as a stable molecule. This effect, in molecular bonding generally, accounts for the *saturation* of the covalent bond and leads to the familiar representation of this bond as an *electron pair* with opposite spins.

For a *homopolar* molecule such as hydrogen it is of course necessary that $c_A^2 = c_B^2$ in the LCAO wavefunction, equation (A 2.16). But in the general *heteropolar* case, e.g. in the HF molecule, the electrons are concentrated preferentially on the more electronegative partner, which is represented by $c_A^2 \neq c_B^2$. There is then some *ionicity* in the bond, which is taken into account in the LCAO theories by including a suitably weighted ionic term, such as $\psi_A(1)\psi_A(2)$, in the wavefunction.

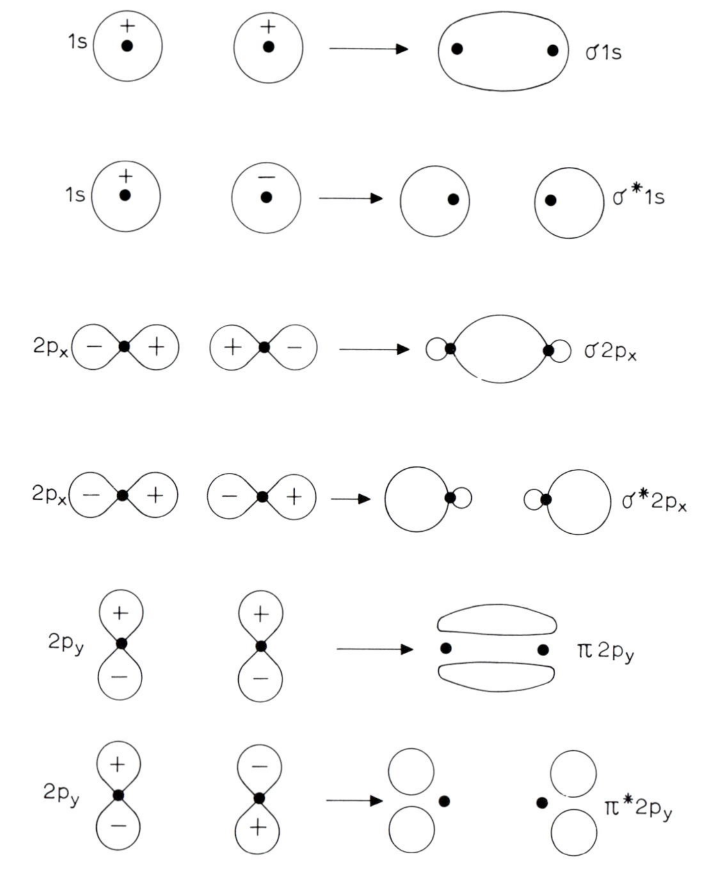

Fig. A 2.3 **The combination of atomic s and p orbitals to form various kinds of molecular orbital**

SYMMETRY OF OVERLAPPING ORBITALS

A requirement for a strong electron-pair bond, in the LCAO picture, is that the bonding atomic orbitals *overlap* considerably. Various effects related to the size and symmetry of orbitals are important in this. In the first place, in covalent bonding it is mainly the outer s, p electrons of atoms that participate because the outer regions of these atomic orbitals extend far from the nucleus and so are the first to overlap when atoms come together.

Since s orbitals are spherically symmetrical they can overlap equally well in any direction. The symmetry effect in this case reduces to the question of the signs of the overlapping atomic functions. As shown in Fig. A 2.3, when the signs are the same a bonding orbital (σ 1s) can form, but if they are opposite a nodal plane divides them and the molecular orbital is antibonding (σ^* 1s). The non-spherical symmetry of the p, d and f orbitals leads to further variations,

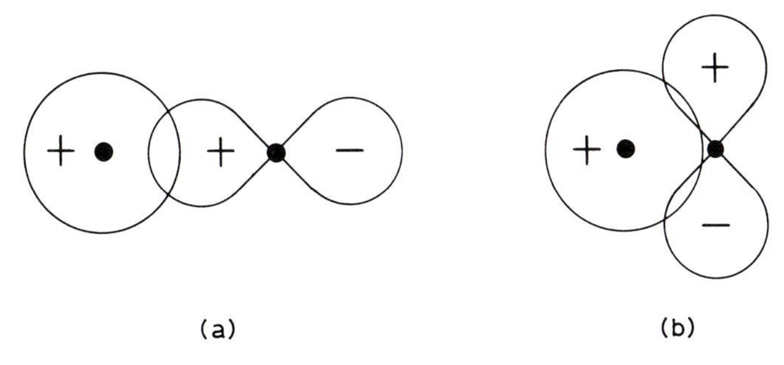

Fig. A 2.4 Overlaps between s and p orbitals: (a) bonding, (b) not bonding

some of which are shown in Fig. A 2.3. Thus, when two nuclei approach along the x axis, the p_x orbital which extends along this axis can give a strong bond ($\sigma 2p_x$) when the overlapping lobes have the same sign, whereas the p_y and p_z orbitals, because of their transverse orientations, overlap little and contribute only slightly to a bond in this molecular configuration (e.g. $\pi 2p_y$).

In the standard notation used in Fig. A 2.3, the symbol σ indicates that the molecular orbital has *circular* symmetry about the molecular (x) axis. Similarly, π means that the sign of the orbital goes through one cycle of alternation along a circuit round the molecular axis, so that the bonding π orbital looks like a meringue, with a nodal plane marking the cream layer (i.e. the plane $y = 0$ for the $\pi 2p_y$ orbital). Although not shown in the diagram, the symbol δ is used for those molecular orbitals formed from d states in which the sign alternates twice, round the circuit. The molecular symbols σ, π and δ obviously derive from the atomic ones, s, p, d. An asterisk on a molecular symbol denotes the antibonding form of the orbital.

Electron-pair bonds between different types of atomic orbitals can form, provided a certain condition is satisfied. The example in Fig. A 2.4(a) shows that a bonding overlap is possible when the p orbital is in the σ orientation relative to the s orbital; but in Fig. A 2.4(b), where the p orbital is in the π orientation, the bonding (+ +) part of the overlap is countered by the antibonding (+ −) part. The condition for bonding is thus that the orientations of the two participating atomic orbitals relative to the molecular axis shall be of the same type: both σ, both π or both δ.

PROMOTION AND HYBRIDIZATION

The ground state of the beryllium atom is $(1s)^2\,(2s)^2$, i.e. with two electrons in 1s and two in 2s. Since the 2s sub-shell is full and there is an energy gap between this and the empty 2p shell, it might be expected that beryllium should be chemically inert, whereas it is an active divalent atom. The reason is that this gap is only small, so that the energy cost incurred in *promoting* the beryllium atom to the *excited* state $(1s)^2\,(2s)^1\,(2p)^1$ can be more than recouped when the two singly occupied orbitals, 2s and 2p, form bonding molecular orbitals with other atoms. The same effect occurs in other atoms. An important example is *carbon*, the ground state of which is $(1s)^2\ (2s)^2\ (2p_x)^1\ (2p_y)^1$, whereas the element is quadrivalent. The promotion of one 2s electron to the $2p_z$ orbital gives four singly occupied orbitals, 2s, $2p_x$, $2p_y$ and $2p_z$, for bonding.

The four carbon bonds, which in substances such as methane (CH_4) and diamond stand out from the atom in the directions of the corners of a tetrahedron (which has the carbon atom at its centre), provide a familiar example of the *hybridization* effect. As we saw in Appendix 1, and have used in the LCAO method, a major feature of quantum mechanics is that wavefunctions are *linearly superposable*. As a result, there is no unique way of describing degenerate wave patterns. Although the s, p, d, f classification has many advantages, a complete atomic shell is spherically symmetric and it is thus a matter of *mere convenience*, for the n shell, which particular pattern of n^2 independent wavefunctions ψ we care to use to represent it. To this extent the classification into s, p, d, f types is purely conventional and any linear combination of these, with the same number of nodes, is equally valid as an atomic orbital. In fact, the combinations of *hybridized orbitals*

$$\left.\begin{aligned}\psi_A &= 2s + 2p_x + 2p_y + 2p_z \\ \psi_B &= 2s + 2p_x - 2p_y - 2p_z \\ \psi_C &= 2s - 2p_x + 2p_y - 2p_z \\ \psi_D &= 2s - 2p_x - 2p_y + 2p_z\end{aligned}\right\} \qquad \text{(A 2.27)}$$

suitably normalized, give four atomic orbitals which stand out strongly from the atom in the four tetrahedral directions and so are particularly suitable for describing the bonding of carbon with four symmetrically arranged partners. Because quantum mechanics (e.g. Schrödinger's equation) is a mathematically *linear* theory, this representation is essentially no more than a convenience. The final result of every *complete* calculation is always the same, whichever full set of atomic orbitals is used. Nevertheless, the calculation can be made much easier by working with orbitals which closely model the actual molecular structure being described.

APPENDIX 3

The free-electron theory of metals

THE FERMI DISTRIBUTION

In the free-electron theory the freedom of the electrons is expressed by the assumption that V = constant = 0 in the Schrödinger equation. The independent-electron assumption is also made, so that the equation reduces to a one-electron form, simply $\nabla^2\psi = -E\psi$ in atomic units. This equation has the *plane-wave* solutions

$$\psi_k(\boldsymbol{r}) = L^{-3/2} \exp(\mathrm{i}\boldsymbol{k}\cdot\boldsymbol{r}) \tag{A 3.1}$$

for a piece of metal shaped as a cube of side L (although the physically significant results of the theory are virtually the same for other shapes, apart from minor surface effects). Here

$$\boldsymbol{k}\cdot\boldsymbol{r} \equiv k_x x + k_y y + k_z z \tag{A 3.2}$$

$\boldsymbol{k}$ being the *wave-vector* ($k = \lambda/2\pi$). The energy is given by

$$E = \frac{\hbar^2 k^2}{2m} \tag{A 3.3}$$

Although perfectly free inside the metal, an electron is imprisoned within the cube by the sharp rise of V to some higher value outside the boundaries. Thus, ψ falls to zero beyond the surface of the metal and is limited to those **k** values which achieve this. To avoid prematurely complicating the theory with this surface effect, it is simpler to bypass the boundary problem altogether. This is done in the *Born–von Karman* cyclic assumption. In one dimension, for example, we simply imagine the metal as a wire of length L which we join, end to end, to form a single closed loop. The requirement then is that the wavefunction ψ of equation (A 3.1), should join smoothly on itself, having made one circuit of the loop, i.e. that $\psi(x + L) = \psi(x)$. In the three-dimensional case the corresponding assumption is that

$$\psi(x, y, z + L) = \psi(x, y + L, z) = \psi(x + L, y, z) = \psi(x, y, z) \tag{A 3.4}$$

From equation (A 3.1) this is possible only if

$$\exp(ik_xL) = \exp(ik_yL) = \exp(ik_zL) = 1 \tag{A 3.5}$$

i.e. if

$$k_x = \frac{2\pi}{L}n_x, \quad k_y = \frac{2\pi}{L}n_y, \quad k_z = \frac{2\pi}{L}n_z \tag{A 3.6}$$

where n_x, n_y and n_z are integers.

Each particular set n_x, n_y, n_z of numbers denotes a quantum state which can hold two electrons of opposite spins. At 0 K all the $\frac{1}{2}N$ states of lowest energy are filled by the N free electrons of the metal, up to a highest filled energy level, the *Fermi level* E_F. The corresponding *Fermi wave-vector* $\mathbf{k}_F$, regarded as a radius from the origin in *k-space*, defines the *Fermi sphere*, i.e. the free-electron version of the *Fermi surface* which is the boundary in k-space of the occupied states.

The integral values of n_x, n_y and n_z mark out in k-space the points of a simple cubic lattice, spaced at intervals $2\pi/L$ along the cube axes. The volume of k-space per point is $(2\pi/L)^3$, and thus the volume of the Fermi sphere, $\frac{4}{3}\pi k_F^3$, is $4\pi^3 N/L^3$. Writing $L^3 = \Omega$, we obtain

$$k_F = \left(\frac{3\pi^2 N}{\Omega}\right)^{1/3} \tag{A 3.7}$$

and the corresponding value of E_F is then obtained from equation (A 3.3).

Since k_F and E_F depend on N/Ω, i.e. on the *density* of free electrons in the metal, these quantities do not change when we join two such cubes to make a larger piece, or an arbitrary number to make an irregularly shaped piece. Typically, in metals k_F corresponds to an electron wavelength of the order of the atomic spacing of the crystal. Such electrons move at high speed, about 1% of the speed of light, so that, unlike a classical gas in which the thermal motion disappears on cooling towards 0 K, the particles of the free-electron gas are intensely active at 0 K. They have large kinetic energies, as indicated by the Fermi energy

$$E_F = \frac{\hbar^2}{2m}\left(\frac{3\pi^2 N}{\Omega}\right)^{2/3} \tag{A 3.8}$$

Typical values for E_F are given in Table A 3.1.

The *density of states* is the number of quantum states $N(E)\,\mathrm{d}E$ per unit volume of metal with energies in the range E to $E + \mathrm{d}E$. Expressing N as a function of E from equation (A 3.8) we have

Table A 3.1 The Fermi energy E_F (eV) in some metals

	Li	Na	K	Cu	Ag	Au	Mg	Al
E_F	4·7	3·2	2·1	7·1	5·5	5·6	7·1	11·6

$$N(E) = \frac{1}{4\pi^2}\left(\frac{2m}{\hbar^2}\right)^{3/2} E^{1/2} \tag{A 3.9}$$

giving a parabolic relation. The *average* kinetic energy of a free electron is then

$$\langle E\rangle = \frac{\int EN(E)\,\mathrm{d}E}{\int N(E)\,\mathrm{d}E} = \frac{\int E^{3/2}\,\mathrm{d}E}{\int E^{1/2}\,\mathrm{d}E} = \frac{3}{5}E_{\mathrm{F}} \tag{A 3.10}$$

THERMAL PROPERTIES OF A FREE-ELECTRON GAS

To acquire such energy by *thermal motion*, a classical gas would need to be heated to a temperature of about $E_{\mathrm{F}}/k_{\mathrm{B}} \simeq 10^4$–$10^5$ K. At ordinary temperatures the available thermal energy (e.g. $k_{\mathrm{B}}T \simeq 0{\cdot}025$ eV at room temperature) is far too small to affect any electrons except those few which happen to be in quantum states near the Fermi surface. The specific heat capacity of the electron gas, unlike that of a classical gas, is thus very small since the Fermi distribution is hardly modified by thermal motion (*see* Fig. A 3.1).

The thermal properties are described by *Fermi–Dirac* statistics. Consider those quantum states in the energy range E to $E + \mathrm{d}E$. It will be convenient here to distinguish between the two spin directions of such states. We now say that there are $2N(E)\,\mathrm{d}E$ states in this energy range, each of which can hold *one* electron. Suppose that there are n electrons in these m states. Because $\mathrm{d}E$ is infinitesimal, the probability of occupation of a state is constant over this energy range. The number of arrangements of the n electrons in the m states is then

$$w = \frac{m!}{n!\,(m-n)!} \tag{A 3.11}$$

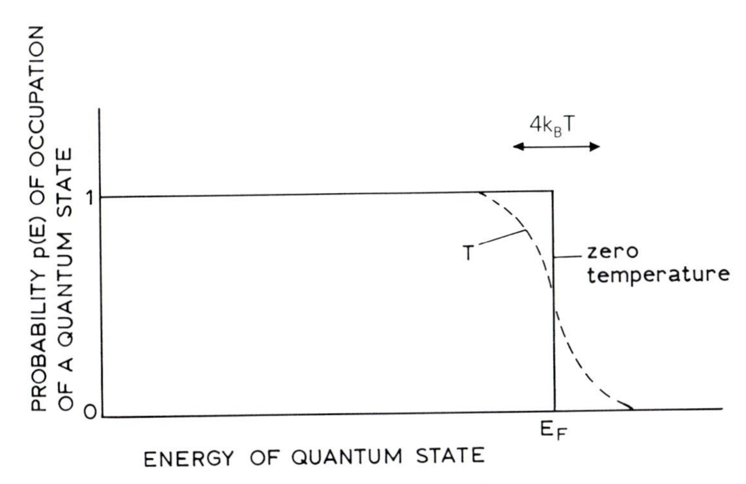

Fig. A 3.1 The Fermi distribution at 0 K and T K

Imagine that the whole energy band is divided up into these infinitesimal energy ranges, containing $m_1, m_2, m_3, \ldots$ states. The total number of arrangements of the entire distribution of electrons is

$$W = w_1 w_2 w_3 \ldots w_i \ldots \tag{A 3.12}$$

where

$$w_i = \frac{m_i!}{n_i!\,(m_i - n_i)!} \tag{A 3.13}$$

Using Stirling's approximation, $\ln x! \simeq x \ln x - x$, we obtain

$$\ln W = \sum_i [m_i \ln m_i - n_i \ln n_i - (m_i - n_i) \ln (m_i - n_i)] \tag{A 3.14}$$

The condition of equilibrium requires $\ln W$ to be a maximum, i.e. $\delta \ln W = 0$ for shifts of small numbers δn of particles from one energy range to another. For such shifts,

$$\sum_i \frac{\partial \ln W}{\partial n_i} \delta n_i = \sum_i \ln \left(\frac{m_i - n_i}{n_i} \right) \delta n_i \tag{A 3.15}$$

This maximization has to be subject to the conditions that both the total number of particles and the total energy remain constant:

$$\sum_i \delta n_i = 0 \tag{A 3.16}$$

$$\sum_i E_i \delta n_i = 0 \tag{A 3.17}$$

These two auxiliary conditions are combined with equation (A 3.15) in the *method of undetermined multipliers* by introducing two multipliers, λ for equation (A 3.16) and μ for equation (A 3.17), and setting the whole to zero:

$$\sum_i \left[\ln \left(\frac{m_i - n_i}{n_i} \right) - \lambda - \mu E_i \right] \delta n_i = 0 \tag{A 3.18}$$

This condition for the maximum is certainly satisfied when the total within the square brackets is zero, i.e. when

$$\frac{m_i - n_i}{n_i} = \exp(\lambda)\exp(\mu E_i) \tag{A 3.19}$$

and this rearranges into the *Fermi–Dirac distribution*

$$p(E) = \frac{1}{\exp[(E - E^*)/k_B T] + 1} \tag{A 3.20}$$

for the probability $p(E)$, i.e. n/m, of the occupation of states with energy E, where $E^* = -\lambda k_B T$ and where the relation

$$\mu = 1/k_B T \tag{A 3.21}$$

follows from the usual thermodynamical arguments.

When $E = E^*$ then $p(E) = \frac{1}{2}$. Thus E^* (called the *chemical potential*) is the energy level of those states with a 'half-chance' of being occupied. Hence $E^* = E_F$ at 0 K. At higher temperatures E^* falls to lower values, although the change is small at room temperature. For energy levels E below E^*, the negative value of $E - E^*$ makes the exponential factor small and thus $p(E) \rightarrow 1$. For higher levels where $E - E^*$ is positive, the exponent can become large and this section of the Fermi–Dirac distribution then resembles the *Maxwell–Boltzmann* distribution:

$$p(E) \rightarrow \exp[-(E - E^*)/k_B T] \tag{A 3.22}$$

when $E \gg E^*$, as seen in Fig. A 3.1. The narrowness of the transition region should be noted. Thus $p(E) \approx 0{\cdot}9$ at $E = E^* - 2k_B T$, and $p(E) \approx 0{\cdot}1$ at $E = E^* + 2k_B T$.

From the Fermi–Dirac distribution the energy and specific heat capacity of the gas can be evaluated at any temperature. When $k_B T \ll E_F$ the order of magnitude can be deduced readily. Only electrons with energies not more than about $k_B T$ below E_F can take up thermal energy in significant amounts. The number of these near-E_F electrons is of the order of $k_B T/E_F$ of the total number. Hence the specific heat capacity is smaller than that of a classical gas (i.e. $\frac{3}{2}Nk_B$) by a factor of the order of $k_B T/E_F$.

APPENDIX 4

Electrons in a periodic field

BLOCH WAVES

The *Bloch* (or *Floquet*) theorem is the cornerstone of the quantum theory of crystals. Not only does it determine the general form of the wavefunction for a crystal, but it does so by an argument of great generality, requiring no more than the periodicity of the field of potential in a perfect crystal, which is applicable to all crystalline matter.

Consider a one-dimensional 'crystal' as in Appendix 3, with a cyclic boundary condition, i.e. a closed ring of N ($\gg 1$) identical atoms of equal spacing a. Let x measure position round the ring from some arbitrary origin. Each atom exerts the same potential on an electron, i.e. if the potential is $V(x)$ when the electron is at x and $V(x + a)$ at $x + a$, and so on, then

$$V(x) = V(x + a) = V(x + 2a) = \dots \quad \text{(A 4.1)}$$

so that $V(x)$ in the one-dimensional, one-electron Schrödinger equation has the periodicity of the one-dimensional lattice.

We would then naturally expect the solutions of this equation also to have the periodicity of the lattice, i.e. to be functions of the form

$$u(x + a) = u(x) \quad \text{(A 4.2)}$$

which repeat identically at each lattice site. This is too restrictive, however. The translational symmetry merely requires that

$$\psi(x + a) = C\psi(x) \quad \text{(A 4.3)}$$

where C is a constant to be determined. With this, each increase of x by the amount a multiplies the wavefunction by C. Thus, for N such increases

$$\psi(x + Na) = C^N\psi(x) \quad \text{(A 4.4)}$$

However, this completes one circuit of the ring and brings us back to the starting point; therefore

$$\psi(x + Na) = \psi(x), \qquad \text{i.e.} \quad C^N = 1 \quad \text{(A 4.5)}$$

so that C is one of the N roots of unity:

$$C = \exp(2\pi i n/N), \quad n = 0, 1, 2, \ldots, N-1 \tag{A 4.6}$$

We define

$$k = 2\pi n/Na \tag{A 4.7}$$

so that

$$C = \exp(ika) \tag{A 4.8}$$

and thus find, using equation (A 4.2), that equation (A 4.3) is satisfied by

$$\psi(x) = u(x)\exp(ikx) \tag{A 4.9}$$

which follows from

$$\begin{aligned}\psi(x+a) &= u(x+a)\exp[ik(x+a)] \\ &= u(x)\exp(ikx)\exp(ika) = C\psi(x)\end{aligned} \tag{A 4.10}$$

The *Bloch wavefunction*, equation (A 4.9), can be generalized to three dimensions:

$$\psi(\boldsymbol{r}) = u(\boldsymbol{r})\exp(i\boldsymbol{k}\cdot\boldsymbol{r}) \tag{A 4.11}$$

where $\boldsymbol{r}$ is the vectorial position of the electron and $\boldsymbol{k}$ is the wave-vector, as in equation (A 3.2). When u = constant, the Bloch function reduces to the *plane wave* solution (A 3.1). Thus $u(\boldsymbol{r})$ reflects the extent to which the periodic field $V(\boldsymbol{r})$ deviates from the free-electron field V = constant. From molecular theory, we expect $V(\boldsymbol{r})$ to approximate to the single-atom potential (e.g. Hartree–Fock) of the atom nearest the point $\boldsymbol{r}$; and $u(\boldsymbol{r})$ similarly to approximate to an atomic orbital of that atom (*see* Fig. 1.3).

The Bloch wave, although modulated, is otherwise the same as a free-electron wave. It extends throughout the whole crystal and, in its time-dependent form (*see* equation (A 1.34))

$$\Psi(\boldsymbol{r}, t) = u(\boldsymbol{r})\exp[i(\boldsymbol{k}\cdot\boldsymbol{r} - Et/\hbar)] \tag{A 4.12}$$

it represents an electron in a stationary state moving with uniform velocity through the crystal. The atomic fields $V(\boldsymbol{r})$ do not deflect the electron from its state of uniform motion because the wavelets 'scattered' from each atom superpose coherently, in this perfectly periodic structure, and so reconstitute the same undeflected total wave. All this interaction of the electron with the field of the perfect crystal has already been fully taken into account in the Schrödinger equation, which leads to the Bloch wave as the solution for the perfect lattice.

MOTION OF AN ELECTRON

For a more detailed discussion of the motion of an electron, equation (A 4.12) is not the ideal form of wavefunction to use because, having a single value of E, it represents an electron with a sharp value of momentum and thus, by the Heisenberg principle (equation (A 1.1)), with a uniform spread across the entire crystal. It is better to use a wavefunction Ψ which represents a more localized electron. By the Heisenberg principle this means a wavefunction with a spread of $\boldsymbol{k}$ values around the mean. Such a function is constructed by adding together a number of functions of the type (A 4.12), but with slightly different $\boldsymbol{k}$ values and with a weighting factor attached to each, so that the whole sum represents a *wave packet* (e.g. a gaussian distribution) which gives localized 'humps' both in k-space and real space for the position of the electron.

The theory of the motion of a wave packet goes back to de Broglie's pioneering contributions to quantum mechanics. The velocity $\mathbf{v}$ of the centre of a wave packet, i.e. the *group velocity* of the waves, is in the present notation

$$\mathbf{v} = \frac{\mathrm{d}\omega}{\mathrm{d}\boldsymbol{k}} = \frac{1}{\hbar}\frac{\mathrm{d}E}{\mathrm{d}\boldsymbol{k}} \qquad \text{(A 4.13)}$$

where ω is the wave frequency (equation (A 1.35)). For a free electron, for which $E = (\hbar^2/2m)\,k^2$ and $\boldsymbol{p} = \hbar\boldsymbol{k}$, equation (A 4.13) reduces simply to

$$\mathbf{v} = \boldsymbol{p}/m \qquad \text{(A 4.14)}$$

the classical relation between velocity, momentum and mass. In recognition of this, $\hbar\boldsymbol{k}$ is termed the *crystal momentum*, although in a periodic lattice field it is not a true momentum, as we shall see. Suppose that an external electric field $\mathscr{E}$ is applied along the x axis, and consider components of motion along this axis. The work δE done by this field on the electron in time δt is $\delta E = e\mathscr{E}v\,\delta t$. However, $\delta E = (\mathrm{d}E/\mathrm{d}k)\,\delta k = \hbar v\,\delta k$. Equating these gives

$$\hbar\frac{\mathrm{d}k}{\mathrm{d}t} = e\mathscr{E} \qquad \text{(A 4.15)}$$

Thus, in the crystal, the external force $e\mathscr{E}$ on the electron is equal to $\hbar(\mathrm{d}k/\mathrm{d}t)$, whereas for a free electron it is $m(\mathrm{d}v/\mathrm{d}t)$.

The acceleration of the electron is given by

$$\frac{\mathrm{d}v}{\mathrm{d}t} = \frac{1}{\hbar}\frac{\mathrm{d}^2E}{\mathrm{d}k\,\mathrm{d}t} = \frac{1}{\hbar}\left(\frac{\mathrm{d}^2E}{\mathrm{d}k^2}\right)\frac{\mathrm{d}k}{\mathrm{d}t} = \frac{e\mathscr{E}}{\hbar^2}\left(\frac{\mathrm{d}^2E}{\mathrm{d}k^2}\right) \qquad \text{(A 4.16)}$$

Since for a free electron $\mathrm{d}v/\mathrm{d}t = e\mathscr{E}/m$, this expression has a similar form if an *effective mass* m^* is defined:

$$m^* = \frac{\hbar^2}{(\mathrm{d}^2E/\mathrm{d}k^2)} \qquad \text{(A 4.17)}$$

This m^* is, however, not a constant, as we shall see. It depends on direction through the crystal and so should be generalized to a *mass tensor*,

$$m_{ij}^* = \frac{\hbar^2}{(d^2 E/dk_i \, dk_j)} \qquad \text{(A 4.18)}$$

but even in a single direction its value depends on k and becomes *negative* in certain ranges, i.e. where the curvature of E as a function of k becomes negative.

BANDS

All these deviations from the behaviour of free electrons arise from the periodicity of the lattice field in which the crystal electron moves. They are implicit in the Bloch function, which is a sequence of identical u functions, one at each lattice site, each rotated in the complex $\exp(ikx)$ plane by the same standard increment, relative to its predecessor, for a given k. The rotation is zero when $k = 0$, which is the ground Bloch state in the limit $Na \rightarrow \infty$, and the bending of the wavefunction is then the least possible, for a given shape of u. The energy of this Bloch state is at its lowest if the u function is s-like, as in Fig. 1.3. When k is increased, the Bloch wavelength $\lambda = 2\pi/k$ decreases and the energy increases in consequence. A limit is reached, however, when $k = \pm \pi/a$, because the misorientation of u from one site to the next, due to the rotation, is then as large as it can be (i.e. π) so that if $\psi(0) = +u$ then $\psi(a) = -u$. Further increase of k beyond π/a simply starts to bring $\psi(a)$ back towards the alignment of $\psi(0)$, as the angle of rotation increases beyond π towards 2π, so that the misorientation is here diminished.

Thus, in terms of misorientation of the u function between neighbouring sites, the sequence ends at $k = \pm \pi/a$ and thereafter merely repeats itself. Physically, it is misorientation that matters since this concerns the relation of the u functions to one another at neighbouring sites. There is thus an upper limit to the energy of a Bloch function, for a given s-like lattice factor u, which is reached when $k = \pm \pi/a$. All the energy levels of the various stationary states for this particular u fall within a *band*, between the lower limit at $k = 0$ and the upper limit at $k = \pm \pi/a$. A state which might be thought to have a *shorter* wavelength than the corresponding limiting value, $\lambda = 2a$, is in fact simply a wave going in the opposite direction with a *longer* wavelength. Furthermore, $k = 2\pi/a$ gives $\exp(ika) = \exp(2\pi i) = +1$, so that $\exp(ikx)$ in this case merely repeats itself identically whenever x is increased by a. This $\exp(ikx)$ is then a *periodic* function of x and so can be absorbed within the periodic function $u(x)$. Thus, for any arbitrary value of k it is always possible to add or subtract $2\pi/a$ without changing the form of the Bloch function; and any general k can always be reduced by this means to a value lying in the range $\pm \pi/a$.

For a simple treatment of the band, in the Schrödinger equation

$$E\psi = -\frac{\hbar^2}{2m}\frac{d^2\psi}{dx^2} + V\psi \qquad \text{(A 4.19)}$$

we can express $d^2\psi/dx^2$ approximately as a finite difference between the values of ψ at the sites $x + a$, x and $x - a$,

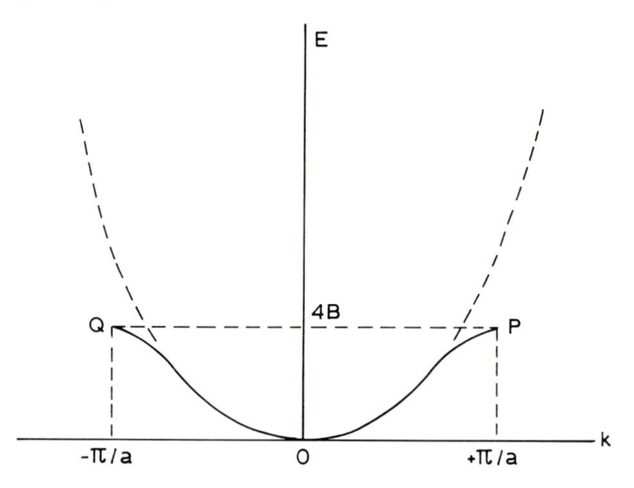

Fig. A 4.1 An energy band for a Bloch function

$$\frac{d^2\psi}{dx^2} \propto \psi(x+a) + \psi(x-a) - 2\psi(x) \qquad \text{(A 4.20)}$$

thus obtaining

$$E\psi(x) = A\psi(x) - B[\psi(x+a) + \psi(x-a)] \qquad \text{(A 4.21)}$$

and so

$$E = A - B\frac{\psi(x+a) + \psi(x-a)}{\psi(x)}$$

$$= A - B[\exp(\mathrm{i}ka) + \exp(-\mathrm{i}ka)] = A - 2B\cos ka \qquad \text{(A 4.22)}$$

This gives the E, k relation shown in Fig. A 4.1. Further calculations show that the bandwidth $4B$ diminishes as the deviation from the free-electron field increases, i.e. as the amplitude of the periodic $V(x)$ increases. In the limit of V functions like deep wells separated by high energy walls, so that an electron can tunnel from one site to the next only infrequently, $4B$ narrows towards zero. In this limit the energy band contracts into a single degenerate energy level as the atoms become isolated from one another.

Near the bottom of the band the E, k relation resembles the free-electron parabola,

$$E = \frac{\hbar^2 k^2}{2m^*} \qquad \text{(A 4.23)}$$

but with an effective mass that depends on the bandwidth. In the upper half of the band the curvature d^2E/dk^2 is negative, so the electron behaves as if it had negative mass (or, equivalently, positive charge) in this region. The reason is that, as k approaches the band limits $\pm\pi/a$, *Bragg reflection* plays an increasingly significant role in the motion of the electron.

This is also seen in the velocity. From equations (A 4.13) and (A 4.22) we have

$$\hbar v = 2aB \sin ka \qquad \text{(A 4.24)}$$

which is zero at both $k = 0$ and $k = \pm\pi/a$, and has a maximum at $k = \pm\pi/2a$. The velocity is zero at the top of the band (which shows that crystal momentum is not ordinary momentum) because at this critical value of k, waves travelling in opposite directions have equal influence, and in this limit the Bloch wave is converted from a *travelling* to a *standing* wave:

$$\psi = u(x)[\exp(ikx) + \exp(-ikx)] = 2u(x)\cos kx \qquad \text{(A 4.25)}$$

which is directly related to the Bragg condition for diffraction, since the *Bragg law*

$$\lambda = 2d \sin\theta \qquad \text{(A 4.26)}$$

then reduces to $k = \pi/a$, with $\lambda = 2\pi/k$, $d = a$ and $\theta = \frac{1}{2}\pi$. For this critical value of k successive wavelets reflected off different atoms interfere constructively to make a backward-going wave which is as strong as the forward one. The superposition of these two waves gives the standing wave. Physically, the electron shuttles forwards and backwards so that its mean velocity in either direction is zero.

To calculate the *number of quantum states* in an energy band, we note that since each Bloch wave must join smoothly to itself round the ring, of length Na, the wavelength must satisfy $n\lambda = Na$ where n is an integer. Hence $k(= \pm 2\pi/\lambda) = \pm 2\pi n/Na$, where the $\pm$ signs indicate the two directions for k around the ring. The number of quantum states, up to the limiting k $(= \pm\pi/a)$, which corresponds to $n = \frac{1}{2}N$, is thus $2n$, i.e. equal to the number N of crystal sites in the ring.

THE RECIPROCAL LATTICE

The generalization of the above to three dimensions leads to the concept of *Brillouin zones* in *k-space*. The *reciprocal lattice* of crystallography is useful here, because k-space is reciprocal space, and because the reciprocal lattice provides a natural way to express Bragg's law, which governs the form of the Brillouin zones. Consider a crystal in which the atomic positions mark out a *Bravais lattice*, i.e. are located at the points

$$\mathbf{l} = l\mathbf{a} + m\mathbf{b} + n\mathbf{c} \qquad \text{(A 4.27)}$$

where l, m and n are integers, and $\mathbf{a}$, $\mathbf{b}$ and $\mathbf{c}$ are vectors of the basic crystal cell. Imagine that superposed on this lattice is a plane wave, e.g. of X-rays or electrons, $\exp(\mathrm{i}\boldsymbol{k}\cdot\boldsymbol{r})$, of some arbitrary wavelength and direction of propagation. Then for certain special values of $\boldsymbol{k}$, which we shall denote by $\boldsymbol{g}$, the periodicity of the wave coincides with that of the crystal in this particular direction. The set of all such special $\boldsymbol{g}$ values of $\boldsymbol{k}$ is the *reciprocal lattice* of this Bravais lattice. The condition for the coincidence of the two periodicities,

$$\exp[\mathrm{i}\boldsymbol{g}\cdot(\boldsymbol{l}+\boldsymbol{r})] = \exp(\mathrm{i}\boldsymbol{g}\cdot\boldsymbol{r}) \tag{A 4.28}$$

is equivalent to

$$\exp(\mathrm{i}\boldsymbol{g}\cdot\boldsymbol{l}) = 1 \tag{A 4.29}$$

from which the set of reciprocal lattice vectors $\boldsymbol{g}$ is defined in terms of the vectors $\boldsymbol{l}$ of the real lattice. Two useful properties of the reciprocal lattice are that the vector $\boldsymbol{g}(hkl)$ from the origin to the reciprocal lattice point with coordinates hkl is normal to the planes in the real lattice with Miller indices (hkl); and the length g of this reciprocal lattice vector is inversely related to the spacing d of these planes:

$$g = 2\pi n/d \tag{A 4.30}$$

where n is an integer.

Consider two plane waves, $\exp(\mathrm{i}\boldsymbol{k}'\cdot\boldsymbol{r})$ and $\exp(\mathrm{i}\boldsymbol{k}\cdot\boldsymbol{r})$, of the same wavelength, i.e.

$$|\boldsymbol{k}'| = |\boldsymbol{k}| \tag{A 4.31}$$

If

$$\boldsymbol{k}' = \boldsymbol{k}+\boldsymbol{g} \tag{A 4.32}$$

then, using the Bloch theorem (equation (A 4.11)),

$$\begin{aligned}\psi_{\boldsymbol{k}'}(\boldsymbol{r}+\boldsymbol{l}) &= \exp(\mathrm{i}\boldsymbol{k}'\cdot\boldsymbol{l})\,\psi_{\boldsymbol{k}'}(\boldsymbol{r})\\ &= \exp[\mathrm{i}(\boldsymbol{k}+\boldsymbol{g})\cdot\boldsymbol{l}]\,\psi_{\boldsymbol{k}'}(\boldsymbol{r})\\ &= \exp(\mathrm{i}\boldsymbol{g}\cdot\boldsymbol{l})\exp(\mathrm{i}\boldsymbol{k}\cdot\boldsymbol{l})\,\psi_{\boldsymbol{k}'}(\boldsymbol{r})\\ &= \exp(\mathrm{i}\boldsymbol{k}\cdot\boldsymbol{l})\,\psi_{\boldsymbol{k}'}(\boldsymbol{r})\end{aligned} \tag{A 4.33}$$

so that the state $\psi_{\boldsymbol{k}'}$ behaves as if it has the wave-vector $\boldsymbol{k}$, in Bloch's theorem. Physically, the phases of the two waves keep in step through the crystal, and the crystal is thereby unable to recognize any distinction between them. This is of course the condition for Bragg reflection. Thus, referring to Fig. A 4.2, where equations (A 4.31) and (A 4.32) apply, we have $g = 2k\sin\theta$; and hence, from equation (A 4.30) and the relation $k = 2\pi/\lambda$, we obtain Bragg's law, equation (A 4.26).

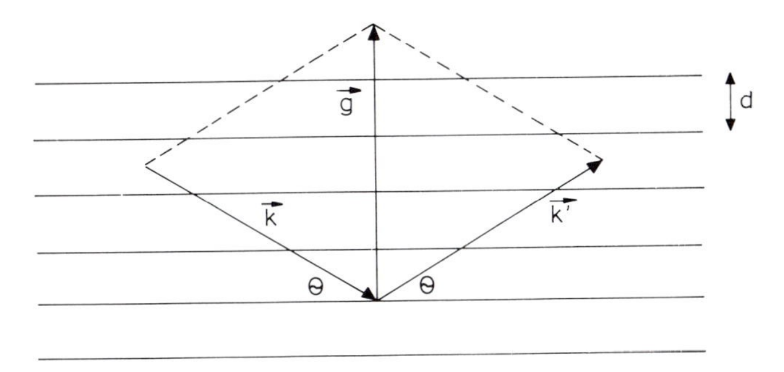

Fig. A 4.2 **Bragg reflection**

BRILLOUIN ZONES

For an energy band in one dimension we saw that it was sufficient to use k values lying within the range $\pm \pi/a$. We do the same in three dimensions by taking advantage of equations (A 4.32) and (A 4.33) to use a *reduced representation* in which the smallest possible k values are used, i.e. those that lie nearest the origin of the reciprocal lattice. To find in this representation the Bragg reflection conditions equivalent to $k = \pm \pi/a$ in one dimension, we construct in the reciprocal lattice a *Wigner–Seitz* cell by drawing connecting lines from the origin to nearby lattice sites and then bisecting each such line by a plane. The Wigner–Seitz cell is the volume inside all these planes. The boundary of this cell defines the condition for Bragg reflection, as shown in Fig. A 4.3, and because of this the Wigner–Seitz cell in reciprocal space is the *Brillouin zone* of the crystal represented by this particular reciprocal lattice.

To count the number of states in a Brillouin zone we note first that the definition of $\boldsymbol{k}$ in the Bloch function (equations (A 4.7), (A 4.9) and (A 4.11)) depends only on the number of crystal sites, *not* on the amplitude of V. Hence, for this purpose we can assume the limiting case, V = constant and u = constant,

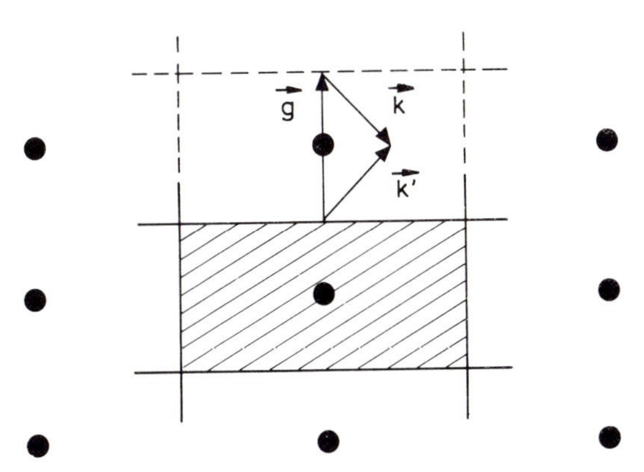

Fig. A 4.3 **A Brillouin zone constructed as the Wigner–Seitz cell of a reciprocal lattice**

and so use the results of free-electron theory. In particular, from Appendix 3 we have a volume $8\pi^3/\Omega$ of k-space per quantum state for a crystal of volume Ω. From Fig. A 4.3 the volume of the Brillouin zone is clearly equal to the volume of a unit cell of the reciprocal lattice, which is known from standard crystallographic theory to be $8\pi^3/\omega$, where ω is the volume of the unit cell of the Bravais lattice, i.e. $N\omega = \Omega$. The number of quantum states in the Brillouin zone is thus $(8\pi^3 N/\Omega)/(8\pi^3/\Omega)$, i.e. N. Thus, as in our earlier one-dimensional case, *the number of quantum states in the Brillouin zone is equal to the number of unit cells in the crystal*.

The zone described above, which represents the most widely spaced reflecting planes in the crystal, is the *first Brillouin zone*, i.e. the zone bounded by the set of points in k-space that can be reached from the origin without crossing any bisecting plane. There are of course other reflecting planes in a crystal, of closer spacing, and corresponding to these there are other bisecting planes and *higher zones*. The *nth Brillouin zone* is the zone bounded by the set of points that can be reached from the origin by crossing $n - 1$ bisecting planes. Examples are shown in Fig. A 4.4 of the first few zones for the simple square lattice.

The higher zones are seen to be fragmented pieces separated from each other by the lower zones inside them. But we recall from equations (A 4.32) and (A 4.33) that the addition or subtraction of a reciprocal lattice vector makes no physical difference to a wavefunction in the crystal. We can thus translate the separate pieces of the higher zones into the first zone, by suitably applying the reciprocal vectors, $\pm g_x$ and $\pm g_y$ in Fig. A 4.4. As we see from Fig. A 4.5, each translated higher zone exactly fills the first zone. This is a general result and means that each zone occupies the same volume of k-space. It further follows that, not just the first zone, but every zone contains N quantum states for a crystal of N unit cells, each of which can of course hold two electrons of opposite spins.

Next, we consider lattices *with a basis*, i.e. crystals in which there is more than one atom per lattice point. In pure elements there are two examples of these:

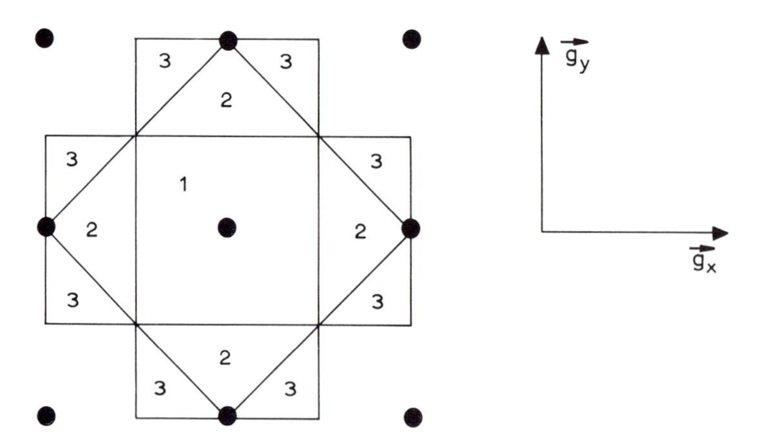

Fig. A 4.4 The first three Brillouin zones of the simple square lattice

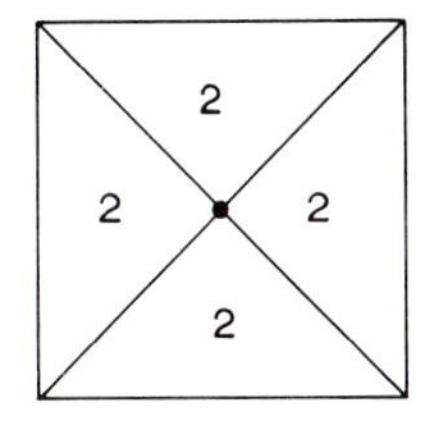

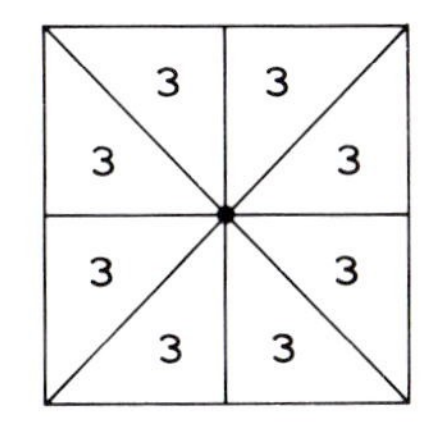

Fig. A 4.5 How the second and third Brillouin zones fit into the first, for the simple square lattice

those crystal structures such as close-packed hexagonal (c.p.h.) in which the basis is intrinsic, and those such as body-centred cubic (b.c.c.) and face-centred cubic (f.c.c.) which are strictly Bravais lattices but are more conveniently envisaged as cubic lattices with a basis, b.c.c. having two atoms per cubic cell and f.c.c. four.

The concept of the *structure factor* from diffraction theory is useful here. If the incident and diffracted beams have wave-vectors $\boldsymbol{k}$ and $\boldsymbol{k}'$ respectively, and

$$\boldsymbol{q} = \boldsymbol{k}' - \boldsymbol{k} \tag{A 4.34}$$

then the structure factor for a system of N atoms, sited at positions $\boldsymbol{l}$ relative to an origin, is

$$S(\boldsymbol{q}) = \frac{1}{N}\sum_{l} \exp(-\mathrm{i}\boldsymbol{q}\cdot\boldsymbol{l}) \tag{A 4.35}$$

This expression sums the various scattered wave amplitudes from the individual atoms, in the direction of $\boldsymbol{k}'$, taking account of their phase differences through the exponential factor. Thus the incident wave differs at the point $\boldsymbol{l}$ by the factor $\exp(\mathrm{i}\boldsymbol{k}\cdot\boldsymbol{l})$ from what it is at the origin; and the diffracted wave similarly differs by the factor $\exp(\mathrm{i}\boldsymbol{k}'\cdot\boldsymbol{l})$.

Taking the b.c.c. unit cell as an example, we have one atom at the origin and the other at

$$\boldsymbol{l} = \tfrac{1}{2}(\boldsymbol{a} + \boldsymbol{b} + \boldsymbol{c}) = \frac{a}{2}[111] \tag{A 4.36}$$

where a is the length of the cube edge and $\boldsymbol{a}$, $\boldsymbol{b}$ and $\boldsymbol{c}$ are the (cubic axis) unit lattice vectors. This is a simple cubic lattice, with a basis, and its reciprocal lattice is also cubic, with a cube edge of length $2\pi/a$. An arbitrary vector $\boldsymbol{g}$ in this reciprocal lattice has the form

$$\boldsymbol{g} = n_1\boldsymbol{a}^* + n_2\boldsymbol{b}^* + n_3\boldsymbol{c}^* \tag{A 4.37}$$

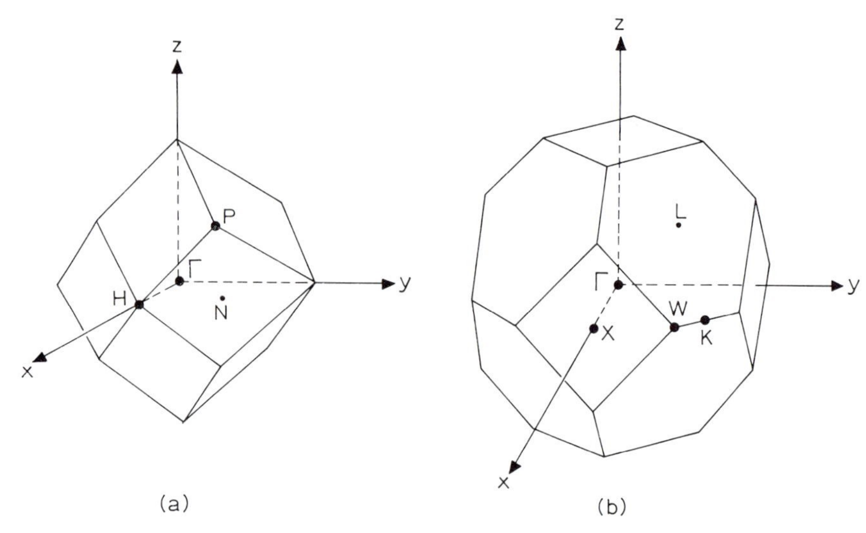

Fig. A 4.6 The first Brillouin zones of (a) the b.c.c. and (b) the f.c.c. lattice. The cubic axes are x, y, z; the centre of each zone is denoted by $\boldsymbol{\Gamma}$; the notation H, N, P and K, L, W, X is widely used to denote points of high symmetry on the surfaces of the zones

where n_1, n_2 and n_3 are integers, and $\boldsymbol{a}^*$, $\boldsymbol{b}^*$ and $\boldsymbol{c}^*$ are the reciprocal lattice vectors, which are related to $\boldsymbol{a}$, $\boldsymbol{b}$ and $\boldsymbol{c}$ as follows:

$$\left.\begin{aligned} &\boldsymbol{a}^* \cdot \boldsymbol{a} = \boldsymbol{b}^* \cdot \boldsymbol{b} = \boldsymbol{c}^* \cdot \boldsymbol{c} = 2\pi \\ &\boldsymbol{a}^* \cdot \boldsymbol{b} = \boldsymbol{a}^* \cdot \boldsymbol{c} = \boldsymbol{b}^* \cdot \boldsymbol{a} = \boldsymbol{b}^* \cdot \boldsymbol{c} = \boldsymbol{c}^* \cdot \boldsymbol{a} = \boldsymbol{c}^* \cdot \boldsymbol{b} = 0 \end{aligned}\right\} \qquad \text{(A 4.38)}$$

Substituting these relations into equation (A 4.35), we obtain

$$\begin{aligned} NS(\boldsymbol{g}) &= 1 + \exp^{[-i\pi(n_1 + n_2 + n_3)]} = 1 + (-1)^{(n_1 + n_2 + n_3)} \\ &= \begin{cases} 2 & (\text{for } n_1 + n_2 + n_3 = \text{even}) \\ 0 & (\text{for } n_1 + n_2 + n_3 = \text{odd}) \end{cases} \end{aligned} \qquad \text{(A 4.39)}$$

Interference between waves scattered by the two atoms of the basis thus leads to *missing reflections* from (100) and similar planes. The first Brillouin zone for the b.c.c. structure is thus the *dodecahedron* formed from reflections from (110) planes (*see* Fig. A 4.6(a)). This zone holds two quantum states per unit cell but, since there are two atoms per unit cell in the b.c.c. structure, it holds one quantum state per atom, the same as for a Bravais lattice.

The corresponding analysis for the f.c.c. unit cell shows that (100) and (110) reflections do not occur. The first zone is then a *truncated octahedron* formed from (111) and (200) reflections (*see* Fig. A 4.6(b)). It holds four quantum states per cell, i.e. one per atom, since there are four atoms per cell.

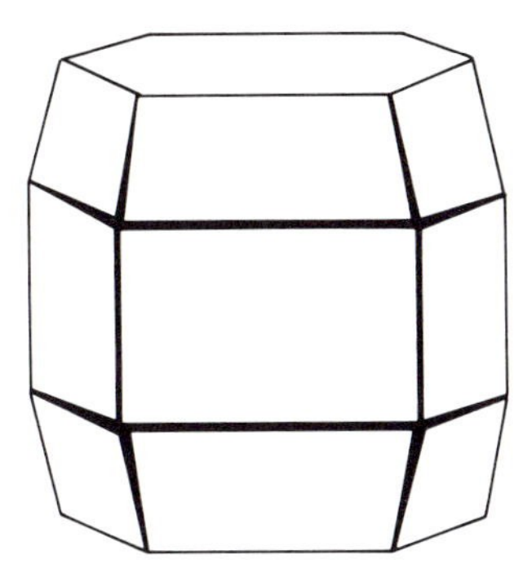

Fig. A 4.7 The composite c.p.h. zone

The first Brillouin zone of the c.p.h. structure consists of a hexagonal prism which contains one quantum state per atom, but there is no energy discontinuity across its hexagonal faces because the reflections from alternate basal planes in the crystal cancel. We may thus consider either the second zone (and the first and second together contain two quantum states per atom) or the *Jones zone*, which is the smallest region totally enclosed by planes of energy discontinuity. As shown in Fig. A 4.7, it is a composite of first-zone and second-zone faces and can hold

$$2 - \frac{3}{4}\left(\frac{a}{c}\right)^2\left[1 - \frac{1}{4}\left(\frac{a}{c}\right)^2\right] \tag{A 4.40}$$

electrons per atom, for a crystal of axial ratio c/a. For ideal close-packing, i.e. $c/a = 1{\cdot}63$, the value is $1{\cdot}745$.

THE NEARLY-FREE-ELECTRON THEORY

Whereas the Brillouin zone of a crystal can be deduced exactly from simple crystallographic arguments, the calculation of the energy levels in the zone is much more difficult. Many approximate methods have been developed, of which the two main ones are the *nearly-free-electron theory* (NFE) and the *tight-binding theory*, which we outline in this section and the next, using the one-dimensional problem for simplicity.

The idea in NFE theory is to start with the free-electron theory as a basis and to introduce a weak periodic variation of V as a *perturbation*. We thus take as a *zero-order* solution the simple plane wave functions

$$\psi_k^{(0)} = A \exp{(\mathrm{i}kx)} \tag{A 4.41}$$

in one dimension, where the superscript zero indicates the order of the approximation and A is the normalization factor. The corresponding energy levels, given by equation (A 3.3), are denoted by $E_k^{(0)}$. We now try to improve the values of ψ and E by the use of perturbation theory, writing them as

$$\left.\begin{aligned} \psi &= \psi^{(0)} + \psi^{(1)} + \text{higher-order terms} \\ E &= E^{(0)} + E^{(1)} + \text{higher-order terms} \end{aligned}\right\} \tag{A 4.42}$$

in which each successive term is an order of magnitude smaller than its predecessor. In the Schrödinger equation

$$(\mathrm{H}^{(0)} + V)\psi = E\psi, \qquad \mathrm{H}^{(0)} = -\frac{\hbar^2}{2m}\frac{\mathrm{d}^2}{\mathrm{d}x^2} \tag{A 4.43}$$

we use the free-electron equation $\mathrm{H}^{(0)}\psi^{(0)} = E^{(0)}\psi^{(0)}$ to remove the zero-order terms and so obtain for the first-order terms (including V)

$$E_k^{(1)}\psi_k^{(0)} = V\psi_k^{(0)} + (\mathrm{H}^{(0)} - E_k^{(0)})\psi_k^{(1)} \tag{A 4.44}$$

The fact that, in the free-electron theory, the wavefunctions $\psi_k^{(0)}$ form a complete orthonormal set (*see* Appendix 1) allows any other wavefunction to be written as a sum of them, i.e.

$$\psi_k^{(1)} = \sum_{k'} a_{k'k}\psi_{k'}^{(0)} \tag{A 4.45}$$

with coefficients $a_{k'k}$. Substituting this into the first-order equation gives

$$E_k^{(1)}\psi_k^{(0)} = V\psi_k^{(0)} + \sum_{k'} a_{k'k}(E_{k'}^{(0)} - E_k^{(0)})\psi_{k'}^{(0)} \tag{A 4.46}$$

The orthonormal properties of $\psi^{(0)}$ have further value here. We multiply through this equation by the complex conjugate $\psi_k^{(0)*}$ and integrate over the whole volume of the metal. The final term of course vanishes when $k' = k$, but it also vanishes when $k' \neq k$ because of the orthogonality property (equation (A 1.24)). Furthermore, the normalization condition gives

$$\int \psi_k^{(0)*} E_k^{(1)}\psi_k^{(0)}\,\mathrm{d}v = E_k^{(1)} \int \psi_k^{(0)*}\psi_k^{(0)}\,\mathrm{d}v = E_k^{(1)} \tag{A 4.47}$$

The *first-order* change in the energy levels as a result of the perturbation is thus

$$E_k^{(1)} = \int \psi_k^{(0)*} V\psi_k^{(0)}\,\mathrm{d}v = \langle k|V|k\rangle \tag{A 4.48}$$

where we have introduced Dirac's bra–ket notation (*see* Appendix 1), with

$$\langle k| = \psi_k^{(0)*} \quad \text{and} \quad |k\rangle = \psi_k^{(0)} \tag{A 4.49}$$

This confirms our result obtained with perturbation theory in Appendix 2, i.e. that *the first-order change in energy is simply the average of the perturbation operator over the unperturbed quantum state of the system.*

To obtain $\psi^{(1)}$ we multiply through equation (A 4.46) by $\psi_{k'}^{(0)*}$, integrate over v again, and use the orthonormal properties

$$\int \psi_{k'}^{(0)*} \psi_k^{(0)} \, \mathrm{d}v = \delta_{k'k} \tag{A 4.50}$$

to obtain

$$a_{k'k} = \frac{\langle k' | V | k \rangle}{E_k^{(0)} - E_{k'}^{(0)}} \tag{A 4.51}$$

when $k' \neq k$. By normalizing ψ_k it can be shown that $a_{kk} = 0$. The improved wavefunction

$$\psi_k = \psi_k^{(0)} + \sum_{k'(\neq k)} a_{k'k} \psi_{k'}^{(0)} \tag{A 4.52}$$

is a weighted sum of the zero-order functions. From perturbation theory we might expect that $\psi_k \simeq \psi_k^{(0)}$ and the coefficients $a_{k'k}$ all to be small. We shall see, however, that the most significant aspect of NFE theory stems from the *failure* of this assumption near the Brillouin zone boundaries. Equation (A 4.52) can, however, be used to find the energy levels of states in other parts of the zones.

This first-order refinement of ψ enables us to repeat the calculation of the energy change to obtain the *second-order* refinement $E^{(2)}$. This calculation, similar to the above, gives

$$E_k^{(2)} = \sum_{k'(\neq k)} \frac{|\langle k' | V | k \rangle|^2}{E_k^{(0)} - E_{k'}^{(0)}} \tag{A 4.53}$$

The above perturbation theory is inapplicable when there is degeneracy, i.e. when two or more zero-order states have the same energy, since equation (A 4.51) then yields an infinite quantity. The method in this case is first to make new zero-order states by linearly combining the degenerate wavefunctions. It is precisely this situation that exists for a state $\boldsymbol{k}$ at the limit of a Brillouin zone since, by adding or subtracting a reciprocal lattice vector $\boldsymbol{g}$, another state $\boldsymbol{k}' = \boldsymbol{k} - \boldsymbol{g}$ can be found in an exactly equivalent position on the opposite side of the zone, and with exactly the same energy (e.g. the states at P and Q in Fig.A 4.1). When $\boldsymbol{k}$ is at or near the zone boundary there are, instead of equation (A 4.52), *two* major contributors to ψ_k, i.e. the two states $\psi_k^{(0)}$ and $\psi_{k-g}^{(0)}$ which represent forward and backward waves, in accordance with Bragg reflection. We thus take, in place of equation A 4.52,

$$\psi_k = a_k \exp(\mathrm{i}kx) + a_{k-g} \exp[\mathrm{i}(k-g)x] \quad \text{(A 4.54)}$$

with equal or nearly equal coefficients. We substitute this into the Schrödinger equation, from which we then obtain two equations by the standard procedure of multiplying through by the complex conjugates of each of the two terms in equation (A 4.54), integrating over (unit) v and applying the orthonormality conditions. In these two equations terms of the type $\int \exp(-\mathrm{i}k'x) V \exp(\mathrm{i}kx)\, \mathrm{d}v$ can be simplified by using the periodicity of V. We can thus express V as a *Fourier sum* of plane waves:

$$V(x) = \sum_g V_g \exp(\mathrm{i}gx) \quad \text{(A 4.55)}$$

where V_g is the amplitude of the component with the reciprocal lattice vector $\mathbf{g}$. Then

$$\begin{aligned} \int \exp(-\mathrm{i}k'x) V \exp(\mathrm{i}kx)\,\mathrm{d}v &= \int \exp(-\mathrm{i}k'x)\left[\sum_g V_g \exp(\mathrm{i}gx)\right]\exp(\mathrm{i}kx)\,\mathrm{d}v \\ &= \sum_g V_g \int \exp[\mathrm{i}(k+g-k')x]\,\mathrm{d}v \\ &= \begin{cases} V_g & (\text{when } k'-k = g) \\ 0 & (\text{otherwise}) \end{cases} \end{aligned} \quad \text{(A 4.56)}$$

The two equations then become

$$\left.\begin{aligned} (E_k^{(0)} - E)a_k + V_g a_{k-g} &= 0 \\ (E_{k-g}^{(0)} - E)a_{k-g} + V_{-g} a_k &= 0 \end{aligned}\right\} \quad \text{(A 4.57)}$$

and they can be solved for E:

$$E^{\pm} = \tfrac{1}{2}(E_k^{(0)} + E_{k-g}^{(0)}) \pm \tfrac{1}{2}[(E_k^{(0)} - E_{k-g}^{(0)}) + 4|V_g|^2]^{1/2} \quad \text{(A 4.58)}$$

The two energies E^- and E^+ correspond to the two different E values on the inner and outer faces of the Brillouin zone boundary at the point k, and come from the two combinations of the forward and backward waves, as discussed in Chapter 1 (*see* equations (1.18) and (1.19)). At the zone boundary, where $k = \frac{1}{2}g$, we have

$$E_k^{(0)} = E_{k-g}^{(0)} = E_{g/2}^{(0)} \quad \text{(A 4.59)}$$

and thus

$$E^{\pm} = E_{g/2}^{(0)} \pm |V_g| \quad \text{(A 4.60)}$$

The *band gap* ΔE (i.e. AB in Fig. 1.5) at this point on the zone boundary is given by

$$\Delta E = 2|V_g| \qquad \text{(A 4.61)}$$

Typically, $\Delta E \simeq$ 1–3 eV for the simple (non-transition) metals.

The NFE theory may appear unconvincing in that it assumes a weak periodic variation of V, whereas in fact V goes to $-\infty$ at the centre of each atom of the crystal. However, the *pseudopotential* theory outlined in Chapter 2 has shown that the approximation is an excellent one for the outermost electrons (s, p) of non-transition metals.

THE TIGHT-BINDING THEORY

The tight-binding theory starts from the opposite simplifying assumption to that of NFE theory: it assumes that the periodic variation of V is extremely strong. Its starting model is not a free-electron gas, but a crystal of such large spacing that each atom is electrically neutral and virtually isolated from all others by a thick barrier of height V. The approximation is then to assume that each Bloch electron exists in a state made of *atomic orbitals* slightly modified by the other atoms. We can think of such an electron as being localized for long periods of time on one atom, but occasionally moving forward to the next by penetrating, through the *tunnel effect* (*see* Appendix 1), the high potential barrier between the atoms. Not surprisingly, the tight-binding method, which is closely related to the valence-bond LCAO method of molecular theory (*see* Appendix 2), is particularly suitable for dealing with small overlaps between atomic orbitals and narrow energy bands, e.g. the d states in transition metals.

Suppose that $\phi(x)$ is a normalized s state wavefunction of an electron in the field $U(x)$ of a free atom and that the energy of this state is E_0. In a linear crystal of N atoms of spacing a, the potential at any point x is

$$V(x) = \sum_n U(x - na) \qquad \text{(A 4.62)}$$

where $U(x - na)$ is the potential at x due to the atom at the lattice point na. The atom nearest x of course makes the main contribution to $V(x)$, as shown in Fig. A 4.8.

The zero-order wavefunction $\psi(x)$ in the crystal is simply the sum of the unchanged atomic orbitals:

$$\psi_k(x) = \sum_n c_n \phi(x - na) \qquad \text{(A 4.63)}$$

The coefficients c_n are given directly as exp $(ikna)$ by the Bloch theorem and so the Bloch function in the tight-binding approximation is

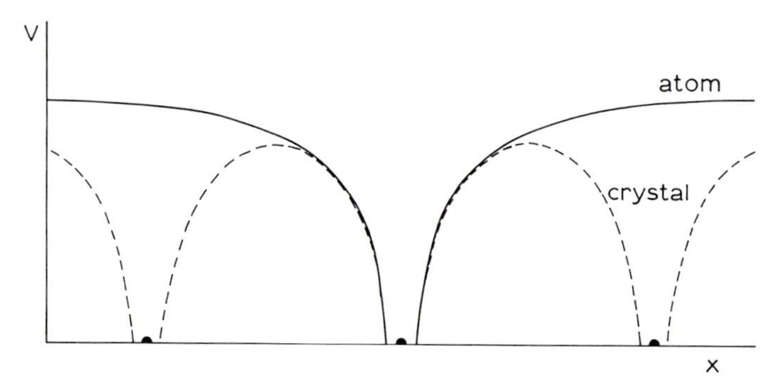

Fig. A 4.8 **The potential of an electron in a free atom and in a crystal**

$$\psi_k(x) = \sum_n \phi(x - na) \exp(ikna) \tag{A 4.64}$$

This is a line of unchanged atomic orbitals, each multiplied by a wave factor which changes the orientation from one orbital to the next by an amount that depends on the wavenumber k.

In the limit of infinitely tight binding, the energy of an electron in this Bloch state is of course E_0, i.e. when the atoms are isolated the bandwidth is zero. More generally, the energy E_k of the state ψ_k can be found by using equations (A 1.30) and (A 1.31) with the Hamiltonian operator

$$\mathrm{H} = -\frac{\hbar^2}{2m}\frac{\mathrm{d}^2}{\mathrm{d}x^2} + V \tag{A 4.65}$$

Since

$$\left(-\frac{\hbar^2}{2m}\frac{\mathrm{d}^2}{\mathrm{d}x^2} + U\right)\phi = E_0\phi \tag{A 4.66}$$

we have

$$\mathrm{H}\phi = E_0\phi + \mathrm{H}^{(1)}\phi \tag{A 4.67}$$

where $\mathrm{H}^{(1)}$, i.e.

$$\mathrm{H}^{(1)} = V - U \tag{A 4.68}$$

is the perturbation term in the Hamiltonian.

We assume that the overlaps of the atomic orbitals are so small that the normalization factor in equation (A 1.30) can be set to unity; the matrix elements in which we are interested are then

$$\langle k'|\mathrm{H}^{(1)}|k\rangle = \int \psi_{k'}^{*}\mathrm{H}^{(1)}\psi_{k}\,\mathrm{d}x \tag{A 4.69}$$

Because of the periodicity of the lattice potential, $\mathrm{H}(x) = \mathrm{H}(x + na)$ and thus

$$\langle k'|\mathrm{H}^{(1)}|k\rangle = \sum_{n}\sum_{n'} \exp(\mathrm{i}kna)\exp(-\mathrm{i}k'n'a)\int \phi^{*}(x - n'a)\mathrm{H}^{(1)}\phi(x - na)\,\mathrm{d}x \tag{A 4.70}$$

Let $m = n' - n$, and move the origin to the point na so that

$$\langle k'|\mathrm{H}^{(1)}|k\rangle = \sum_{n} \exp[\mathrm{i}(k - k')na]\sum_{m} \exp(\mathrm{i}k'ma)\int \phi^{*}(x - ma)\mathrm{H}^{(1)}\phi(x)\,\mathrm{d}x \tag{A 4.71}$$

If $k' \neq k$ the various values n in the first sum sample the sinusoidal wave $\exp[\mathrm{i}(k - k')na]$ evenly over its whole cycle, and the mutual cancellation of the positive and negative parts of this cycle leads to a zero sum. But when $k' = k$, each term is unity and so the sum is N. Thus the matrix element simplifies to

$$\langle k|H^{(1)}|k\rangle = N\sum_{n} \exp(\mathrm{i}kna)\int \phi^{*}(x - na)\mathrm{H}^{(1)}\phi(x)\,\mathrm{d}x \tag{A 4.72}$$

Since $\phi^{*}(x - na)$ and $\phi(x)$ are localized atomic orbitals, at least one of them is almost zero if $n > 1$. Hence we consider only the two terms $n = 0$ and $n = 1$:

$$\begin{aligned} N\int \phi^{*}(x)\mathrm{H}^{(1)}\phi(x)\,\mathrm{d}x &= \alpha \\ N\int \phi^{*}(x - a)\mathrm{H}^{(1)}\phi(x)\,\mathrm{d}x &= \beta \end{aligned} \tag{A 4.73}$$

and obtain

$$\begin{aligned} E_k &= E_0 + \langle k|\mathrm{H}^{(1)}|k\rangle \\ &= E_0 + \alpha + \beta\exp(\mathrm{i}ka) \end{aligned} \tag{A 4.74}$$

The first matrix element, α, changes the average energy level and the second, β, splits this level into a band, in accordance with the various values of k allowed by the Bloch theorem. The matrix elements obviously become stronger as the lattice constant is reduced and the orbitals overlap more. The band thus

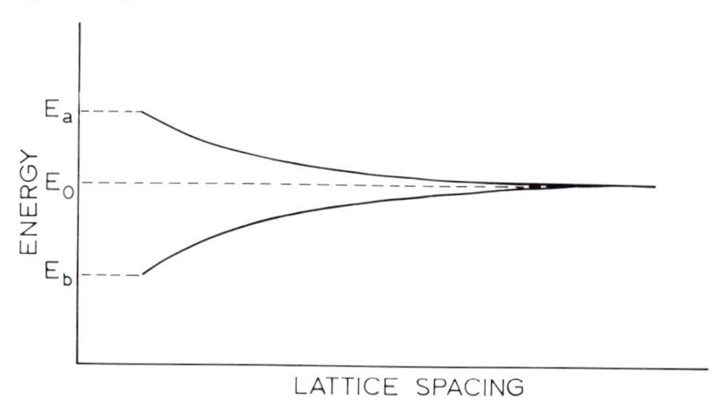

Fig. A 4.9 The formation of a narrow energy band

develops as in Fig. A 4.9, where $\alpha = E_0 - \frac{1}{2}(E_a + E_b)$. It is an N-atom version of the two-atom molecular theory presented in Appendix 2. The highest level E_a represents a completely antibonding combination of the atomic orbitals (compare Fig. A 2.2), while the lowest represents a completely bonding combination; and the $N - 2$ intermediate levels represent intermediate cases in which some neighbouring atoms have the bonding configuration of their orbitals and others have the antibonding configuration.

The extension to three dimensions is straightforward. In a simple cubic lattice the six nearest neighbours are located at the positions $\pm a, 0, 0$; $0, \pm a, 0$ and $0, 0, \pm a$. The three-dimensional generalization of equation (A 4.74) then gives the energy as

$$E_{\mathbf{k}} = E_0 + \alpha + 2\beta(\cos k_x a + \cos k_y a + \cos k_z a) \qquad \text{(A 4.75)}$$

and the bandwidth is

$$E_a - E_b = 12|\beta| \qquad \text{(A 4.76)}$$

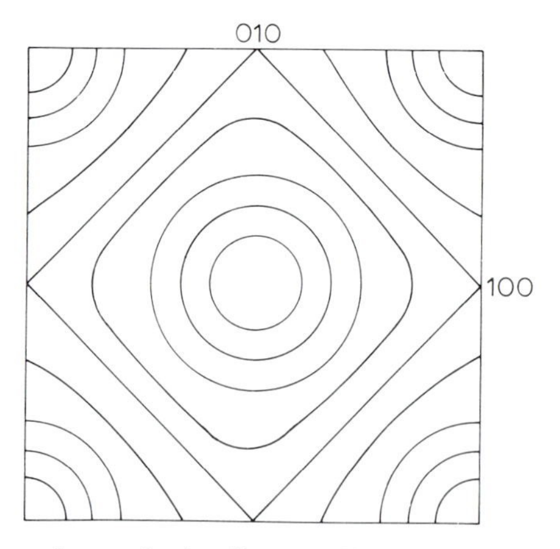

Fig. A 4.10 Section through the first Brillouin zone of the simple cubic lattice, in the tight-binding approximation

Similar expressions can be derived for the b.c.c. and f.c.c. lattices.

Figure A 4.10 shows the energy contours in the first zone of the simple cubic lattice. They are, not unexpectedly, very different from the near-circles of the NFE theory, except near the origin.

BANDS FOR s, p AND d STATES

By working with atomic orbitals, the tight-binding theory gives physical meaning to the periodic lattice function $u(\boldsymbol{r})$ in the Bloch state (equation (A 4.11)). Each $u(\boldsymbol{r})$ generates its own first Brillouin zone and corresponding energy band, and so we expect one such zone for each of the 1s, 2s, 3s, . . . atomic orbitals, three for each of the 2p, 3p, . . . orbitals, five for each of 3d, 4d, . . . orbitals, and so on, with the $u(\boldsymbol{r})$ function of each zone closely resembling in each atom the atomic orbital from which the zone is named. This proves to be the case up to a point, but there are complications due to the *different symmetries* of these orbitals and also due to *hybridization* when the energy bands overlap.

In order to get a qualitative idea of the effect of symmetry, suppose, in our one-dimensional model, that $u(x)$ is a $2p_x$ atomic orbital. A sequence of such orbitals along the x coordinate of the line of atoms can then be arranged to give good bonding, according to the molecular principles outlined in Appendix 2, as in Fig. A 4.11(a). This sequence obviously has the lowest energy of any Bloch state, but it corresponds to the case where $k = \pm\pi/a$ because each successive orbital has the opposite orientation to that of its predecessor. The *lowest* energy of this p band thus occurs at the Bragg limits. If we progressively reduce k from these limits towards zero, more and more antibonding orientations of neighbouring orbitals appear in the sequence until, at the limit $k = 0$, all the states are aligned in the same orientation, as in Fig. A 4.11(b), which is clearly the most antibonding combination and so has the highest energy in the band (*see* Fig. A 4.12).

We now consider hybridization. If the bandwidths of two different $u(\boldsymbol{r})$ zones are large compared with the difference in the energy levels of the two parent atomic orbitals, it is no longer possible to recognize the bands separately and identify them unambiguously with their parent orbitals. Suppose that there are two such orbitals, $\phi_{(0)}$ and $\phi_{(1)}$, which when not hybridized give the two corresponding Bloch functions, $\psi_{(0)k}(x)$ and $\psi_{(1)k}(x)$, in equation (A 4.64). The

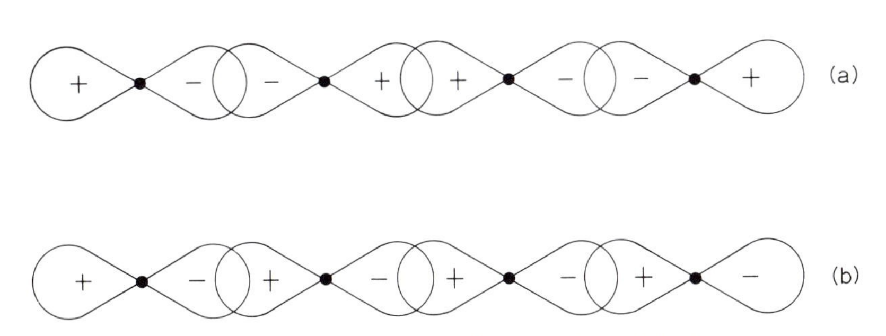

Fig. A 4.11 Orbitals for p states at (a) $k = \pm\pi/a$ and (b) $k = 0$

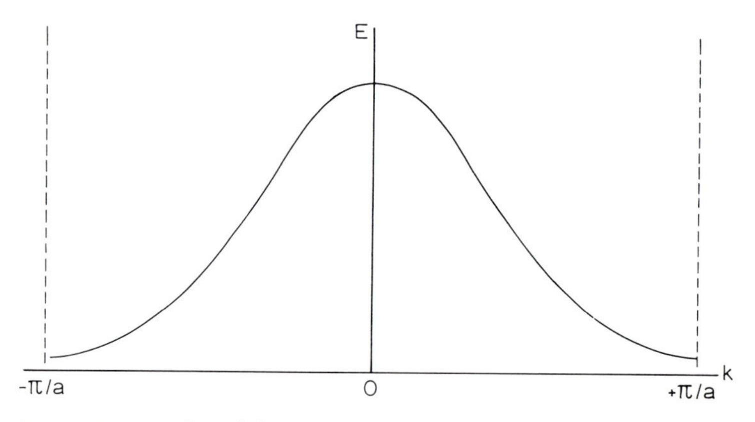

Fig. A 4.12 Energy band for p-like Bloch states

above tight-binding analysis then has to be generalized to include the possibility that $H^{(1)}$ couples together these two functions; i.e. the matrix element of equation (A 4.72) has to be generalized to

$$\langle pk|\mathrm{H}^{(1)}|qk\rangle = N\sum_n \exp(\mathrm{i}kna)\int \phi_p^*(x-na)\,\mathrm{H}^{(1)}\phi_q(x)\,\mathrm{d}x \qquad \text{(A 4.77)}$$

where p and q each represent the two values, 0 and 1, corresponding to the different orbitals. For each term of our previous matrix element, equation (A 4.72), there are now four, corresponding to $p = q = 0$, $p = q = 1$, $p = 0$ and $q = 1$, and $p = 1$ and $q = 0$. The analysis of the energy levels is then a more elaborate repetition of that given above (Mott and Jones 1936). It leads, for overlapping s and p states in a simple cubic lattice, to the energy levels E_k on the k_x axis, where

$$2E_k = H_0 + H_1 \pm [(H_0 - H_1)^2 + 16\beta^2 \sin^2 ak_x]^{1/2} \qquad \text{(A 4.78)}$$

$$\beta = \int \phi_{(0)}^*(x+a)\,\mathrm{H}^{(1)}\phi_{(1)}(x)\,\mathrm{d}v \qquad \text{(A 4.79)}$$

and H_0 and H_1 are the respective energies of non-hybridized s and p Bloch functions. The most interesting case is where the two bands of these functions *cross*, as k_x increases from 0 to π/a, as shown in Fig. A 4.13. The dashed curves are the overlapping E, k curves for the bands before hybridization, and the continuous curves show the effect of hybridization, as in equation (A 4.78). The lower dashed band starts out as s-like when k is small, but its energy rises steeply at high k values because the symmetrical s-like form of its $u(\boldsymbol{r})$ state fits badly onto the rapidly alternating carrier wave, exp $(\mathrm{i}kx)$, when k is large. There are lower states in this region, coming down from the upper dashed band,

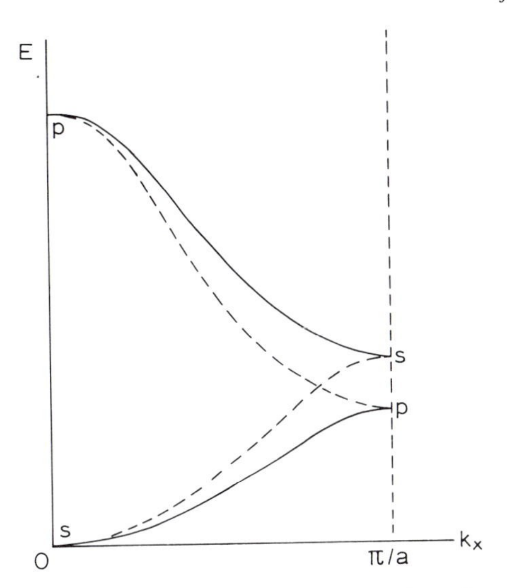

Fig. A 4.13 The effect of hybridization on overlapping bands with s and p symmetry at k = 0

because the p-like form of the $u(\mathbf{r})$ function of these fits well (*see* Fig. A 4.11(a)) on the carrier wave there. The line of lowest energy thus starts at $k = 0$, with s-like states, and then rises only slightly, to the p-like states at $k = \pi/a$. Along this line the Bloch functions in the first band thus gradually change from s-like to p-like as k increases. Correspondingly, the upper band starts off as p-like at $k = 0$ and then falls, more gradually than a pure p-like band, to become s-like at $k = \pi/a$.

We see that in this case hybridization has the major effect, in addition to changing the s or p character of the states as k increases, of *splitting* overlapping, non-hybridized bands into two, non-overlapping hybridized bands which are separated by an energy gap. In effect, the hybridization 'pushes' the energy levels of the overlapping states apart, so opening up a stateless gap between them.

REFERENCE

Mott, N. F., and Jones, H., *The Theory of the Properties of Metals and Alloys*, Oxford University Press (1936), p. 72.

APPENDIX 5

Combination of overlapping states

To understand why, in the tight-binding approximation, two neighbouring atomic states do not combine well if their energies are very different, we can use the LCAO method from Appendix 2. Consider two normalized atomic orbitals, ψ_A on atom A and ψ_B on B. Suppose that they combine to form the state

$$\psi = c_A\psi_A + c_B\psi_B \tag{A 5.1}$$

We can treat c_A and c_B as parameters whose values are to be adjusted so as to give the best possible wavefunction ψ, i.e. that of lowest energy. The energy is

$$E = K\int \psi^* \mathrm{H}\psi \, \mathrm{d}v \tag{A 5.2}$$

where $K^{-1} = \int \psi^*\psi \, \mathrm{d}v$ is the normalization factor. We substitute equation (A 5.1) in this and define

$$\begin{aligned} E_A &= \int \psi_A^* \mathrm{H}\psi_A \, \mathrm{d}v, \qquad E_B = \int \psi_B^* \mathrm{H}\psi_B \, \mathrm{d}v \\ E_{AB} &= \int \psi_A^* \mathrm{H}\psi_B \, \mathrm{d}v = \int \psi_B^* \mathrm{H}\psi_A \, \mathrm{d}v \end{aligned} \tag{A 5.3}$$

to obtain

$$E = K(c_A^2 E_A + c_B^2 E_B + 2c_A c_B E_{AB}) \tag{A 5.4}$$

Apart from a minor normalization adjustment, E_A and E_B represent the energies of the separate atomic states ψ_A and ψ_B.

In the *method of variations* we seek the minimum of E through the conditions

$$\frac{\partial E}{\partial c_A} = 0 \quad \text{and} \quad \frac{\partial E}{\partial c_B} = 0 \tag{A 5.5}$$

which, when applied to equation (A 5.4) (and noting that ψ_A and ψ_B are normalized), give

$$\left.\begin{aligned}(E_A - E) + c(E_{AB} - ES) &= 0\\(E_{AB} - ES) + c(E_B - E) &= 0\end{aligned}\right\} \tag{A 5.6}$$

where $c = c_B/c_A$ and

$$S = \int \psi_A \psi_B \, dv \tag{A 5.7}$$

Equations (A 5.6) are known as *secular equations* (from an astronomical analogy). More generally, when ψ is represented as a linear combination of n functions, $c_A\psi_A, c_B\psi_B, \ldots$, there are n secular equations, arising from n conditions of the type (A 5.5), and these simultaneous equations can be solved in the form of a *secular determinant* to give the energy E.

We eliminate c between these two secular equations to obtain

$$(E - E_A)(E - E_B) - (E_{AB} - ES)^2 = 0 \tag{A 5.8}$$

Under tight-binding conditions, where the spatial overlap of the combining states is small, there is generally no volume element in which ψ_A and ψ_B are both large, so that S is small and E_{AB} is small compared with E_A or E_B. Taking advantage of this, and assuming that $E_A \ll E_B$, we can simplify equation (A 5.8) into the approximate solutions

$$\left.\begin{aligned}E_- &= E_A - \frac{(E_{AB} - E_A S)^2}{E_B - E_A}\\E_+ &= E_B + \frac{(E_{AB} - E_B S)^2}{E_B - E_A}\end{aligned}\right\} \tag{A 5.9}$$

We obtain two values, E_+ and E_-, for E and notice that E_+ lies *above* the higher atomic level E_B, whereas E_- lies *below* the lower level, E_A. Also, since $E_B - E_A$ is large, E_- and E_+ differ only slightly from the atomic energy levels E_A and E_B, i.e. the energy levels are *hardly altered* by the overlap of ψ_A and ψ_B. Moreover, the fact that $E_- \simeq E_A$ shows that c is small in the combination with energy E_-, so that this state is nearly pure ψ_A. Similarly, c is large in the combination with energy E_+, so this state is nearly pure ψ_B. Thus there is little combination of ψ_A and ψ_B when E_A and E_B are very different.

APPENDIX 6

Fourier methods

FOURIER TRANSFORMS

If a function $f(x)$ is sufficiently well-behaved mathematically*, it can be expressed as a *Fourier integral*:

$$f(x) = \int_{-\infty}^{+\infty} f(q) \exp(-iqx) \, dq \qquad \text{(A 6.1)}$$

where $f(q)$ is its *Fourier transform*, given by the *inverse Fourier transformation*

$$f(q) = \frac{1}{2\pi} \int_{-\infty}^{+\infty} f(x) \exp(iqx) \, dx \qquad \text{(A 6.2)}$$

Sometimes these two expressions are written symmetrically by introducing $(2\pi)^{-1/2}$ into each, rather than $(2\pi)^{-1}$ into the inverse one.

The Fourier integral is a generalization of the familiar *Fourier series* in which a *periodic function* (e.g. equation (A 4.1)) is expressed as a sum of sines and cosines. Since $\exp(-iqx) = \cos qx - i \sin qx$, equation (A 6.1) expresses $f(x)$ as a *weighted sum* of sines and cosines, summed over a continuous range of the wavenumber q. Each wave contributes to this sum an amount given by its weighting factor $f(q)$, i.e. its Fourier transform, and with a phase given by the argument of its exponent, $\exp(-iqx)$. Each weighted wave $f(q) \exp(-iqx)$ is a *Fourier component* (with a wavenumber q) of $f(x)$. Unlike the Fourier series, the Fourier integral can represent *non-periodic* $f(x)$ functions, or, equivalently, functions whose period is infinite.

The generalization to more than one dimension is straightforward. In three dimensions, with position vector $\boldsymbol{r}$ and wave-vector $\boldsymbol{q}$, the expressions become

$$f(\mathbf{r}) = \int_{-\infty}^{+\infty} f(\boldsymbol{q}) \exp(-i\boldsymbol{q} \cdot \boldsymbol{r}) \, d\boldsymbol{q} \qquad \text{(A 6.3)}$$

*Smoothly varying functions are generally of this well-behaved type; the general conditions are defined in Titchmarsh, E. C., *Introduction to the theory of Fourier integrals*, Oxford University Press (1937).

Table A 6.1 Some examples of standard Fourier transforms

1	$\exp(-x^2/2)$	$(2\pi)^{-1/2} \exp(-q^2/2)$
2	$\begin{cases} b & (x^2 < a^2) \\ 0 & (x^2 > a^2) \end{cases}$	$\dfrac{b \sin qa}{\pi q}$
3	$\exp(-gx)$	$\dfrac{g}{\pi(g^2 + q^2)}$
4	1	$\delta(q)$
5	$\dfrac{1}{r}\exp(-gr)$	$\dfrac{1}{2\pi^2(g^2 + q^2)}$

In examples 1 to 4 the function is $f(x)$ and the transform is $f(q)$, as given by equations (A 6.1) and (A 6.2). In example 5 they are $f(\mathbf{r})$ and $f(\mathbf{q})$, as in equations (A 6.3) and (A 6.4).
In example 4 $\delta(q)$ is the *Dirac delta function*, which represents a sharp point at $q = 0$. It is zero everywhere except at $q = 0$, where $\int \delta(q)\, \mathrm{d}q = 1$.

$$f(\mathbf{q}) = \frac{1}{(2\pi)^3} \int_{-\infty}^{+\infty} f(\mathbf{r}) \exp(\mathrm{i}\mathbf{q} \cdot \mathbf{r})\, \mathrm{d}\mathbf{r} \qquad \text{(A 6.4)}$$

where $\mathrm{d}\mathbf{q}$ and $\mathrm{d}\mathbf{r}$ are volume elements in q-space and real space, respectively.

Many problems of mathematical physics are conveniently solved by first working them out for one Fourier component and then generalizing, through the Fourier integral and weighting factor, to the complete solution $f(x)$. Catalogues of standard transforms are available; some examples are given in Table A 6.1.

APPLICATION TO THE PSEUDOPOTENTIAL THEORY

The weakness of the pseudopotential (*see* Chapter 2) justifies the use of the NFE theory (*see* Appendix 4), in which the effects of the pseudopotential are dealt with as a slight perturbation of the energy levels and plane wave functions $\psi^0 \propto \exp(\mathrm{i}\mathbf{k} \cdot \mathbf{r})$ of the free-electron theory. If $V(\mathbf{r})$ is the pseudopotential, the characteristic quantities in this treatment are the *matrix elements*

$$\langle k' | V | k \rangle = \int \psi_{k'}^{*} V \psi_k \, \mathrm{d}\mathbf{r} \qquad \text{(A 6.5)}$$

since, for example, the change in the energy level of the state with wave-vector $\mathbf{k}$

is given to the first order by $\langle k|V|k\rangle$ (*see* equation (A 4.48)). We take plane waves for ψ and define

$$W(q) = \langle k'|V|k\rangle \qquad \text{(A 6.6)}$$

where

$$\boldsymbol{q} = \boldsymbol{k}' - \boldsymbol{k} \qquad \text{(A 6.7)}$$

This matrix element $W(q)$ is also the *scattering amplitude*, i.e. the square of its magnitude is proportional to the probability that an electron, travelling through the crystal with wave-vector $\boldsymbol{k}$, will in unit time be deflected by the perturbation $V(\boldsymbol{r})$ into another state $\boldsymbol{k}'$ with the same energy. This interpretation of equation (A 6.5) in terms of scattering can be deduced by repeating the NFE analysis of Appendix 4, but with the operator $-i\hbar\,\partial/\partial t$ in place of E in equation (A 4.43) (cf. *see* equation (A 1.32)). The argument proceeds as before, except that the coefficients in equation (A 4.45) now contain time factors of the type $\exp(-i\omega t)$, where $\omega = \omega_k = E_k^0/\hbar$. The scattering is indicated by the change of these coefficients with time.

This interpretation in terms of scattering links the pseudopotential theory to that of X-ray scattering. We recall that X-ray (and electron) scattering is described in terms of a *form factor*, which gives the scattering from a single atom, and a *structure factor* (equation (A 4.35)), which combines the scatterings from the individual atoms into a total scattered wave. With plane wave functions ψ the matrix element for the scattering $\boldsymbol{k} \rightarrow \boldsymbol{k}'$ becomes

$$W(q) = \int V(r) \exp(-i\boldsymbol{q}\cdot\boldsymbol{r})\, d\boldsymbol{r} \qquad \text{(A 6.8)}$$

which we recognize as a Fourier transform of $(2\pi)^3\, V(r)$ with the wave-vector $\boldsymbol{q}$. Next, we express V as a sum of the pseudopotentials of the individual atoms:

$$V(\boldsymbol{r}) = \sum_{l} v(\boldsymbol{r} - \boldsymbol{l}) \qquad \text{(A 6.9)}$$

where $v(\boldsymbol{r} - \boldsymbol{l})$ is the contribution to $V(\boldsymbol{r})$ at the point $\boldsymbol{r}$ from the atom at $\boldsymbol{l}$. Substituting into equation (A 6.8), we obtain

$$W(q) = w(q)\,S(q) \qquad \text{(A 6.10)}$$

where

$$w(q) = \frac{1}{\Omega_a}\int v(r) \exp(-i\boldsymbol{q}\cdot\boldsymbol{r})\, d\boldsymbol{r} \qquad \text{(A 6.11)}$$

is the *pseudopotential form factor*, Ω_a being the volume per atom, and $S(q)$ is the *structure factor* (equation (A 4.35)).

The pseudopotential scattering amplitude $W(q)$ is thus converted into the product of a form factor and a structure factor, exactly as in the theory of diffraction. This greatly simplifies the theory. Once the form factor of a given atom has been determined, by measurement or theory, it can then be used in all appropriate applications. The analysis of metallic structures on this basis is then reduced to the determination of the structure factor, a well-established problem in the standard theory of X-ray analysis.

This diffraction theory is general and can be applied to any arrangement of the atoms. For a crystal, the *lattice periodicity*

$$V(\boldsymbol{r} + \boldsymbol{l}) = V(\boldsymbol{r}) \qquad \text{(A 6.12)}$$

(*see* equations (A 4.1) and (A 4.27)) reduces the Fourier integral to a sum,

$$V(\boldsymbol{r}) = \sum_{g} V_g \exp(\mathrm{i}\boldsymbol{g} \cdot \boldsymbol{r}) \qquad \text{(A 6.13)}$$

where V_g is the amplitude of that Fourier component of $V(\boldsymbol{r})$ which has the reciprocal lattice vector $\boldsymbol{g}$. From the analysis leading to equation (A 4.56), we then deduce, from equations (A 6.8) and (A 6.13), that

$$W(q) = \begin{cases} C\,V_g & (\text{when } \boldsymbol{q} = \boldsymbol{g}) \\ 0 & (\text{otherwise}) \end{cases} \qquad \text{(A 6.14)}$$

These two values express the fact that diffraction occurs in a perfect lattice when and only when the difference between the incident and diffracted wave-vectors is exactly equal to a reciprocal lattice vector.

APPLICATION TO THE PLASMA THEORY

We can now confirm the statement, made at the end of Section 3.3, that when the plasma oscillations are extracted from the Coulomb interaction, between electrons, what then remains of this interaction is short-ranged. We follow the treatment given by Raimes (1961). Equation (A 6.3) and the standard Fourier transforms allow us to express e^2/r as a spectrum of wavenumber components:

$$\frac{e^2}{r} = \frac{e^2}{2\pi^2} \int_{-\infty}^{+\infty} \frac{1}{k^2} \exp(-\mathrm{i}\boldsymbol{k} \cdot \boldsymbol{r})\, \mathrm{d}\boldsymbol{k} \qquad \text{(A 6.15)}$$

From Section 3.3 we conclude that the small-k end of this spectrum, i.e.

$$k \leqslant k_c \approx \lambda_{\mathrm{TF}}^{-1} \qquad \text{(A 6.16)}$$

where λ_{TF} is given by equation (3.7), may be represented by the plasma oscillations. Thus the total interaction e^2/r between two electrons can be regarded as consisting of a contribution to the general plasma oscillations, plus a remaining interaction

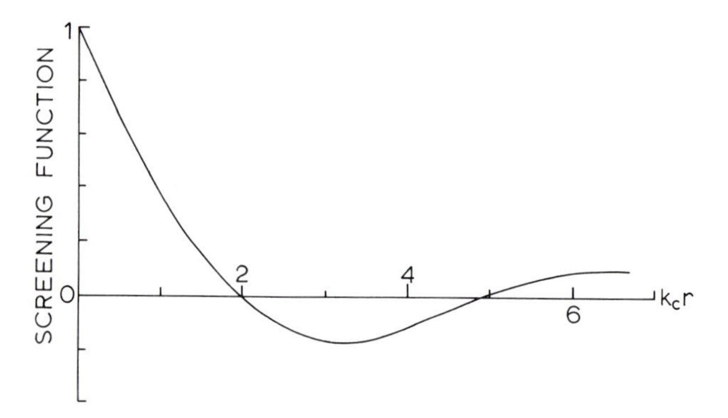

Fig. A 6.1 **The screening function**

$$I_R = \frac{e^2}{r} - \frac{e^2}{2\pi^2}\int_{-k_c}^{+k_c} \frac{1}{k^2} \exp(-i\boldsymbol{k}\cdot\boldsymbol{r})\, d\boldsymbol{k} \tag{A 6.17}$$

where the integral is over the small-k end of the spectrum. The integration over $\boldsymbol{k}$ can be split into an integral over $k = |\boldsymbol{k}|$ and an integration, at constant k, over the spherical surface spanned by the vector $\boldsymbol{k}$. Consider, for fixed k, those vectors $\boldsymbol{k}$ at an angle θ to $\boldsymbol{r}$. We have $\boldsymbol{k}\cdot\boldsymbol{r} = kr\cos\theta$. The area of the elementary annulus centred on the axis $\boldsymbol{r}$, spanned by the range θ to $\theta + d\theta$, is $2\pi k^2 \sin\theta\, d\theta$, i.e. $-2\pi k^2\, d(\cos\theta)$. The θ integral, from 0 to π, is thus

$$\frac{2\pi}{kr}\int_{-kr}^{+kr} \exp(-ikr\cos\theta)\, d(kr\cos\theta) = \frac{4\pi}{kr}\sin kr \tag{A 6.18}$$

The second term in I_R then becomes

$$\frac{e^2}{r}\frac{2}{\pi}\int_0^{k_c r} \frac{\sin kr}{kr}\, d(kr) = \frac{e^2}{r}\frac{2}{\pi}\,\mathrm{Si}\,(k_c r) \tag{A 6.19}$$

where Si (x) is the standard, tabulated, *sine-integral function* which ranges from 0 when $x = 0$ to $\frac{1}{2}\pi$ when $x \to \infty$. The interaction I_R is thus of the form

$$I_R = \frac{e^2}{r}F(k_c r) \qquad \text{where } F(k_c r) = 1 - \frac{2}{\pi}\,\mathrm{Si}\,(k_c r) \tag{A 6.20}$$

The *screening function*, $F(k_c r)$, is shown in Fig. A 6.1. We see that it produces a rapid reduction of I_R as r increases to $2k_c^{-1}$, and a weak oscillation thereafter. More accurate treatments give a greater damping of the interaction when $r > 2k_c^{-1}$.

REFERENCE

Raimes, S., *The Wave Mechanics of Electrons in Metals*, North-Holland, Amsterdam (1961).

APPENDIX 7

Band-structure energy

The weakness of the pseudopotential enables us to calculate the band energy of the electrons for simple NFE metals (*see* Chapter 5) by using perturbation theory (*see* Appendix 4). The zero-order energy of an electron of wavenumber k is simply the free-electron value $\hbar^2 k^2/2m$. The first-order improvement to this is obtained by adding the energy $\langle k|V|k\rangle$, as given by equation (A 4.48). This is basically the energy $E_{el} + E_{ec}$ from Chapter 5, which can be improved by including the exchange correlation term E_x (equation (5.6)). No band-structure (Bragg reflection or Brillouin zone) effects appear in the theory to this level of approximation, and to obtain a first indication of these it is necessary to improve the energy to the second order, as in equation (A 4.53).

In this expression we put $\boldsymbol{k}' = \boldsymbol{k} + \boldsymbol{q}$ and use equations (A 6.6) and (4.9) to express the matrix element $\langle k'|V|k\rangle$ in terms of the structure factor $S(q)$ and form factor $w(q)$, as explained in Appendix 6. The second-order contribution to the energy is then of the form

$$\frac{1}{N}\sum_{q}\sum_{k<k_F}\frac{|S(q)w(q)|^2}{\boldsymbol{k}^2-(\boldsymbol{k}+\boldsymbol{q})^2} = \sum_{q}|S(q)w(q)|^2\chi(q) \tag{A 7.1}$$

where

$$\chi(q) = \frac{1}{N}\sum_{k<k_F}\frac{1}{\boldsymbol{k}^2-(\boldsymbol{k}^2+\boldsymbol{q}^2)} \tag{A 7.2}$$

and is known as the *perturbation characteristic*. The sum is taken as an integral over all occupied states inside the Fermi surface which, to this order of approximation, can be regarded as spherical. This leads again to the logarithmic expression of equations (3.3) and (3.20):

$$\chi(q) = -\frac{3Zm}{4\hbar^2 k_F^2}\left[1+\frac{(1-\alpha^2)}{2\alpha}\ln\left|\frac{1+\alpha}{1-\alpha}\right|\right] \tag{A 7.3}$$

where $\alpha = q/2k_F$ and Z is the number of electrons per ion (see e.g. Heine and

Weaire 1970, Harrison 1980, Ziman 1964). From this the *band-structure energy*, E_{BS} in equation (5.15) can be expressed as

$$E_{BS} = \sum_{q} |S(q)|^2 \mathbf{\Phi}(q) \tag{A 7.4}$$

where

$$\mathbf{\Phi}(q) = |w(q)|^2 \chi(q) \beta(q) \tag{A 7.5}$$

In these expressions we have separated the part that depends on the arrangement of the ions, i.e. on the structure factor $S(q)$, from that which is independent of this structure and so need be evaluated only once for a given metal. The additional factor $\beta(q)$ takes account of terms which, in the general calculation of equation (A 7.1), are counted twice when we sum the interaction of every electron with every other electron. In the important range of q, it is sufficient for qualitative discussions of general physical effects to take $\beta(q) \simeq 1$.

The general form of $\chi(q)$ is, because of its logarithmic function, the same as that of E_x in Fig. 3.1. The factor which multiplies $\chi(q)$ to give $\mathbf{\Phi}(q)$ is determined mainly by the form of the pseudopotential component, $w(\mathbf{q})$, as shown in Fig. 4.1. This characteristically goes from negative to positive as q increases, and we denote by q_0 the point at which it passes through zero. As a result, $\mathbf{\Phi}(q)$ has the general form shown in Fig. A 7.1; the rise that occurs where $q \simeq 2k_F$ is due to the rapid increase of $\chi(q)$ in this region. We note that, since $\chi(q)$ is negative, then so also are $\mathbf{\Phi}_{BS}(q)$ and E_{BS}.

It is remarkable that E_{BS} can be re-expressed as a (spherically symmetrical)

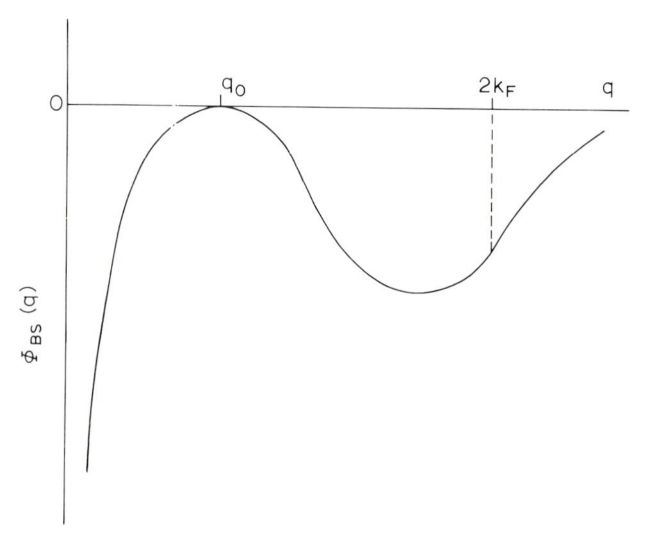

Fig. A 7.1 Typical form of $\mathbf{\Phi}_{BS}(q)$ when $q_0 < 2k_F$

ion–ion interaction potential, similar in this respect to the E_E interaction. The reason is that E_{BS} comes from the response of the electron gas to the electrostatic field of the ions, and this response takes the form of *screening*. It thus acts as a 'counter-Coulomb' field which completely neutralizes each Coulomb ionic field at large distances from the source ion of the field, and so imitates the Coulomb field in being representable as a *central force* – acting indirectly via the electron gas, of course – between each pair of ions. In effect, each ion provides a screened pseudopotential field, such as that shown in Fig. 2.7, in which the other ions sit; and each of these other ions, through its own pseudopotential charge, reacts electrostatically to this field.

The form of this ion–ion potential can be deduced from E_{BS} as given by equation (A 7.4). Since the structure factor is, from equation (4.3),

$$S(q) = \frac{1}{N}\sum_{l} \exp(-\mathrm{i}\boldsymbol{q}\cdot\boldsymbol{l}) \tag{A 7.6}$$

where the sum is over the positions $\boldsymbol{l}$ of the ions, we can write equation (A 7.4) as

$$E_{BS} = \frac{1}{N^2}\sum_{q}\sum_{l_i}\sum_{l_j} \exp[-\mathrm{i}\boldsymbol{q}(\boldsymbol{l}_i - \boldsymbol{l}_j)]\boldsymbol{\Phi}_{BS}(q) \tag{A 7.7}$$

where $\boldsymbol{l}_i$ and $\boldsymbol{l}_j$ are the sites of ions i and j. The sum over $\boldsymbol{q}$ can be converted into a volume integral over k-space, as in Appendices 3 and 4. Since one quantum state has a volume $8\pi^3/N\boldsymbol{\Omega}_a$ in k-space, for a crystal of N atoms of atomic volume $\boldsymbol{\Omega}_a$, we have

$$\sum_{q} = \frac{N\boldsymbol{\Omega}_a}{8\pi^3}\int \mathrm{d}\boldsymbol{q} \tag{A 7.8}$$

We use this and apply it to *one term* of equation (A 7.7), the one which represents the contribution to E_{BS} of the interaction of the ions i and j, separated by the vector $\boldsymbol{r} = \boldsymbol{l}_i - \boldsymbol{l}_j$. This term is then

$$\frac{\boldsymbol{\Omega}_a}{8\pi^3 N}\int \exp(-\mathrm{i}\boldsymbol{q}\cdot\boldsymbol{r})\boldsymbol{\Phi}(\boldsymbol{q})\,\mathrm{d}\boldsymbol{q} = \frac{\boldsymbol{\Omega}_a}{8\pi^3 N}\boldsymbol{\Phi}_{BS}(\boldsymbol{r}) \tag{A 7.9}$$

where $\boldsymbol{\Phi}_{BS}(\boldsymbol{r})$ is the Fourier transform of $\boldsymbol{\Phi}_{BS}(\boldsymbol{q})$, as given by equation (A 6.3). But $\boldsymbol{\Phi}_{BS}(\boldsymbol{r})$ represents an interaction via the band-structure between ions i and j, and it is a function of their distance apart, r. The double sum over $\boldsymbol{l}_i$ and $\boldsymbol{l}_j$ in equation (A 7.7) is twice the sum of all such interaction terms (since each ion appears as both i and j in the double sum). We see that E_{BS} may thus be represented as a sum of ion–ion pairwise interactions, each of which is a function of the interionic distance r.

Band-structure energy

The theory of band-structure energy describes the response of the metallic electrons to the (unscreened) pseudopotentials of the ions, and it is thus essentially also the theory of the *screening* of these pseudopotentials by the electron gas, as outlined in Chapters 2, 3 and 4. Hence, over large distances $\mathbf{\Phi}_{BS}(r)$ acts like a Coulomb potential, equal and opposite to the repulsive Coulomb ionic potential. It is therefore convenient to combine these two into a single *effective ion–ion interaction*

$$\mathbf{\Phi}(r) = \frac{Z^2e^2}{r} + \mathbf{\Phi}_{BS}(r) \qquad \text{(A 7.10)}$$

the form of which is shown in Fig. 5.5. The Friedel oscillations (*see* Chapter 2) which this develops at large r and which arise from the integration of the logarithmic term of equation (A 7.3) up to the surface of the Fermi distribution (*see* Chapter 3), lead to a limiting form of $\mathbf{\Phi}(r)$ at large r:

$$\mathbf{\Phi}(r) = C\frac{\cos(2k_Fr)}{(2k_Fr)^3} \qquad \text{(A 7.11)}$$

where

$$C \simeq \frac{9\pi Z^2|w(2k_F)|^2}{E_F} \qquad \text{(A 7.12)}$$

REFERENCES

Harrison, W. A., *Electronic Structure and the Properties of Solids*, Freeman, San Francisco (1980).
Heine, V., and Weaire, D., *Solid State Phys.*, 24, 249 (1970).
Ziman, J. M., *Principles of the Theory of Solids*, Cambridge University Press (1964).

APPENDIX 8

Perturbation theory of the d band

Following the Friedel theory outlined in Section 6.2, consider the pth d state ($p = 1, 2, 3, 4, 5$) of an atom i. In a free atom this state has an energy level E_0 given by Schrödinger's equation

$$\mathrm{H}_0 |ip\rangle = E_0 |ip\rangle \qquad \text{(A 8.1)}$$

where

$$\mathrm{H}_0 = \mathrm{T} + V_i \qquad \text{(A 8.2)}$$

$$\mathrm{T} = -\frac{\hbar^2}{2m}\left(\frac{\partial^2}{\partial x^2} + \frac{\partial^2}{\partial y^2} + \frac{\partial^2}{\partial z^2}\right) \qquad \text{(A 8.3)}$$

and V_i is the field of potential provided by atom i for an electron in this state. Inside the metal, the tight-binding approximation gives the wavefunction $\psi(E)$ of an electron in the energy level E as

$$\psi(E) = \sum_{i,p} a_{ip} |ip\rangle \qquad \text{(A 8.4)}$$

where the coefficients are normalized, i.e.

$$\sum_{i,p} |a_{ip}|^2 = 1 \qquad \text{(A 8.5)}$$

In the metal an electron on atom i experiences, not only the V_i of this atom, but also weakly the fields V_l of the other atoms. We assume that only the nearest neighbours (and next-nearest neighbours, in the b.c.c. structure) have a significant effect and that it is small enough to be treated as a perturbation. Thus in the solid H_0 is replaced by

$$\mathrm{H} = \mathrm{H}_0 + \mathrm{H}_s = \mathrm{H}_0 + \sum_{l \neq i} V_l \qquad \text{(A 8.6)}$$

for an electron on atom i. Standard perturbation theory (*see* equation (A 4.48)) then gives the change in the energy levels of the d states upon formation of the solid as a sum of matrix elements such as

$$\langle ip | \mathrm{H_s} | jq \rangle \equiv \int \psi_p^* \mathrm{H_s} \psi_q \, \mathrm{d}v \qquad \text{(A 8.7)}$$

where the integral is over the volume of the system. Since $\mathrm{H_s}$ is a sum of V_l, each such matrix element is itself a sum of *component matrix elements* such as

$$\langle ip | V_l | jq \rangle \equiv \int \psi_p^* V_l \psi_q \, \mathrm{d}v \qquad \text{(A 8.8)}$$

which represents the energy of overlap of, in general, three different atoms, i, j and l; i.e. the contribution from each volume element $\mathrm{d}v$ is proportional to the triple product of the two wave amplitudes $|ip\rangle$ and $|jq\rangle$ and the potential V_l, in this volume element. It is convenient here to approximate this difficult three-centre product by the two-centre one used by Slater and Koster (1954). We assume that in every case i, j and l do not refer to more than two different atoms. If $i = j$ then l is a neighbour (nearest or next-nearest) of i, and if $i \neq j$ then i and j are neighbours and V_l is either V_i or V_j.

The coefficients

$$\alpha_{ip} = \langle ip | \sum_l V_l | ip \rangle \qquad \text{(A 8.9)}$$

and

$$\beta_{ip}^{jq} = \langle ip | V_i | jq \rangle \qquad \text{(A 8.10)}$$

are then defined, as in equations (6.2) and (6.3), and substituted along with $\psi(E)$ and H from equations (A 8.4) and (A 8.6) into the Schrödinger equation

$$\mathrm{H}\psi(E) = E\psi(E) \qquad \text{(A 8.11)}$$

which is, at atom i,

$$(E_0 - E + \mathrm{H_s}) \left[a_{ip} | ip \rangle + \sum_{j,q} a_{jq} | jq \rangle \right] = 0 \qquad \text{(A 8.12)}$$

where $j \neq i$ in the sum. Following the standard method of perturbation theory (e.g. as in the derivation of equation (A 4.51), we multiply from the left by $\langle ip|$, integrate over the volume and use the orthonormality relations (equation (A 1.24)) to obtain

$$(E_0 - E + \langle ip|\mathrm{H_s}|ip\rangle)a_{ip} + \sum_{j,q} a_{jq}\langle ip|\mathrm{H_s}|jq\rangle = 0$$

which is

$$(E_0 - E + \alpha_{ip})a_{ip} + \sum_{j,q} a_{jq}\beta_{ip}^{jq} = 0 \qquad \text{(A 8.13)}$$

with $j \neq i$. This equation for a_{ip} and the various a_{jq} is one of a set of simultaneous equations of this type which can be solved to give all these coefficients. The wavefunction $\psi(E)$ of the metal is thus determined from those of the component atoms, e.g. $|ip\rangle$, in the free state.

The energy level E of this $\psi(E)$ is given by the standard expression (*see* equation (A 4.48))

$$E = \langle\psi|\mathrm{H}|\psi\rangle \qquad \text{(A 8.14)}$$

for a normalized ψ. We obtain

$$E = E_0 + E_s \qquad \text{(A 8.15)}$$

where

$$E_s = \langle\psi|\mathrm{H_s}|\psi\rangle = \sum_{j,q} a_{jq}^*\langle jq|\mathrm{H_s}|\sum a_{ip}|ip\rangle$$

$$= a_{ip}^*\langle ip|\mathrm{H_s}|\sum a_{ip}|ip\rangle + \sum_{j,q} a_{jq}^*\langle jq|\mathrm{H_s}|\sum a_{ip}|ip\rangle$$

$$= \sum_{i,p} a_{ip}^* a_{ip}\alpha_{ip} + \sum_{j,q}\sum_{i,p} a_{jq}^* a_{ip}\beta_{ip}^{jq} \qquad \text{(A 8.16)}$$

where $j \neq i$ in the sum over j. This gives equation (6.6).

REFERENCE

Slater, J. C., and Koster, G. F., *Phys. Rev.*, **94,** 1498 (1954).

APPENDIX 9

The moment distribution

To understand the development of equations (6.7)–(6.11), consider the expression

$$\sum_k \langle k | (\mathrm{H} - E_0)^m | k \rangle \tag{A 9.1}$$

where we have written the quantum states of the d band as Bloch states in a crystal, with wavenumbers k, using the Dirac bra–ket notation to represent these states as $\langle k|$ and $|k\rangle$. The Hamiltonian operator is H, and E_0 is the energy of an electron in a d state of the free atom. We can apply the operator $(\mathrm{H} - E_0)^m$ one step at a time. From the Schrödinger equation we obtain for the first step

$$(\mathrm{H} - E_0)|k\rangle = (E - E_0)|k\rangle \tag{A 9.2}$$

where E is the eigenvalue energy for the operation of H on $|k\rangle$. Applying $(\mathrm{H} - E_0)$ a second time gives

$$\begin{aligned}(\mathrm{H} - E_0)^2|k\rangle &= (\mathrm{H} - E_0)(E - E_0)|k\rangle \\ &= (E - E_0)(\mathrm{H} - E_0)|k\rangle = (E - E_0)^2|k\rangle \end{aligned} \tag{A 9.3}$$

and hence

$$\begin{aligned}\sum_k \langle k|(\mathrm{H} - E_0)^m|k\rangle &= \sum_k (E - E_0)^m \langle k|k\rangle \\ &= \sum_k N_k (E - E_0)^m \end{aligned} \tag{A 9.4}$$

where $N_k = \langle k|k\rangle$ is the number of states with wave-vector k (*see* equations (A 1.18) and (A 1.31)). Replacing the sum by an integral and N_k by $Nn(E)\,\mathrm{d}E$, where N is the number of atoms (or Bravais lattice sites) and $n(E)$ is the density of states per atom, we obtain

$$\sum_k \langle k|(\mathrm{H} - E_0)^m|k\rangle = \int Nn(E)(E - E_0)^m \,\mathrm{d}E = NM_m \tag{A 9.5}$$

where M_m is the mth moment as defined in equation (6.7), and the integral is over the d band. According to the tight-binding approximation the set of all the $|k\rangle$ states is equivalent to the set of all atomic states $|ip\rangle$ defined in Section 6.2 (equation (6.5)). Hence, to this approximation,

$$M_m = N^{-1} \sum_{i,p} \langle ip|(\mathrm{H} - E_0)^m|ip\rangle \tag{A 9.6}$$

For the next step it is convenient to use a mathematical device known as the *identity operator*. We note first that the orthonormality condition (equation (A 1.24)) for the integral, over the system, of the product of orthonormal wavefunctions, $|ip\rangle$ and $|jq\rangle$, is in Dirac notation

$$\langle ip|jq\rangle = \delta \tag{A 9.7}$$

where $\delta = 1$ when both $i = j$ and $p = q$, otherwise $\delta = 0$. We now take the wavefunction $|ip\rangle$ and multiply it from the left by $\sum |jq\rangle\langle jq|$, where the sum is over all j and q. This gives, using equation (A 9.7),

$$\sum_{j,q} |jq\rangle\langle jq|ip\rangle = \sum_{j,q} |jq\rangle\delta = |ip\rangle \tag{A 9.8}$$

and hence

$$\sum_{j,q} |jq\rangle\langle jq| = 1 \tag{A 9.9}$$

This is an identity operator. To apply it to equation (A 9.6) we spread out the m products of $\mathrm{H} - E_0$ and insert the identity operator between each successive pair of them. Applying the operator to the right-hand end of this sequence gives

$$\sum_{j,q} |jq\rangle\langle jq|(\mathrm{H} - E_0)|ip\rangle = \sum_{j,q} |jq\rangle\beta_{jq}^{ip} \tag{A 9.10}$$

where β_{jq}^{ip} is defined as in equation (6.3). Repeating this process m times and then using equation (A 9.7) on the term $\langle ip|rt\rangle$, we obtain

$$M_m = N^{-1} \sum \beta_{ip}^{rt} \ldots \beta_{ko}^{jq} \beta_{jq}^{ip} \tag{A 9.11}$$

where the summed term is a product of m coefficients β and the sum is over all indices, subject to the restriction $i \neq j \neq k \neq \ldots$. This is equation (6.11).

APPENDIX 10

Brillouin zones of superlattices

As we saw in Section 4.4 and Appendix 4, the theory of Brillouin zone boundaries follows closely that of X-ray crystal diffraction. For example, the b.c.c. structure is regarded as a simple cubic lattice, of spacing a, but with a *two-atom basis* consisting of one atom at the cube corner (000) and the other at the body centre $(a/2, a/2, a/2)$. In the pattern of simple cubic reflections all those which reflect anti-phase waves, from the interleaved planes passing through these two points, are *missing*; the resulting depleted pattern of Bragg reflections gives the characteristic X-ray signature of the b.c.c. structure and, when marked out in reciprocal space, it gives a Brillouin zone as in Fig. A 4.6. In this example the reflections from (100) planes are missing, and the first zone is the dodecahedron formed from (110) reflections.

The reflections from planes such as (100) are missing not merely because the path difference between the reflections from the interleaved planes is in this case

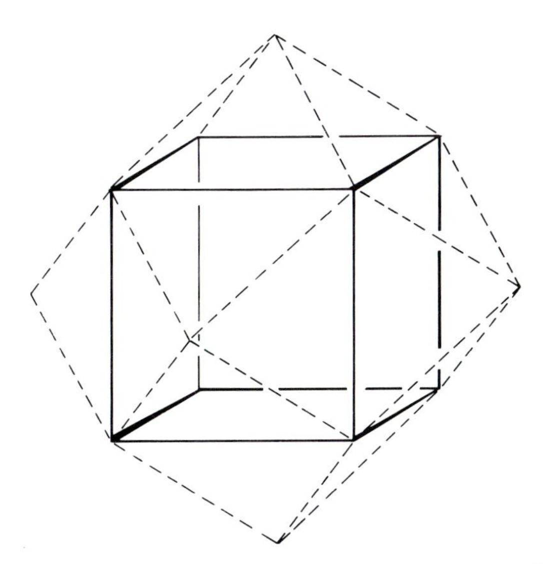

Fig. A 10.1 The cubic superlattice zone inscribed in the first Brillouin zone of the b.c.c. structure

half a wavelength: the amplitudes of the two out-of-phase waves are exactly equal, since the planes have identical atomic form factors, and hence give perfect mutual cancellation. In a superlattice, however, there is generally an inequality of amplitudes because the different atoms on alternate planes have different form factors or pseudopotentials. For example, the b.c.c. superlattice is the CsCl or ordered β brass structure, with one species of atoms at the cube corners and the other at the cube centres. The intensities of the reflections vary as the square of the atomic form factors (*see* equation (4.20)), and so weak *superlattice reflections* then appear in the positions of the missing reflections of the b.c.c. structure, of intensity proportional to the difference of the squared form factors (*see* equation (8.4)). Correspondingly, extra boundaries appear in the Brillouin zone structure. For C_sCl the first of these boundaries forms a cube in k-space, representing (100) reflections with a $\mathbf{g}$ in the $\langle 100 \rangle$ direction of length $2\pi/a$. This cube is shown inscribed in the b.c.c. dodecahedral zone in Fig. A 10.1.

This cubic superlattice zone is important not only in ordered alloys such as CuZn and HgMg, and ionic crystals such as CsCl; it also has physical significance in antiferromagnetic crystals such as b.c.c. chromium because the two sets of atoms, ordered according to their opposite spins, appear as physically distinct species to an electron of given spin passing through the metal.

APPENDIX 11

Bound electron pairs

Consider, in a free-electron gas, two electrons with wavenumbers k_1 and k_2, and positions r_1 and r_2. Using free-electron wavefunctions (equation (A 3.1)), we form a two-electron wavefunction for this pair of electrons of the type

$$\psi(1, 2) = \exp(ik_1 r_1) \exp(ik_2 r_2) \quad \text{(A 11.1)}$$

ignoring normalization. For simplicity, suppose that $k_1 = -k_2 = k$, so that the wavefunction becomes

$$\psi = \exp[ik(r_1 - r_2)] \quad \text{(A 11.2)}$$

Now imagine that there is an attraction between these two electrons. We take the Hamiltonian for them as

$$\mathrm{H} = \mathrm{H}^{(0)} + \mathrm{H}^{(1)} \quad \text{(A 11.3)}$$

where $\mathrm{H}^{(1)}$ represents this interaction and $\mathrm{H}^{(0)}$ is the free-electron Hamiltonian

$$\mathrm{H}^{(0)} = \frac{p_1^2}{2m} + \frac{p_2^2}{2m} = \frac{p^2}{m} \quad \text{(A 11.4)}$$

p being the momentum operator (*see* equations (A 1.3)), and we look for a solution of the *two-electron* Schrödinger equation

$$\mathrm{H}\phi = E\phi \quad \text{(A 11.5)}$$

of the form

$$\phi(r_1, r_2) = \sum_k c_k \exp[ik(r_1 - r_2)] \quad \text{(A 11.6)}$$

where the coefficients c_k have to be determined.

For this, we use the orthonormal properties of the free-electron wavefunctions (equation (A 1.24)). We operate on $\phi(r)$ with $\mathrm{H} - E$, then multiply from the left by $\exp[-ik'(r_1 - r_2)]$ and integrate over the whole range of $r_1 - r_2$ so as to eliminate, by equation (A 1.24), all terms except that for which $k = k'$. The $H^{(0)}$ operation on equation (A 11.6) gives

$$H^{(0)}\phi = \sum_k \frac{\hbar^2 k^2}{m} c_k \exp[ik(r_1 - r_2)] \qquad \text{(A 11.7)}$$

and hence, from equation (A 1.24),

$$\int \exp[-ik'(r_1 - r_2)H^{(0)}\phi \, d(r_1 - r_2) = E_{k'}c_{k'} \qquad \text{(A 11.8)}$$

where

$$E_{k'} = \frac{\hbar^2 k'^2}{m} \qquad \text{(A 11.9)}$$

The $-E\phi$ term similarly gives

$$-\int \exp[-ik'(r_1 - r_2)]E\phi \, d(r_1 - r_2) = -Ec_{k'} \qquad \text{(A 11.10)}$$

The electron–electron interaction $H^{(1)}$ is expected to scatter the two electrons, from initial states k and $-k$ into a similar pair of states k'' and $-k''$, of the same total momentum. The scattering amplitude (*see* equation (A 6.6)) for this can be written in the usual Dirac matrix element notation (*see* equation (A 1.31)) as $\langle k'', -k'' | H^{(1)} | k, -k \rangle$, where for example $| k, -k \rangle$ is the ket form of the initial two-electron state, $k, -k$. The pair of states $k'', -k''$ have the wavefunction given by equation (A 11.2), with k'' in place of k, and with an amplitude given by this matrix element. As before, when we multiply this wavefunction by $\exp[-ik'(r_1 - r_2)]$ and integrate over the whole volume, the orthonormality condition removes all terms except that with $k'' = k'$, and we thus obtain $\sum_k c_k \langle k', -k' | H^{(1)} | k, -k \rangle$ as the contribution of the $H^{(1)}$ term.

Putting the three contributions together and remembering that $(H - E)\phi = 0$ we obtain with, for convenience, interchanged symbols k and k':

$$(E_k - E)c_k + \sum_{k'} c_{k'} \langle k, -k | H^{(1)} | k', -k' \rangle = 0 \qquad \text{(A 11.11)}$$

A similar equation can be obtained for each of the c_k in the wavefunction of the electron pair (equation (A 11.6)), so that in principle this set of *secular equations* (*see* Appendix 5) can be solved to give the wavefunction. To continue, we assume that

$$\langle k, -k | H^{(1)} | k', -k' \rangle = -V \qquad \text{(A 11.12)}$$

where V is a positive constant (with respect to k and k') for all wave-vectors that lie in a thin shell on the (spherical) Fermi surface; and is zero outside this shell. To link up with the BCS theory, we define the shell as lying within the energy limits $E_F - \hbar\omega$ to $E_F + \hbar\omega$. We also assume that the free-electron states exp (ikr), which are used to construct the wavefunction of equation (A 11.6), are chosen from the energy range E_F to $E_F + \hbar\omega$. The sum in equation (A 11.11) is over the group of all those initial two-electron states k', $-k'$ within the energy range $2E_F$ to $2(E_F + \hbar\omega)$ that can by interaction scatter into the two-electron state k, $-k$. We can thus represent this group of electron-pair states as a density of states, and so replace the sum by an integral. This density of states is usually expressed as the density of single-electron states of one spin in the normal state, and written as $N(E')$. With these changes, equation (A 11.11) becomes

$$(E_k - E)c_k = V\int_{2E_F}^{2(E_F + \hbar\omega)} N(E')c_{k'}\,\mathrm{d}E' \qquad \text{(A 11.13)}$$

None of the terms on the left-hand side of this equation appear on the right-hand side. Hence we have a constant C;

$$(E_k - E)c_k = C \qquad \text{(A 11.14)}$$

Using this to substitute for both c_k and $c_{k'}$, equation (A 11.13) becomes

$$V\int_{2E_F}^{2(E_F + \hbar\omega)} \frac{N(E')\,\mathrm{d}E'}{E_{k'} - E} = 1 \qquad \text{(A 11.15)}$$

Because the shell is thin ($\pm\,\hbar\omega$) we can over this range approximate $N(E')$ by its value $N(E_F)$ at the Fermi level, and so take $N(E_F)$ outside this integral. The integration can then be made and gives

$$\frac{1}{N(E_F)V} = \int_{2E_F}^{2(E_F + \hbar\omega)} \frac{\mathrm{d}E'}{E' - E} = \ln\frac{2\hbar\omega + \Delta}{\Delta} \qquad \text{(A 11.16)}$$

where

$$E = 2E_F - \Delta \qquad \text{(A 11.17)}$$

It follows that

$$\Delta = \frac{2\hbar\omega}{\exp(1/N(0)V) - 1} \qquad \text{(A 11.18)}$$

where we have written $N(E_F)$ in its usual form $N(0)$ as used in the theory of superconductivity. In the limit of *weak coupling*, where $N(0)V \ll 1$, this expression simplifies to

$$\Delta \simeq 2\hbar\omega \exp(-1/N(0)V) \tag{A 11.19}$$

It follows that the energy of the whole electron system can be reduced if two such electrons are excited into the shell just above the Fermi level and then allowed to interact, with energy $-V$, to form a pair. The Fermi distribution of a free-electron gas is thus unstable with respect to attractive interactions between the electrons, which enable bound electron pairs to form.

APPENDIX 12

The energy gap of a superconductor

Consider a system of electrons in *pair states*, such as $\boldsymbol{k}$ and $-\boldsymbol{k}$, which we will simply denote as k. Let ε_k be the energy, measured relative to the Fermi energy, of an electron in a *free-electron state* of wavenumber k, before the interaction $-V$ (equation (A 11.12)) is 'switched on'. Let ρ_k be the probability that the two states of the pair k are occupied, and $1 - \rho_k$ the probability that this pair is empty. Then, before $-V$ is switched on, the energy of the set of pairs relative to that of the normal metal is $2\sum \varepsilon_k \rho_k$, summed over all pairs, the factor 2 entering because there are two electrons in each filled pair.

Now switch on $-V$ and consider the interaction which scatters the two electrons from k to k', or vice versa. The transition $k \rightarrow k'$ requires k to be full and k' to be empty, initially. Hence its probability is proportional to $\rho_k(1 - \rho_{k'})$. Similarly, the probability of $k' \rightarrow k$ is proportional to $\rho_{k'}(1 - \rho_k)$. The whole process is symmetrical between these two and, as a result, the probability for its contribution to the energy of the system of pair states is given by $[\rho_k(1 - \rho_{k'}) \rho_{k'}(1 - \rho_k)]^{1/2}$.

The pair state k has similar interactions with all other pair states like k'. Hence the total energy W of the superconducting state in its ground state at zero temperature, relative to that of the metal in the normal state, is

$$W = 2\sum_k \varepsilon_k \rho_k - V \sum_k \sum_{k'} [\rho_k(1 - \rho_{k'})\rho_{k'}(1 - \rho_k)]^{1/2} \qquad \text{(A 12.1)}$$

We want to choose ρ_k so that W is as low as possible. Minimizing W with respect to ρ_k leads to

$$\rho_k = \frac{1}{2}\left[1 - \frac{\varepsilon_k}{(\varepsilon_k^2 + \varepsilon_0^2)^{1/2}}\right] \qquad \text{(A 12.2)}$$

where

$$\varepsilon_0 = V \sum_{k'} [\rho_{k'}(1 - \rho_{k'})]^{1/2} \qquad \text{(A 12.3)}$$

i.e.

$$\varepsilon_0 = \frac{V}{2} \sum_k \frac{\varepsilon_0}{(\varepsilon_k^2 + \varepsilon_0^2)^{1/2}} \tag{A 12.4}$$

We can now proceed as in Appendix 11 and turn the sum into an integral over an energy range ε that corresponds to the range of k, measuring the number of states in the neighbourhood of the Fermi energy ($\varepsilon = 0$) once more by means of the density $N(0)$ of free-electron states of one spin at the Fermi level. Then equation (A 12.4) becomes

$$\frac{1}{N(0)V} = \int_0^{\hbar\omega} \frac{d\varepsilon}{(\varepsilon^2 + \varepsilon_0^2)^{1/2}} \tag{A 12.5}$$

which gives

$$\varepsilon_0 = \frac{\hbar\omega}{\sinh\,[1/N(0)V]} \tag{A 12.6}$$

Substituting this back through the above equations, we obtain for the ground state energy of the superconducting state

$$W = -\frac{2N(0)(\hbar\omega)^2}{\exp\,[2/N(0)V] - 1} \simeq -2N(0)(\hbar\omega)^2 \exp\,[-2/N(0)V] \tag{A 12.7}$$

when $N(0)V \ll 1$.

The contribution of one electron pair, e.g. k, to W is given by

$$2\left\{\varepsilon_k\rho_k - V\sum_{k'} [\rho_{k'}(1 - \rho_{k'})]^{1/2}\right\} \tag{A 12.8}$$

For k sufficiently near the Fermi surface the first term becomes negligible and we are left with the second one, which represents the energy required to break up this pair by removing one of its electrons to some free-electron state, so producing two unpaired electrons. The finite value of this term is the basis of the energy gap which, at zero temperature, separates the superconducting ground state from the lowest excited state. Substituting from equations (A 12.3) and (A 12.6), we see that the magnitude of the gap is

$$2\varepsilon_0 = \frac{2\hbar\omega}{\sinh\,[1/N(0)V]} \simeq 4\hbar\omega \exp\,[-1/N(0)V] \tag{A 12.9}$$

so that

$$\varepsilon_0 = \Delta \tag{A 12.10}$$

where Δ is given by equations (A 11.18) and (A 11.19).

APPENDIX 13

The layered perovskite structure of superconducting oxides

This structure has been described by Robinson (1987). Figure A 13.1 shows a unit cell of a perovskite oxide, which for convenience we will write as $XCuO_3$, where atoms X occupy cube centres, the copper atoms are at the cube corners and the oxygen atoms occupy mid-points along the cube edges. Each copper ion is sited at the centre of an octahedron of six oxygen ions.

To go from this to the layered structure of a superconducting oxide of composition $AB_2Cu_3O_7$, in which A is yttrium (or some rare earth) and B is an alkaline earth, the first step is to make a vertical three-cell stack of these perovskite units, placing e.g. a barium atom at the body centre of each of the top and bottom cubes and an yttrium atom at the centre of the middle cube. This produces a tetragonal cell, of composition $YBa_2Cu_3O_9$. It is convenient to regard this as a multi-sandwich of seven horizontal two-dimensional layers of oxide: starting from the bottom, there is first a copper layer (CuO_2), followed in sequence by layers of barium, copper, yttrium, copper, barium and copper.

Next, we reduce the oxygen content to O_7. This is done by removing all the oxygen atoms from the yttrium layer, and half those in the top and bottom copper layers. Oxygen atoms are removed from the copper layers in such a way as to reduce these layers to parallel lines of -Cu-O-Cu-O- atoms threading through the structure along one of the originally cubic axes. The structure

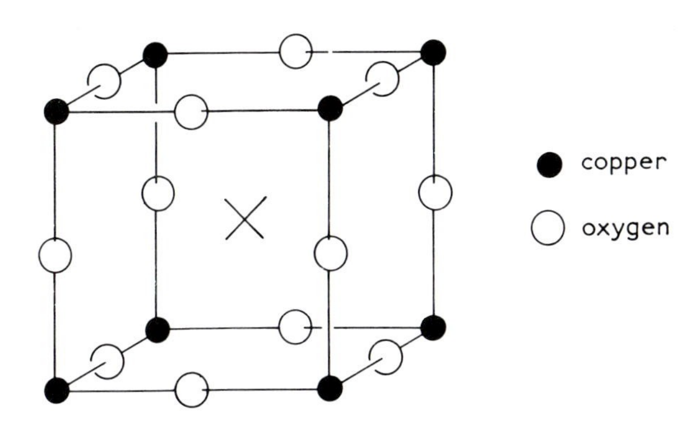

Fig. A 13.1 The perovskite structure of $XCuO_3$

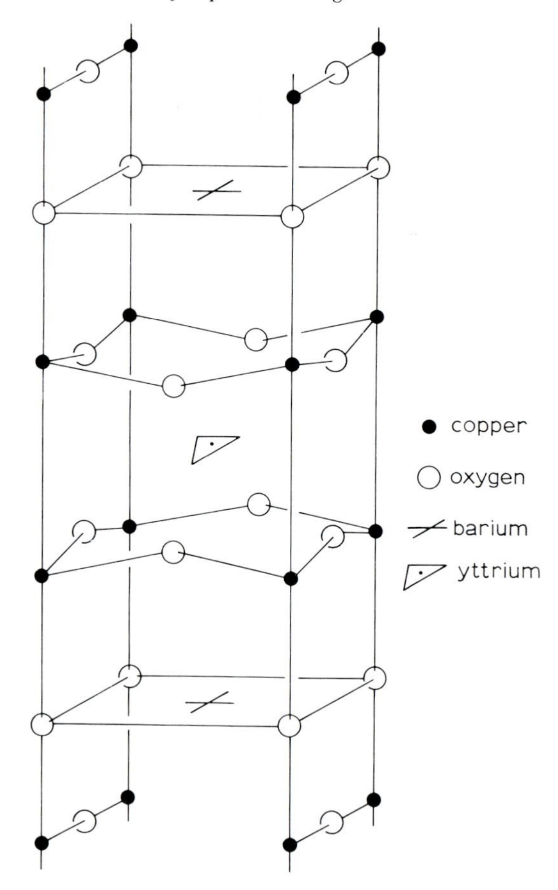

Fig. A 13.2 The layered perovskite structure of $YBa_2Cu_3O_7$

thereby distorts slightly, becoming orthorhombic, and the two CuO_2 layers next to the yttrium one become 'dimpled' through a small displacement of their oxygen ions towards the oxygen-free yttrium layer. The structure is shown in Fig.A 13.2.

The organization of the top and bottom layers into parallel copper–oxygen lines is a form of low-temperature ordering which leads to a small inequality in the two horizontal orthogonal cell dimensions. Above 650°C the oxygen atoms in these two layers become distributed equally in both directions between the copper atoms, and the two horizontal cell dimensions then become equal. On cooling into the ordered state, various domains of the crystal make different choices of the two axes for their copper–oxygen lines, and the ensuing dimensional changes generate elastic strains between the domains.

REFERENCE

Robinson, A. L., *Science*, **236**, 1063 (1987).

INDEX

P

Q

R

S